Biotechnological Approaches
for
Sustainable Development

THE AUTHOR

Prof. T.Pullaiah obtained his M.Sc. (1973) and Ph.D. (1976) degrees in Botany from Andhra University. He was a Post Doctoral Fellow at Moscow State University, Russia during 1976-78. He traveled widely in Europe and USA and visited Universities and Botanical Gardens in about 17 countries. Professor Pullaiah joined Sri Krishnadevaraya University as Lecturer in 1979 and became Professor in 1993. He held several positions in the University, which include Dean, Faculty of Life Sciences, Head of the Department of Botany, Chairman, BOS in Botany, Head of the Department of Sericulture, Co-ordinator and Chairman, BOS in Biotechnology, Vice Principal and Principal, S.K. University College. He retired from active service on 31[51] May 2011.

He was selected by UGC as UGC-BSR Faculty Fellow and is working in Sri Krishnadevaraya University. Prof. Pullaiah has published 53 books, 295 research papers and 35 popular articles. His books include *Flora of Andhra Pradesh* (5 volumes), *Flora of Eastern Ghats* (vol. 1-4), *Biodiversity in India* (vol. 1-6), *Encyclopaedia of World Medicinal Plants* (5 volumes), *Indian Medicinal Plants* (2 volumes), *Herbal Antioxidants* (3 volumes) *Taxonomy of Angiosperms* (presently in 3rd edition), *Plant Development, Plant Reproduction, Plant Tissue Culture, Flora of Guntur district. Flora of Kurnool, Flora of Anantapur district, Flora of 'Zamabad, Flora of Ranga Reddi district* etc. He is Principal Investigator of 20 Major Research Projects totaling more than a Crore of Rupees funded by DBT, DST, CSIR, UGC, BSI, WWF, GCC etc. Under his guidance 52 students obtained their Ph.D. degrees and 35 students their M.Phil. degrees. He is recipient of Best Teacher Award from Government of Andhra Pradesh, Prof. P. Maheswari Gold Medal and Prof. G. Panigrahi Memorial Lecture of Indian Botanical Society and Prof. Y.D. Tyagi Gold Medal of Indian Association for Angiosperm Taxonomy. He is President of Indian Botanical Society and President of Indian Association for Angiosperm Taxonomy. He was Member of Species Survival Commission of International Union for Conservation of Nature and Natural Resources (IUCN).

Biotechnological Approaches for Sustainable Development

Editor

T. Pullaiah

Department of Botany
Sri Krishnadevaraya University
Anantapur-515003, A.P. India

2015

Regency Publications

A Division of

Astral International (P) Ltd

New Delhi 110 002

Published by   : **Regency Publications**
                         A Division of
                         Astral International Pvt. Ltd.
                         – ISO 9001:2008 Certified Company –
                         House No. 96, Gali No. 6,
                         Block-C, 30ft Road, Tomar Colony, Burari
                         New Delhi-110 084
                         E-mail: info@astralint.com
                         Website: www.astralint.com

Sales Office   : 4760-61/23, Ansari Road, Darya Ganj
                         New Delhi-110 002   Ph. 011-23245578, 23244987

Laser Typesetting   : **Rajender Vashist**
                         Delhi - 110 059

Printed at   : **Replika Press Pvt. Ltd.**

PRINTED IN INDIA

Preface

Biotechnology is expanding at phenomemal speed and there is every necessity to catch it. Although internet gives all the information it is difficult and time consuming for the investigator to search for the information. Many experts are working for many years and acquiring knowledge on the topic of their research. There is every necessicity for the experts to bring together all the information that they accumulated over the years. Thi book brings together articles by different experts on diverse topics like probiotics, molecular markers, DNA sequencing, medicinal plants micropropagation, in vitropropagation of pteridophytes and phytoremediation of Uranium.

I request readers and researchers to give their suggestions for improvement of these articles and topics for future edited books on Biotechnology.

T. Pullaiah

Contents

Preface   v

List of Contributors   ix

1. Probiotics - Not just for your gut only: Impact on diseases   1
   beyond gut
   –Pongali B. Raghavendra

2. Molecular Markers: An Array of Technological Advancements   21
   and Their Applications in Crop Improvement
   —A. Chandra Sekhar, P. Chandra Obul Reddy, T. Krishnaveni,
   P. Ramesh and K.V.N. Ratnakar Reddi

3. *In-silico* re-evaluation of DNA Sequences for Drawing   55
   Phylogeny of Orchids with Special Emphasis on *Dendrobium*
   Sw.
   –Pritam Chattopadhyay and Nirmalya Banerjee

4. An Effective Nutritional Requirements and an *in vitro*   83
   Approach for Asymbiotic Seed Germination and Seedling
   Growth of Some Terrestrial Orchids
   –Madhumita Majumder and Nirmalya Banerjee

5. *In vitro* Approaches in Medicinal Plants–A Viable Strategy to   141
   Strengthen the Resource Base of Plant Based Systems of
   Medicines
   –A.A. Waman, P. Bohra, B.N. Sathyanarayana and B.G.
   Hanumantharaya

6. Somaclonal Variation in Plant Tissue Culture and its Role in   157
   Crop Improvement
   –*P. Chandramati Shankar and H. Fathima Nazneen*

7. *In vitro* Multiplication of *Pronephrium Triphyllum* (Sw.) Holttum   169
   - An Endangered Fern
   –*M. Johnson and V.S. Manickam*

8. Effect of Plant Growth Regulators on *in vitro* Raised   179
   Gametophytes of *Phlebodium aureum* l.
   –*M. Johnson and V.S. Manickam*

9. Antiulcer Activity of an Isolated Compound (KR–1) from   189
   Kaempferia rotunda Linn. Leaf in Rats
   –*Prasanta Kumar Mitra, Tanaya Ghosh, Prasenjit Mitra and*
   Gayatri Mitra

10. Comparative Assessment of *in Vitro* Antioxidant and   201
    Antiproliferative Activity in *Morinda citrifolia* Fruit and
    Commercial Juice
    –*K. Rajaram and P. Suresh Kumar*

11. Remediation of Uranium Contaminated Soils: Conventional   211
    and Emerging Technologies
    –*Sravani Konduru, Chandra Obul Reddy Puli,*
    Chandra Sekhar Akila, Varakumar Pandit,
    Krishna Kumar Guduru, Jayanna Naik Banavath

Index   251

List of Contributors

A. Chandra Sekhar, Department of Biotechnology, School of Life Sciences, Yogi Vemana University, Kadapa-516 003 (A.P., India).

A.A. Waman, Plant Tissue Culture Laboratory, Department of Horticulture, University of Agricultural Sciences, GKVK Campus, Bengaluru-560 065, India.ajit.hort595@gmail.com

B.G. Hanumantharaya, Plant Tissue Culture Laboratory, Department of Horticulture, University of Agricultural Sciences, GKVK Campus, Bengaluru-560 065, India.

B.N. Sathyanarayana, Plant Tissue Culture Laboratory, Department of Horticulture, University of Agricultural Sciences, GKVK Campus, Bengaluru-560 065, India.

Chandra Obul Reddy Puli, Department of Plant Sciences, School of Life Sciences, Yogi Vemana University, Vemanapuram Kadapa 516 003 A.P. India.

Chandra Sekhar Akila, Department of Biotechnology, School of Life Sciences, Yogi Vemana University, Vemanapuram Kadapa 516 003 A.P., India

Gayatri Mitra' Biochem Academy, Saktigarh, Siliguri -734 005, West Bengal, India.

H. Fathima Nazneen, Department of Biotechnology, School of Life Sciences, Yogi Vemana University, Vemanapuram, Kadapa-516 003, Andhra Pradesh, India.

Jayanna Naik Banavath, Department of Plant Sciences, School of Life Sciences, Yogi Vemana University, Vemanapuram Kadapa 516 003 A.P., India.

K. Rajaram, Department of Biotechnology, Bharathidasan Institute of Technology, Anna University, Tiruchirappalli-620 024, Tamil Nadu, India.

Krishnakumar Guduru, Department of Plant Sciences, School of Life Sciences, Yogi Vemana University, Vemanapuram Kadapa-516 003 A.P., India

K.V.N. Ratnakar Reddi, Department of Biotechnology, School of Life Sciences, Yogi Vemana University, Kadapa-516 003 (AP., India).

M. Johnson, Centre for Plant Biotechnology, Department of Botany, St. Xavier's College (Autonomous), Palayamkottai-627 002, Tamil Nadu, India. ptcjohnson@gmail.com

Madhumita Majumder, Department of Botany, Visva-Bharati, Santiniketan-731235, West Bengal, India.

Nirmalya Banerjee, Cytogenetics and Plant Biotechnology Laboratory, Department of Botany, Visva-Bharati University, Santiniketan-731 235, WB, India E-mail: nirmalya_b@rediffmail.com

P. Bohra, Plant Tissue Culture Laboratory, Department of Horticulture, University of Agricultural Sciences, GKVK Campus, Bengaluru-560065, India.

P. Chandra Obul Reddy, Department of Plant Sciences, School of Life Sciences, Yogi Vemana University, Kadapa-516 003, (A.P., India).

P. Chandramati Shankar, Department of Biotechnology, School of Life Sciences, Yogi Vemana University, Vemanapuram, Kadapa-516 003, Andhra Pradesh., India. pchandra20@gmail.com

P. Suresh Kumar, Department of Biotechnology, Bharathidasan Institute of Technology, Anna University, Tiruchirappalli-620 024, Tamilnadu, India, E.mail: drsureshbiotech2003@gmail.com

Pongali B. Raghavendra, Department of Physiology and Division of Human Pathology, Michigan State University, East Lansing, MI 48823, USA Email: pbraghav@msu.edu / raghavbiot@gmail.com

Prasanta Kumar Mitra, Department of Biochemistry, North Bengal Medical College, Sushrutanagar-734 012, Dist. Darjeeling, West Bengal, India. dr_pkmitra@rediffmail.com

Prasenjit Mitra, Biochem Academy, Saktigarh, Siliguri-734 005, West Bengal, India.

Pritam Chattopadhyay, Cytogenetics and Plant Biotechnology Laboratory, Department of Botany, Visva-Bharati University, Santiniketan-731 235, WB, India.

Sravani Konduru, Department of Plant Sciences, School of Life Sciences, Yogi Vemana University, Vemanapuram, Kadapa-516 003, A.P., India.

T. Krishnaveni, P. Ramesh, Department of Biotechnology, School of Life Sciences, Yogi Vemana University, Kadapa-516 003 (AP., India).

Tanaya Ghosh, Department of Biochemistry, North Bengal Medical College, Sushrutanagar-734 012, Dist. Darjeeling, West Bengal, India. biochemacademy@rediffmail.com

V.S. Manickam, Centre for Biodiversity and Biotechnology, Department of Botany, St. Xavier's College (Autonomous), Palayamkottai-627 002, Tamil Nadu, India.

Varakumar Pandit, Department of Plant Sciences, School of Life Sciences, Yogi Vemana University, Vemanapuram Kadapa-516 003, A.P., India.

Chapter 1

Probiotics - Not just for your gut only: Impact on diseases beyond gut

Pongali B. Raghavendra

Introduction

Gut microbial community has profound effects in the development of both the intestinal mucosal and systemic immune systems. A healthy gut microbiota is essential to promote host health and well-being, overgrowth of the bacterial population results in a variety of detrimental conditions, and different strategies are employed by the host to prevent this outcome. Gut microbiota has evolved with humans as a mutualistic partner, but dysbiosis in a form of altered gut metagenome and collected microbial activities, in combination with classic genetic and environmental factors, may promote the development of various disorders. Further different recent study findings are emerging for the role of intestinal microflora and its influence in the development of metabolic disorders such as obesity, diabetes etc., allergy, periodontal diseases, bone related disorders and few more. Probiotics are viable non–pathogenic micro-organisms which, when consumed, exert a positive influence on host health or physiology. There are growing evidences relating for the efficacy of specific or with several probiotic strains in human gastrointestinal diseases. In addition, studies suggest probiotics have the potential to alleviate diet-induced obesity and modulate genes associated with metabolism and inflammation in the liver and adipose tissue, regulate allergic inflammation locally and systemically, anti-dental caries activity of probiotic treatment, cholesterol lowering therapies, increase in circulating 25-hydroxyvitamin D in response to oral probiotic supplementation, femoral and vertebral bone formation increase in response to oral probiotic use. The advent of probiotic treatment appears to be a promising pharmaco-nutritional approach to reverse the host metabolic alterations linked to dysbiosis observed several disorders beyond to gut. More molecular and cellular study approaches are needed to unravel the modulations in the gut microbial community or counteract the development of related disorders.

Gut Microflora

Collection of microorganisms resident in the gastrointestinal tract is termed as the microflora. Gut microflora contains a variety of different bacteria and fungi of which there are typically approximately 400 different types of microorganisms with a total population of $\sim 10^{14}$ (that is 100,000,000,000,000 bacteria) throughout the length of the intestinal tract (Björkstén *et al.*, 2001; Guarner and Malagelada, 2003; Sears, 2005; Steinhoff 2005; Savage, 1977). The gut microflora is a complex collection of microorganisms which are distributed throughout the whole length of the gut. Within particular regions the organisms may be found in three niches:

(a) associated with gut wall. This can either take the form of direct attachment to the epithelium or entrapment in the mucous layer of the epithelium

(b) attachment to food particles

(c) suspension in the liquid phase of the gut contents

The composition of the microflora varies in different regions of the intestine and is dependent on factors such as pH. The microflora which develops in the human intestinal tract is characteristic for that species which has evolved a symbiotic association with the host (Lebba *et al.*, 2012). However, there are a few microorganisms which are only found in the human infant gut for example *Bifidobacterium infantis* (Bettelheim *et al.*, 1974). Before birth, the foetus is contained within a sterile environment. Within hours of birth, the baby acquires a complex collection of microorganisms which populate its oral cavity. Within days the full length of the gastrointestinal tract will be colonised with microorganisms. The digestive tract is colonised by microorganisms which the newly born infant comes into contact with namely from the mother, father etc, milk and the immediate environment (Buddington and Sangild, 2011; Guarino *et al.*, 2012).

Bifidobacterium as well as Lactobacilli will dominate the gastrointestinal tract of infants until weaning whereby the types of microorganisms colonising the gastrointestinal tract will resemble an adult (Mikami *et al.*, 2009). At birth, babies have a very weak immune system. That system needs to be developed for survival of the infant. Breast milk which contains nearly all the constituents necessary for growth and enhancement of the immune system also contains beneficial bacteria which attach to the baby's intestinal wall. It is these beneficial bacteria which allow digestion in the gut to occur. As the child grows, additional microorganisms are ingested from other food sources. Due to the fact that the microflora is not fully developed until after weaning, is one of the reasons why young children are often more susceptible to illness than adults (Conroy *et al.*, 2009; Salminen and Isolauri, 2008; Mackie *et al.*,1999; Kononen, 2000; Adlerberth, 2008).

Major Influencing Factors for Gut Dysbiosis

A frequent disorder of intestinal function is dysbiosis, i.e., the overgrowth of pathogenic bacteria in the intestine. This microflora plays critical roles in the digestion and absorption of nutrients, in the synthesis of vitamins (B and K groups) and fatty acids, in the detoxification of ingested chemicals, but also in the regulation of the immune system. Alterations in the composition of the gut microflora may have serious consequences for the host health. Factors that can affect the microflora include antibiotic use, stress, and diet and genetic factors.

Antibiotic use is a common cause of major alterations. Dosage, length of administration, spectrum of activity will determine the impact on the microbial flora (Wynne *et al.*, 2004; Vedantam and Hecht, 2003; Beaugerie and Petit, 2004; Carman *et al.*, 2004).

Psychological stress can also affect the composition of the flora, including a significant decrease in beneficial bacteria (*Lactobacilli* and *Bifidobacteria*) and an increase in pathogenic *E. coli*. Stress may affect bacterial growth by significantly reducing the mucosal production of mucopolysaccharides and mucins, which are important for inhibiting the adherence of pathogenic organisms, and by decreasing the production of immunoglobulin A (IgA), which play a crucial role in their elimination (Hart and Kamm, 2002; Alverdy *et al.*, 2005; Lewis and Mckay, 2009). Neurochemicals produced upon psychological stress can also directly enhance the growth of pathogenic organisms: norepinephrine stimulates the growth of *Y. enterocolitica, P. aeruginosa,* and gram-negative bacteria such as *E. coli* (Freestone *et al.*, 2007a, b).

Diet, is another factor that may have an impact on the human intestinal flora (Mainous *et al.*, 1994). Some diets promote the growth of beneficial microorganisms, while others promote harmful microfloral activities. For instance, diets rich in sulfur compounds (dairy products, eggs, certain vegetables, dried fruits) promote the growth of sulfate-reducing bacteria (Bernardez *et al*, 1994). Globally it appears that populations consuming the typical Western diet have more anaerobic bacteria, less enterococci, and fewer yeasts than populations consuming a vegetarian or high complex-carbohydrate diet.

Genetic factors seem to be key, that predispose to the development of Irritable Bowel Syndrome (IBS): immediate relatives of an individual with IBS are 4 to 20 times more likely to develop the disease than the general population (Hahm Ki Baik, 2012). The identification of an exact genetic mutation has been so far elusive however recent results suggest an implication of genes relevant to inflammation in general (cytokines), as well as genes coding for Toll-like receptors (notably TLR-4, that recognizes LPS) (Shibolet and Podolsky, 2007; Sanderson and Walker, 2007; Henckaerts *et al.*, 2007).

Gastrointestinal Disorders

Gastrointestinal infections caused by *Helicobacter pylori*, traveller's diarrhoea, rotavirus diarrhoea, antibiotic-associated diarrhoea (AAD) and *Clostridium difficile*-induced diarrhoea are common disorders and have major effects on GI tract. Inflammatory bowel disease and irritable bowel syndrome, two idiopathic conditions where alterations in the normal microflora have been implicated as responsible for initiation. In most of these cases the commensal bacterial species are diminished and the list of disorders include, such as acute pancreatitis, necrotizing enterocoloitis, pouchitis, ulcerative colitis, crohn's disease, diverticular colitis, lactose intolerance and celiac diseases.

Concept of Probiotics

Probiotics are defined as live microbial feed supplements, which beneficially affect the host animal by improving its intestinal microbial balance (Resta, 2009). Metchnikoff in early 1900's hypothesized: "When people have learnt how to cultivate a suitable flora in the intestines of children as soon as they are weaned from the breast, the normal life may extend to twice my 70 years." The most common species of bacteria in probiotic products are listed below in **Table 1.1.**

Table 1.1. Most common species of bacteria in probiotic foods

Lactobacillus acidophilus/johnsonii/gasseri
Lactobacillus casei
Lactobacillus paracasei
Lactobacillus reuteri
Lactobacillus rhamnosus
Lactobacillus plantarum
Bifidobacterium animalis/lactis
Bifidobacterium bifidum
Bifidobacterium breve
Bifidobacterium adolescentis
Pediococcus acidilactici
Saccharomyces boulardii

Probiotics have been shown to work by the following mechanisms:

(i) **Competition for Nutrients:** Within the gut, beneficial as well as pathogenic microorganisms will be utilizing the same types of nutrients. Thus there will be a general competition for these nutrients to grow and reproduce. The more the gut is flooded with beneficial microorganisms, the more competition is created between beneficial and pathogenic microorganisms (Wilson and Perini, 1988).

(ii) **Competition for Adhesion sites:** Adhering to adhesion sites along the wall of the gut is an important colonisation factor and many intestinal pathogens rely on adhesion to the gut wall to prevent them being swept away by peristaltic of food along the intestinal tract. An important function of these probiotic bacteria is to prevent or limit the growth and colonisation of potentially pathogenic bacteria such as *E. coli, Salmonella, Listeria, Campylobacter* and *Clostridia* within the gut. These pathogenic bacteria are known to cause major disturbances within the gut thus preventing efficient digestion and nutrient absorption within the gut and may result in diarrhoea or vomiting. Where the gut microflora is well balanced, the beneficial microorganisms colonised within the gut can hence help to reduce the risk of pathogenic challenge (Alander *et al.*, 1997; Kleeman and Klaenhammer, 1982; Conway *et al.*, 1987; Goldin *et al.*, 1992; Mack *et al.*, 1999; Davidson and Hirsch 1976; Rigothier *et al.*, 1994; Bernet *et al.*, 1994).

(iii) **Stimulation of Immunity:** Probiotics have been shown to ensure the optimum microflora balance in order to stimulate and maintain the natural immune system of the host. These enhanced immune effects help to prevent illness when probiotics are used regularly (Steidler *et al.*, 2000; Dahan *et al.*, 2003; Petrof *et al.*, 2004; Kaila *et al.*, 1992, 1995; de Vrese *et al.*, 2004; Park *et al.*, 2002; Buts *et al.*, 1990; Christensen *et al.*, 2002.

(iv) **Direct Antimicrobial effect:** This can either operate via bacteriocins which are known to be produced by many species of lactic acid bacteria or by the production of organic acids which can either have a direct effect or operate by reducing the pH of the gut (Wollowski *et al.*, 2001; Shi and Walker, 2004; Agarwal *et al.*, 2003).

(v) **Improvement in Digestion:** Probiotic microorganisms act like and add to the healthy microflora by producing enzymes which aid the breakdown of polysaccharides such as carbohydrates to allow the absorption of the energy obtained from these nutrients by the gut. The microflora also ferments carbohydrates which have not been digested in the upper gut and produces vitamins which supply a secondary source to the host (Jahn *et al.*, 1996).

Probiotics and Sources

While commercial supplements often come to mind when people mention probiotics, traditional fermented foods are teeming with these beneficial bacteria. Fermented vegetables, fermented milk products (clabber, yogurt, cheese, buttermilk), kefir, fermented soy products (natto, miso, tempeh, soy sauce, fermented tofu), and even naturally fermented, unpasteurized beers are some of the most complete probiotics available (Sanders, 2008). Yet, while eating traditional fermented foods is the most natural way to get probiotic benefits, many people find it difficult to do on a consistent basis. An effective alternative is to take a probiotic supplement. The probiotic products available in the global market are listed in the below Table 2.

Apart from dairy sources, there are many, dairy-free foods rich in probiotics and beneficial bacteria. The common ones include, *Sauerkraut* the only form of probiotic-rich fermented cabbage. Furthermore, the process of fermenting cabbage actually creates isothiocyanate – a substance thought to inhibit the formation of cancer and tumors. *Kombucha* is another great source of beneficial bacteria in the form of fermented tea. It is source for vitamin B_{12} and also contains a substance called glucaric acid. Glucaric acid is deeply detoxifying and recent research indicates great promise that glucaric acid is effective in the treatment and prevention of cancer. *SauerrÃ¼ben*, like sauerkraut, is a fermented vegetable widely available from northern Europe. SauerrA¼ben are a great source of vitamin C. *Miso* is high in vitamin K (vitamin K and as well as vitamin B_6). Apart from them it's also a good source of phosphorus, manganese and zinc. Zinc, in particular, is essential for proper immune system function. *Water kefir*, alternatively known as tibicos and Japanese water crystals, is a probiotic beverage similar to kombucha. Water kefir grains are a symbiotic culture of bacteria and yeasts including *Lactobacillus hilgardii*. *Moroccan preserved lemons* are naturally fermented, *Coconut kefir* is a probiotic beverage prepared from young coconut water and a starter culture. Probiotic giber beer, Sour pickles, Coconut milk yogurts are other common ones in market with beneficial bacteria, which play a critical role in body's ability to fully absorb the nutrients and for good health.

Table 1.2: Current status of Probiotic culture at Global level

Source	Manufacturer
L. rhamnosus GG	Valio
L. casel Shirota	Yakult
L. plantarum 299v	Probi AB
L. rhamnosus 271	Probi AB
L. casel DN-114001	Danisco
L. johnsonli La7	Nestle
L. acidophilus NCFM	Nestle
L. casel	Chr Hanse
L. reuteri	BioGala
B. animalis DN-173010	Danisco

Probiotics impact on diseases beyond gut

Allergy

The increasing prevalence of allergic disease has been linked to reduced microbial exposure in early life. Allergy is caused by an immune reaction that is out of all proportion to the antigenic stimuli. Classical allergy is a type I hypersensitivity reaction mediated by the interaction of (mast cells and eosinophils) coated with allergen-specific IgE and a cross-linking allergen. The physiological outcome is inflammation commonly displayed by urticaria, rhinitis, vomiting and diarrhoea, depending on the route of allergen entry. In extreme reactions anaphylactic shock can result that may lead to death (Kalliomaki and Isolauri, 2004; Viljanen *et al.*, 2005; Pouchard *et al.*, 2002; Rosenfeldt, 2004; Mastrandrea *et al.*, 2004; Pohjavuori *et al.*, 2004). Chronic allergic responses most commonly present themselves as asthma and eczema. All these symptoms are the consequence of an imbalanced immune system making an unsuitable response to an environmental or food antigen. On bacterial colonisation of the colon after birth the appropriate microbiological stimuli is essential to redress the balance of the skewed T-helper 2 immune response present in the newborn. This normal interaction between baby and microbes is thought to be compromised in the Western world, with a reduction in bifidobacteria and an increase in clostridial species, particularly in bottle-fed infants. Probiotics have recently been advocated for the prevention and treatment of allergic disease. The Finnish study of Kalliomaki was the first report to describe that the frequency of atopic dermatitis in neonates treated with *Lactobacillus rhamnosus* GG (LGG) was half that of the placebo (Kalliomaki *et al.*, 2003). Evidence found that babies considered at high risk for allergies who were given probiotics for six months following their births, cut their incidence of eczema by 40%. Another study showed that allergy-prone mothers who took probiotics during pregnancy were less likely to have children suffering from eczema. Probiotic bacteria continue to represent the most promising intervention for primary prevention of allergic disease, and well-designed definitive intervention studies should now be a research priority (Kirjaveinen *et al.*, 2003; Rosenfeldt *et al.*, 2003; Kalliomaki and Isolauri, 2004; Kopp and Salfeld, 2009).

Infections

Because probiotics help build up the immune system, it only makes sense that they also help fight infection. A seven-month study of more than 570 children in day care centers found that drinking a probiotic milk reduced the number and severity of respiratory infections and the need for antibiotics (Tagg and Dieksen, 2003; Golledge and Riley, 1996; Hatakka *et al.*, 2001). The intake of the probiotic combination, *L. gasseri* PA 16/8, *Bifidobacterium longum* SP 07/3, and *B. bifidum* MF 20/5, had no effect on the incidence of common cold infections, but significantly shortened duration of episodes, reduced the severity of symptoms, and led to increased numbers of cytotoxic, suppressor, and helperTcell counts (de Vrese *et al.*, 2005). Daily intake of *L. reuteri* was shown to reduce sick leaves related to gastrointestinal or respiratory tract diseases by 60% (Tubelius *et al.*, 2005).

Hyperlipidaemia

Globally, hyperlipidaemia causes almost double the number of deaths as those which are caused by cancer and 10 times as those which are caused by accidents. Probiotics have exhibited their potential for the improvement of the lipid profiles, St. Onge *et al.*, demonstrated the effect of fermented dairy products on the serum cholesterol, especially with selected strains of lactic acid bacteria, by increasing the utilization of cholesterol for the de novo synthesis of the bile acids. Study has reported that *L. casei* NCDC-19 (10^9 CFU*) and *Saccharomyces boulardii* (10^9 CFU) caused a 19 % decrease in the total serum cholesterol, whereas the LDL- cholesterol levels had decreased by 37% (St-Onge *et al.*, 2000) Another study which was conducted as a placebo-controlled study on hypercholesterolaemia- induced pigs (Yorkshire barrows), found that the probiotic fed group had 11.8% reduced total blood cholesterol (De Rodas *et al.*, 1996). In a study, it was demonstrated that 36 male sprague-dawley hypercholesterolaemic rats had 25% reduced serum cholesterol and significantly reduced VLDL, IDL and LDL in comparison to the controls, after receiving a supplementation of *L. acidophilus* ATCC 43121 (2×10^6 CFU/day) (Ibnou-Zekri *et al.*, 2003). In other study it was to observe the antioxidative effects of *Lactobacillus casei* on hyperlipidaemic rats, it was shown that the supplementation of *L. casei* significantly reduced the Malondialdehyde (MDA) levels, whereas the Superoxide Dismutase (SOD) and the glutathione peroxidase levels were increased both in the serum and the liver of these rats (Yong *et al.*, 2010). Thus, we can notice by these studies that probiotics reduce the lipid peroxidation and improve the lipid metabolism *in vivo*. One more study which was done on the *in vitro* cholesterol reduction abilities of probiotics, that the *L. fermentum* KC5b strain was able to maintain viability for two hours at pH 2 in bile acids and it was also able to remove a maximum of 14.8 mg of cholesterol per gram (dry weight) of cells from the culture medium (Pereira and Gibson, 2002).

Diabetes

There are several studies which suggest that individuals who have consumed a high fat diet over long periods have a poor inflammatory status which is related with the onset of diabetes in such people. A new strategy that can be employed in the prevention or delay of diabetes and the subsequent reduction in the incidence of hypertension could be the consumption of probiotics. Administration of dahi (an Indian fermented product) which contained *Lactobacillus acidophilus*, *L. casei* and *L. lactis* to fructose induced diabetic rats decreased the accumulation of glycogen in the liver of the rats. One study has also shown that Bifidobacteria can reduce the intestinal endotoxin levels and improve the mucosal barrier and thus reduce the systemic inflammation and subsequently reduce the incidence of diabetes (Al-Salami *et al.*, 2008a, b). In the same study they demonstrated that the pre-treatment of diabetic and healthy Wistar rats (n=10) with probiotics (75 mg/kg body weight) before the administration of a gliclazide suspension, resulted in an optimum control over hyperglycaemia, as well as, that it showed improved signs and symptoms in those diabetic animals. The authors reported a significant increase in the gliclazide uptake which was induced by probiotics. There are 2 reasons which have been mentioned for this; firstly, the probiotic treatment can restore the activity of the drug transporters and secondly, restoration of the disturbed gut flora which is associated with diabetes. Such findings point towards the beneficial effects of probiotics in treating diabetes, in synergism with other diabetes drugs, thereby reducing the incidence of diabetes related hypertension (Lye *et al.*, 2009; Tabuchi *et al.*, 2003).

Dental Caries

Bacterial interference with probiotic bacteria to support or restore diversity in the oral biofilm is enjoying a growing momentum to prevent and control dental caries (Twetman, 2012). Several randomized controlled clinical trials with microbiological and clinical endpoints have been published in recent years. One study reported on caries development in an animal model. Specific pathogen-free rats were inoculated with *S. mutans* and fed with a diet supplemented with a heat-killed probiotic strain (*L. paracasei* DSMZ16671) or placebo. The animals were sacrificed after 42 days, and a significant caries reduction (prevented fraction, 27%) was displayed in the DSMZ16671 group compared with the controls (Tanzer *et al.*, 2010). The authors concluded that the intervention appeared efficacious and safe. The other article demonstrated that probiotic *L. salivarius* LS1952R possessed an inherent cariogenic activity in rats, although to a somewhat lesser extent when compared with that in those superinfected with both the probiotic strain and *S. mutans* (Matsumoto *et al.*, 2005).

Bone Disorders

Bone is living tissue that provides shape and support for the body, as well as protection for some organs. Bone also serves as a storage site for minerals and provides the medium - marrow - for the development and storage of blood cells. Administration of *L. casei* suppressed cartilage destruction and suppressed mononuclear cells in filtrations into cartilage tissues and inflammation in the osteoarthritis (OA) models (So *et al.*, 2011). Inflammatory bowel disease causes bone loss, suggesting the possibility that even low levels of intestinal inflammation can affect bone health (Shiraji *et al.*, 2012; Redlich and Smolen, 2012; Etzel *et al.*, 2011; Klein, 2011; Chae *et al.*, 2001; Yarilina *et al.*, 2011; Zwerina *et al.*, 2007; Wei *et al.*, 2005; Roggia *et al.*, 2001; Li *et al.*, 2007; Lee *et al.*, 2006). The probiotic *L. reuteri* is known to have anti-inflammatory, specifically anti-TNF α, properties. Therefore a study was examined if adult male mice, moved from pathogen free facilities to standard animal facilities, would obtain gut and bone health benefits from *L. reuteri* treatment. Interestingly study reports that the administration of *L. reuteri* increased femoral and verterbral trabecular bone in the male mice (McCabe *et al.*, 2013). As most of the available treatments for osteoporosis function by inhibiting of bone resorption and have unwanted side effects, it is exciting that *L. reuteri* 6475 appears to impact bone health at least in part by promoting bone formation (Sjögren et al., 2012). Thus one can envision *L. reuteri* 6475 being utilized either alone or in combination with existing therapies to treat osteoporosis.

Selection of good probiotic supplement

- Safety - the microorganisms chosen as the components of a probiotic must be non-pathogenic and non-toxic

- Multistrain - a good quality probiotic contains several species of beneficial microorganisms in order to have an improved overall spectrum of activity within the gut and in a wider range of host species.

- Viability - a probiotic can only work if the microorganisms contained within the probiotic remain viable during storage of the product and through the gut to ensure colonisation of these microorganisms

- Minimum dose - the concentration of a probiotic must be such that inclusion rates provide 10^7 - 10^8 CFU per dose (that is 10 million - 100 million beneficial bacteria per dose.

- Quality assurance - it is essential that a probiotic has not become contaminated with any other microorganism other than the particular probiotic microorganisms chosen at any stage e.g.., fermentation, of the manufacturing process or during storage.

Global probiotic products market was estimated at $24.23 billion in 2011. The leading developers and suppliers of probiotic strains include Danisco (Denmark), Chr. Hansen (Denmark), and BioGaia (Sweden) (Bron and Kleerebezem, 2011; Hickson, 2013). The products of these companies are used by FMCG companies such as Nestle and Attune. Asia-Pacific is currently the largest probiotics market, owing to the Japanese market which introduced the concept to the world. Also, high awareness of the benefits of probiotic yoghurts and fermented milk has helped in increasing penetration of the market in APAC and European nations. The U.S. market is also growing rapidly due to the general affinity of the U.S. population towards the probiotic dietary supplements and the concept of preventive health care (Fontana *et al.*, 2013; Lemon *et al.*, 2012). The probiotic products supplement space available in the global market is listed in **Table 1.3**.

Table 1.3: Total 'Probiotic Therapeutic/Supplement' Space

	B. bifidum	B. breve	B. lactis	B. Longum	L. acidophilus	L. casei	L. helveticus	L. paracasei	L. plantarum	L. reuteri	L. rhamnosus	L. salivarius	LC. lactis	P. acidilactici	S. boulardii
Irritable bowel-related pain	x	x	x		x			x							x
Irritable bowel-related diarrhea					x										x
Antibiotic-associated diarrhea			x		x	x					x				x
Traveler's diarrhea						x									
Chemotherapy-induced diarrhea											x				
Radiation-induced diarrhea															
Viral or microorganism-induced diarrhea											x				x
Gas and flatulence					x										
Gastric/duodenal ulcer					x	x	x	x		x	x				
Chemotherapy-induced enterocolitis										x					
Ulcerative colitis														x	
Lactose intolerant					x	x									
Stress-induced gastrointestinal symptoms					x	x									
Bacterial overgrowth					x	x								x	
Recent use of antibiotic								x			x				
Enhance immune performance	x		x	x	x			x					x	x	
Vaginal bacterial infection					x					x	x				
Antimutagenic effects													x		
Radiation-induced intestinal injury and repair					x										
Common cold	x		x	x				x			x				
Pollen-related allergies			x	x	x										
Acne															x
Alcohol-induced liver injury	x							x							
Support immunity in lactating mothers and their children						x									
Pregnant or lactating						x									
Sleep quality							x								

B-Bifidobacterium, L-Lactobacillus, LC - Lactococcus, S - Saccharomycus, P - Pediococcus

Future Prospects and Conclusion

Emerging multidrug resistant pathogens are the main driving force behind the efforts to find an alternative solution as probiotics. The main goals are to decrease antibiotic consumption and fight the negative effects of antibiotic use. Moreover, it has been hypothesized that the sudden change in the intestinal microbiota that parallels the modern life practices of humans may have contributed to the rise in the incidence of autoimmune diseases. Probiotics have showed to possess antimutagenic, anticarcinogenic and hypocholesterolemic properties. The U.S. Food and Drug Administration accept probiotics as dietary supplement that are not subject to regulations made for other pharmaceuticals. The European authorities are currently considering the required minimal colony forming unit counts of probiotics per preparation. The European Food Safety Authority has recently adopted a qualified presumption of safety approach for microorganism use in foods and feeds; however, no definitive guideline exists for commercially used probiotics. Although many trials shed light on our understanding of the mechanisms of action and the beneficial effects of probiotics, it is very hard to draw an exact conclusion from these trials and meta-analyses because of the heterogeneity of patient populations, probiotic strains, dosages, and commercial preparations. There is still a great need of well-designed, placebo controlled, sufficiently powered studies that will reflect the actual role of probiotics.The ecology shared by the host and gut microflora should now be considered a new player that can be manipulated, using pharmacological and nutritional approaches, to control physiological functions and pathological outcomes. What now remains is to demonstrate in depth the molecular, cellular events; connection between the intestinal microflora and above described gut mediated various diseases.

In conclusion, the present review shows, that intestinal microflora component can produce molecules that are contributory to host health. Indeed, the intestinal microflora can act, *via* the production of specific molecules, not only as an enemy of the host, but also as a new friend to improve the defense mechanisms of the host. Currently probiotic's confer huge benefits to the gut but the future holds huge possibilities that these benefits may be extended to above discussed directly or indirectly gut associated other disorders. Over all this review highlights a promising new area of investigation for the future in the areas beyond to gut.

References

Adlerberth, I. 2008. Factors influencing the establishment of the intestinal microbiota in infancy.Nestle Nutr. Workshop Ser. Pediatr. Programme. **62:**13-29; discussion 29-33.

Agarwal, R., Sharma, N. and Chaudhry, R. 2003. Effects of oral *Lactobacillus* GG on enteric microflora in low-birth-weight neonates. J. Pediatr. Gastroenterol. Nutr. **36:** 397-402.

Alander, M., Korpela, R., Saxelin, M., Vilponnen-Salmela, T., MattilaSandholm, T. and vonWright, A. 1997. Recovery of *Lactobacillus rhamnosus*, GG from human colonic biopsies. Lett. Appl. Microbiol, **24:** 361-364.

Al-Salami, H., Butt, G., Fawcett, J.P., Tucker, I.G., Golocorbin-Kon, S. and Mikov, M. 2008a. Probiotic treatment reduces blood glucose levels and increases systemic absorption of gliclazide in diabetic rats. Eur. J. Drug Metab. Pharmacokinet. **33(2):** 101-6.

Al-Salami, H., Butt, G., Tucker, I., Skrbic, R., Golocorbin-Kon, S. and Mikov, M. 2008b. Probiotic pre-treatment reduces gliclazide permeation (*ex vivo*) in healthy rats but increases it in diabetic rats to the level seen in untreated healthy rats. Arch. Drug Inf. l; **1(1):** 35-41.

Alverdy, J, Zaborina, O. and Wu, L. 2005. The impact of stress and nutrition on bacterial-host interactions at the intestinal epithelial surface. Curr. Opin. Clin. Nutr. Metab. Care. **8(2):** 205-9.

Beaugerie, L. and Petit, J.C.. 2004. Microbial-gut interactions in health and disease. Antibiotic-associated diarrhoea. Best Pract. Res. Clin. Gastroenterol. **18** (2): 337–52.

Bernardez, L.A., de Andrade Lima, L.R., de Jesus, E.B., Ramos, C.L. and Almeida, P.F. 2013. A kinetic study on bacterial sulfate reduction. Bioprocess Biosyst Eng. 1 Epub ahead of print]

Bernet, M .F., Brassart, D ., N eeser, JR. and Servin, A L. 1994. *Lactobacillus acidophilus* LA 1 binds to cultured human intestinal cell lines and inhibits cell attachment and cell invasion by enterovirulent bacteria. Gut **35**: 483-48.

Bettelheim, K.A., Breadon, A., Faiers, M.C., O'Farrell, S.M. and Shooter, R.A. 1974. The origin of O serotypes of *Escherichia coli* in babies after normal delivery. J. Hyg. (Lond) **72 (1)**: 67–70.

Björkstén, B., Sepp, E., Julge, K., Voor, T. and Mikelsaar, M. 2001. Allergy development and the intestinal microflora during the first year of life. J. Allergy Clin. Immunol.**108** (4): 516–20.

Bron, P.A. and Kleerebezem, M. 2011. Engineering lactic acid bacteria for increased industrial functionality. Bioeng. Bugs. **2(2)**: 80-7.

Buddington, R.K. and Sangild, P.T. 2011. Companion animals symposium: development of the mammalian gastrointestinal tract, the resident microbiota, and the role of diet in early life. J. Anim. Sci. **89(5)**: 1506-19.

Buts, J.P., Bernasconi, P., Vaerman, J.P. and Dive, C. 1990. Stimulation of secretory IgA and secretory component of immunoglobulins in small intestine of rats treated with *Saccharomyces boulardii*. Dig. Dis. Sci. **35**: 251-256.

Carman, R.J., Simon, M.A., Fernández, H., Miller, M.A. and Bartholomew, M.J. 2004. Ciprofloxacin at low levels disrupts colonization resistance of human fecal microflora growing in chemostats". Regul. Toxicol. Pharmacol. **40 (3)**: 319–26.

Chae, H. J., S. C. Kim, S. W. Chae, N. H. An, H. H. Kim, Z. H. Lee, and H. R. Kim. 2001. Blockade of the p38 mitogen-activated protein kinase pathway inhibits inducible nitric oxide synthase and interleukin-6 expression in MC3T3-E1E-1 osteoblasts. Pharmacol. Res. **43(3)**: 275-283.

Christensen, H.R., Frokiaer, H. and Pestka, J.J. 2002. *Lactobacilli* differentially modulate expression of cytokines and maturation surface markers in murine dendritic cells. J. Immunol. **168**: 171-178.

Conroy, M.E., Shi, H.N. and Walker, W.A. 2009. The long-term health effects of neonatal microbial flora. Curr. Opin. Allergy Clin. Immunol. **9(3)**: 197-201.

Conway, P.L., Goldin, B.R. and Gorbach, S.L. 1987. Survival of lactic acid bacteria in the human stomach and adhesion to intestinal cells. J. Dairy Sci. **70**: 1-12.

Dahan, S., Dalmasso, G., Imbert, V., Peyron, J.F., Rampal, P. and Czerucka, D. 2003. *Saccharomyces boulardii* interferes with enterohemorrhagic *Escherichia coli*-induced signaling pathways in T84 cells. Infect. Immun. **71**: 766-773.

Davidson, J.N. and Hirsch, D.C. 1976. Bacterial competition as a means of preventing diarrhea in pigs. Infect. Immun.**13**: 1773-1774.

De Rodas, B.Z., Gilliland, S.E. and Maxwell, C.V. 1996. Hypocholesterolemic action of *Lactobacillus acidophilus* ATCC and calcium in swine with hypercholesterolemia induced by diet. J. Dairy Sci. **79**: 2121–28.

De Vrese, M., Rautenberg, P., Laue, C., Koopmans, M., Herremans, T. and Schrezenmeir, J. 2004. Probiotic bacteria stimulate virus-specific neutralizing antibodies following a booster polio vaccination. Eur. J. Nutr. **44(7):** 406-13.

De Vrese, M., Winkler, P., Rautenberg, P., Harder, T., Noah, C., Laue, C., Ott, S., Hampe, J., Schreiber, S., Heller, K. and Schrezenmeir, J 2005. Effect of *Lactobacillus gasseri* PA 16/8, *Bifidobacterium longum* SP 07/3, *B. bifidum* MF 20/5 on common cold episodes: a double blind, randomized, controlled trial. Clin. Nutr. **24(4):** 481-91.

Etzel, J.P., Larson, M.F., Anawalt, B.D., Collins, J. and Dominitz, J.A. 2011. Assessment and management of low bone density in inflammatory bowel disease and performance of professional society guidelines. Inflamm. Bowel Dis. **17(10):** 2122-9.

Fontana, L., Bermudez-Brito, M., Plaza-Diaz, J., Munoz-Quezada, S. and Gil, A. 2013. Sources, isolation, characterisation and evaluation of probiotics. Br. J. Nutr. **109** Suppl 2:S35.

Freestone, P.P., Haigh, R.D. and Lyte, M. 2007a. Specificity of catecholamine-induced growth in *Escherichia coli* O157:H7, *Salmonella enterica* and *Yersinia enterocolitica*. FEMS Microbiol. Lett. **269(2):** 221-8.

Freestone, P.P., Haigh, R.D. and Lyte, M. 2007b. Blockade of catecholamine-induced growth by adrenergic and dopaminergic receptor antagonists in *Escherichia coli* O157:H7, *Salmonella enterica* and *Yersinia enterocolitica*. BMC Microbiol. **30:** 7:8.

Goldin, B.R., Gorbach, S.L., Saxelin, M., Barakat, S., Gualtieri, L. and Salminen, S. 1992. Survival of *Lactobacillus* species (Strain GG) in human gastrointestinal tract. Dig. Dis. Sci. **37:** 121-128.

Golledge, C.L. and Riley, T.V. 1996. "Natural" therapies for infectious diseases. Med. J. Aust, **164:** 94-95.

Guarino, A., Wudy, A., Basile, F., Ruberto, E. and Buccigrossi, V. 2012. Composition and roles of intestinal microbiota in children. J. Matern. Fetal Neonatal Med. 2012 Apr;25 Suppl 1:63-6.

Guarner, F. and Malagelada, J.R. 2003. Gut flora in health and disease. Lancet **361** (9356): 512–9.

Hahm Ki Baik, 2012. High concentrated probiotics improve inflammatory bowel diseases better than commercial concentration of probiotics. J. Food & Drug Anal. **20:** 292-295.

Hart, A. and Kamm, M.A. 2002. Mechanisms of initiation and perpetuation of gut inflammation by stress. Aliment Pharmacol Ther. **16(12):** 2017-28.

Hatakka, K., Savilahti, E., Ponka, A., Meurman, J.H., Poussa, T. and Nase, L. 2001. Effect of long term consumption of probiotic milk on infections in children attending day care centres: double blind, randomized trial. BMJ **322:** 1327-1331.

Henckaerts, L., Pierik, M., Joossens, M., Ferrante, M., Rutgeerts, P. and Vermeire, S. 2007. Mutations in pattern recognition receptor genes modulate seroreactivity to microbial antigens in patients with inflammatory bowel disease. Gut **56(11):** 1536-42.

Hickson, M. 2013. Examining the evidence for the use of probiotics in clinical practice. Nurs. Stand. 20-26; **27(29):** 35-41.

Ibnou-Zekri, N., Blum. S., Schiffrin, E.J. and von der Weid, W.T. 2003. Divergent patterns of colonization and immune response elicited from two intestinal *Lactobacillus* strains that display similar properties *in vitro*. Infect. Immun. **71:**428–36.

Jahn, H.U., Ullrich, R., Schneider, T., Liehr, R.M., Schieferdecker, H.L., Holst, H. and Zeitz, M. 1996. Immunological and trophical effects of *Saccharomyces boulardii* on the small intestine in healthy human volunteers. Digestion **57:** 95-104.

Kaila, M., Isolauri, E., Saxelin, M., Arvilommi, H. and Vesikari, T. 1995. Viable versus inactivated lactobacillus strain GG in acute rotavirus diarrhoea. Arch. Dis. Child **72:** 51-53.

Kaila, M., Isolauri, E., Soppi, E., Virtanen, E., Laine, S. and Arvilommi, H. 1992. Enhancement of the circulating antibody secreting cell response in human diarrhea by a human *Lactobacillus* strain. Pediatr. Res. **32:** 141-144.

Kalliomaki, M.A. and Isolauri, E. 2004. Probiotics and down-regulation of the allergic response. Immunolgy Allergy Clin. North Am. **24:** 739-752.

Kalliomaki, M., Salminen. S., Poussa. T., Arvilommi, H. and Isolauri, E. 2003. Probiotics and prevention of atopic disease: 4-year follow-up of a randomized placebo-controlled trial. Lancet **361:** 1869-1871.

Kirjavainen, P.V., Salminen, S.J. and Isolauri, E. 2003. Probiotic bacteria in the management of atopic disease: underscoring the importance of viability. J. Pediatr. Gastroenterol. Nut. **36:** 223-227.

Kleeman, E.G. and Klaenhammer, T.R. 1982. Adherence of *Lactobacillus* species to human fetal intestinal cells. J. Dairy Sci. **65:** 2063-2069.

Klein, G.L. 2011. Gut-bone interactions and implications for the child with chronic gastrointestinal disease. J. Pediatr. Gastroenterol. Nutr. **53(3):** 250-4.

Könönen, E. 2000. Development of oral bacterial flora in young children. Ann. Med. **32(2):** 107-12.

Kopp, M.V. and Salfeld, P. 2009. Probiotics and prevention of allergic disease. Curr. Opin. Clin. Nutr. Metab. Care **12(3):** 298-303.

Lebba, V., Nicoletti, M. and Schippa, S. 2012. Gut_microbiota and the immune system: an intimate partnership in health and disease. Int. J. Immunopathol. Pharmacol. **25(4):** 823-33

Lee, S.K., Kadono, Y., Okada, F., Jacquin, C., Koczon-Jaremko, B., Gronowicz, G., Adams, D.J., Aguila, H.L., Choi, Y. and Lorenzo, J.A. 2006. T lymphocyte-deficient mice lose trabecular bone mass with ovariectomy. J. Bone Miner. Res. **21(11):** 1704–12.

Lemon, K.P., Armitage, G.C., Relman, D.A. and Fischbach, M.A. 2012. Microbiota-targeted therapies: an ecological perspective. Sci. Transl. Med..**4(137):** 137rv5.

Lewis, K. and McKay, D.M. 2009. Metabolic stress evokes decreases in epithelial barrier function. Ann. N Y Acad. Sci. **1165:** 327-37.

Li, Y., Toraldo, G., Li, A., Yang, X., Zhang, H., Qian, W.P. and Weitzmann, M.N. 2007. B cells and T cells are critical for the preservation of bone homeostasis and attainment of peak bone mass *in vivo*. Blood. **109(9)**: 3839–48.

Lye, H.S., Kuan, C.Y., Ewe, J.A., Fung, W.Y. and Liong, M.T. 2009. The improvement of hypertension by probiotics: effects on cholesterol, diabetes, renin, and phytoestrogens. Int. J. Mol. Sci. **10(9)**: 3755-75.

Mack, D.R.., Michail, S., Wei, S., Macdougal, L. and Hollingsworth, M.A. 1999. Probiotics inhibit enteropathogenic *E. coli* adherence *in vitro* by inducing intestinal mucin gene expression. Am. J. Physiol. **39**: G941-G950.

Mackie, R.I., Sghir, A. and Gaskins, H.R. 1999. Developmental microbial ecology of the neonatal gastrointestinal tract. Am. J. Clin. Nutr. **69(5)**: 1035S-1045S.

Mainous, M.R., Block, E.F. and Deitch, E.A. 1994. Nutritional support of the gut: how and why. New Horiz. **2(2)**: 193-201.

Mastrandrea, F., Coradduzza, G., Sero, G., Minardi, A., Manelli, M., Ardito, S. and Muratire, L. 2004. Probiotics reduce the CD34+ hemopoietic precursor cell increased traffic in allergic subjects. Allerg. Immunol. **36**: 118-122.

Matsumoto, M., Tsuji, M., Sasaki, H., Fujita, K., Nomura, R., Nakano, K., Shintani, S., and Ooshima, T. 2005. Cariogenicity of the probiotic bacterium *Lactobacillus salivarius* in rats. Caries Res. **39(6)**: 479-83.

McCabe, L.R., Irwin, R., Schaefer, L. and Britton, R.A. 2013. Probiotic use decreases intestinal inflammation and increases bone density in healthy male but not female mice. J. Cell Physiol. doi: 10.1002/jcp.24340.

Mikami, K., Takahashi, H., Kimura, M., Isozaki, M., Izuchi, K., Shibata, R., Sudo, N., Matsumoto, H. and Koga, Y. 2009. Influence of maternal bifidobacteria on the establishment of bifidobacteria colonizing the gut in infants. Pediatr. Res. **65(6)**: 669-74.

Pereira, D.I.A. and Gibson, G.R. 2002. Cholesterol assimilation by lactic acid bacteria and bifidobacterial isolated from the human gut. Appl. Environ. Microbiol. **68(9)**: 4689–93.

Park, J.H., Um, J.I., Lee, B.J., Goh, J.S., Park, S.Y., Kim, W.S., and Kim, P.H. 2002. Encapsulated *Bifidobacterium bifidum* potentiates intestinal IgA production. Cell Immunol. **219**: 22-27.

Petrof, E.O., Kojima, K., Ropeleski, M.J., Musch, M.W., Tao, Y., De Simone, C. and Chang, E.B. 2004. Probiotics inhibit nuclear factor-kappa B and induce heat shock proteins in colonic epithelial cells through proteasome inhibition. Gastroenterology **127**: 1474-1487.

Pohjavuori, E., Viljanen, M., Korpela, R., Kiutunen, M., Tittanen, M., Vaarala, O. and Savilahti, E. 2004. *Lactobacillus* GG effect in increasing IFNgamma production in infants with cow's milk allergy. J. Allergy Clin. Immunol. **114**: 131-136.

Pouchard, P., Gosset, P., Grangette, C., Andre, C., Tonnel, A.B., Pestel, J. and Mercenier, A. 2002. Lactic acid bacteria inhibit TH2 cytokine production by mononuclear cells from allergic patients. J. Allergy Clin. Immunol.**110**: 617-623.

Redlich, K. and Smolen, J.S. 2012. Inflammatory bone loss: pathogenesis and therapeutic intervention. Nat. Rev. Drug Discov. **11(3)**: 234-50.

Resta, S.C. 2009. Effects of probiotics and commensals on intestinal epithelial physiology: implications for nutrient handling. J. Physiol. **587**: 4169-4174.

Rigothier, M.C., Maccanio, J. and Gayral, P. 1994. Inhibitory activity of *Saccharomyces* yeasts on the adhesion of *Entamoeba histolytica* trophozoites to human erythrocytes *in vitro*. Parasitol. Res. **80**: 10-15.

Roggia, C., Gao, Y., Cenci, S., Weitzmann, M.N., Toraldo, G., Isaia, G. and Pacifici, R. 2001. Up-regulation of TNF-producing T cells in the bone marrow: a key mechanism by which estrogen deficiency induces bone loss *in vivo*. Proc. Natl. Acad. Sci.USA **98(24)**: 13960–5.

Rosenfeldt, V., Benfeldt, E. and Nielsen, S.D. 2003. Effect of probiotic *Lactobacillus* strains in children with atopic dermatitis. J. Allergy Clin. Immunol. **111**: 389-395.

Rosenfeldt, V., Benfeldt, E., Valerius, N.H. Paerregaard, A. and Michaelsen, K.F. 2004. Effect of probiotics on gastrointestinal symptoms and small intestinal permeability in children with atopic dermatitis. J. Pediatr. **145**: 612-616.

Salminen, S. and Isolauri, E. 2008. Opportunities for improving the health and nutrition of the human infant by probiotics. Nestle Nutr. Workshop Ser. Pediatr. Program. **62**: 223-33; discussion 233-7.

Sanders, M.E. 2008. Probiotics: definition, sources, selection, and uses. Clin. Infect. Dis. **46** Suppl 2: S58-61.

Sanderson, I.R. and Walker, W.A. 2007. TLRs in the Gut I. The role of TLRs/Nods in intestinal development and homeostasis. Am. J. Physiol. Gastrointest. Liver Physiol. **292(1)**: G6-10.

Savage, D.C.1977. Microbial ecology of the gastrointestinal tract". Annual Review of Microbiology **31**: 107–133.

Sears, C.L. 2005. A dynamic partnership: celebrating our gut flora. Anaerobe **11** (5): 247–51.

Shi, H.N. and Walker, A. 2004. Bacterial colonization and the development of intestinal defenses. Can. J. Gastroenterol. **18**: 493-500.

Shibolet, O. and Podolsky, D.K. 2007. TLRs in the Gut. IV. Negative regulation of Toll-like receptors and intestinal homeostasis: addition by subtraction. Am. J. Physiol. Gastrointest. Liver Physiol. **292(6)**: G1469-73.

Shirazi, K.M., Somi, M.H., Rezaeifar, P., Fattahi, I., Khoshbaten, M. and Ahmadzadeh, M. 2012. Bone density and bone metabolism in patients with inflammatory bowel disease. Saudi J. Gastroenterol. **18(4)**: 241-7.

Sjögren, K., Engdahl, C., Henning, P., Lerner, U.H., Tremaroli, V., Lagerquist, M.K., Bäckhed, F. and Ohlsson, C. 2012. The gut microbiota regulates bone mass in mice. J. Bone Miner. Res. **27(6)**: 1357-67.

So, J.S., Song, M.K., Kwon, H.K., Lee, C.G., Chae, C.S., Sahoo, A., Jash, A., Lee, S.H., Park, Z.Y. and Im, S.H. 2011. *Lactobacillus casei* enhances type II collagen/ glucosamine-mediated suppression of inflammatory responses in experimental osteoarthritis. Life Sci. 14; **88(7-8): 358**-66.

Steidler, L., Hans, W., Schotte, L., Neirynck, S., Obermeier, F. and Falk, W. 2000. Treatment of murine colitis by *Lactobacillus lactis* secreting interleukin-10. Science, **289**: 1352-1355.

Steinhoff, U. 2005. Who controls the crowd? New findings and old questions about the intestinal microflora. Immunol. Lett. **99 (1)**: 12–6.

St-Onge, M.P., Farnworth, E.R. and Jones, P.J.H. 2000. Consumption of fermented and nonfermented dairy products: effects on cholesterol concentrations and metabolism. Am. J. Clin. Nutr.; **71**: 674–81.

Tabuchi, M., Ozaki, M., Tamura, A., Yamada, N., Ishida, T., Hosoda, M. and Hosono, A. 2003. Antidiabetic effect of *Lactobacillus* GG in streptozotocin-induced diabetic rats. Biosci. Biotechnol. Biochem. **67**: 1421–1424.

Tagg, J.R. and Dierksen, K.P.2003. Bacterial replacement therapy: adapting "germ warfare" to infection prevention. Trends Biotechnol. **21**: 217-223.

Tanzer, J.M., Thompson, A., Lang, C., Cooper, B., Hareng, L., Gamer, A., Reindl, A. and Pompejus, M. 2010. Caries inhibition by and safety of *Lactobacillus paracasei* DSMZ16671. J. Dent. Res. **89(9)**: 921-6.

Tubelius, P., Stan, V. and Zachrisson, A. 2005. Increasing work-place healthiness with the probiotic *Lactobacillus reuteri*: a randomised, double-blind placebo-controlled study. Environ. Health 2005 Nov 7; **4**:25.

Twetman, S. 2012. Are we ready for caries prevention through bacteriotherapy? Braz. Oral Res. **26** Suppl 1: 64-70.

Vedantam, G. and Hecht, D.W. 2003. Antibiotics and anaerobes of gut origin. Curr. Opin. Microbiol. **6 (5)**: 457–61.

Viljanen, M., Savilahti, E., Haahtela, T., Juntunen-backman, K., Korpela, R., Poussa, T., Tuure, T. and Kuitunen, M. 2005. Probiotics in the treatment of atopic eczema/dermatitis syndrome in infants: a double-blind placebo-controlled trial. Allergy **60**: 494-500.

Wei, S., Kitaura, H., Zhou, P., Ross, F.P. and Teitelbaum, S.L. 2005. IL-1 mediates TNF-induced osteoclastogenesis. J. Clin. Invest. **115(2)**: 282–90.

Wilson, K.H. and Perini, I. 1988. Role of competition for nutrients in suppression of *Clostridium difficile* by the colonic microflora. Infect. Immun. **56**: 2610-2614.

Wollowski, I., Rechkemmer, G. and Pool-Zobel, B.L.2001. Protective role of probiotics and prebiotics in colon cancer. Am J. Clin. Nutr.**73**: 451S-455S.

Wynne, A.G., McCartney, A.L., Brostoff, J., Hudspith, B.N. and Gibson, G.R. 2004. An *in vitro* assessment of the effects of broad-spectrum antibiotics on the human gut microflora and concomitant isolation of a *Lactobacillus plantarum* with anti-Candida activities. Anaerobe **10 (3)**: 165–9.

Yarilina, A., Xu, K., Chen, J. and Ivashkiv, L.B. 2011. TNF activates calcium-nuclear factor of activated T cells (NFAT)c1 signaling pathways in human macrophages. Proc. Natl. Acad. Sci. USA **108(4):** 1573–8.

Yong, Z., Ruiting, Du, Lifeng, W. and Heping, Z. 2010. The antioxidative effects of probiotic *Lactobacillus casei* on the hyperlipidemic rats. Eur. Food Res. Technol. **231:** 151–58.

Zwerina, J., Redlich, K., Polzer, K., Joosten, L., Kronke, G., Distler, J., Hess, A., Pundt, N., Pap, T., Hoffmann, O., Gasser, J., Scheinecker, C., Smolen, J.S., van den Berg, W. and Schett, G. 2007. TNF-induced structural joint damage is mediated by IL-1. Proc. Natl. Acad. Sci. USA **104(28):** 11742–11747.

Chapter 2

Molecular Markers: An Array of Technological Advancements and Their Applications in Crop Improvement

A. Chandra Sekhar, P. Chandra Obul Reddy, T. Krishnaveni,
P. Ramesh and K. V. N. Ratnakar Reddi

Usually most of the breeding programmes engage, the synthesis of new and improved genotypes that relies upon the processes of recombination and segregation that occur in the progenies of heterozygous individuals. A major challenge for the plant breeder, for that matter for any breeder - is to identify the desirable recombinant phenotype among the segregating population. The advent of molecular markers and their use for genetic map construction has solved the pains of a breeder to select the desired individual of choice with in the segregation population. Genetic markers were originally used in genetic mapping to determine the order of genes along chromosomes. Advances in molecular markers and marker mapping technologies has shown tremendous thrusting effect and penetrated its way into all areas of modern biology, from molecular genetic approach of systematics, evolution, to genomics assisted breeding and from transgenics to developmental biology. Different approaches (including association studies) have recently been adopted for the functional characterization of allelic variation in plants and to identify sequence motifs affecting phenotypic variation. Development and application of molecular markers derived from genes, commonly called genic markers / functional markers, is gaining momentum in plant genetics and breeding. Here we describe a variety of molecular markers that were being used for various purposes including diversity analysis, molecular mapping to association studies.

Introduction

In most plant breeding programmes, the synthesis of new and improved genotypes relies upon the processes of recombination and segregation that occur in the progenies of heterozygous individuals. A major challenge for the plant breeder is to identify the desirable recombinant phenotype among the segregating population. Various scientists have suggested marker based selection strategies to improve both the speed and precision of plant breeding programmes. In a plant breeding context, many morphological markers have undesirable effects on plant phenotype and their use in crop improvement has been limited. Developments in the electrophoretic separation of proteins and in the exploitation of recombinant DNA technology have dramatically increased the number of genetic markers available for use in plant breeding. Molecular genetics has reached to the new heights via molecular markers use in various fields of biological sciences. Molecular markers represent one of the most powerful tools for the analysis of whole genomes. Molecular marker technology has developed rapidly over the last decade, that include a variety of markers viz., morphological markers, Isozyme (or) Biochemical markers, RFLPs (Restriction fragment length polymorphism), RAPDs (Random amplified polymorphic DNA), SSRs (Simple sequence repeats), AFLPs (Amplified fragment length polymorphisms), SCARs (Sequence characterized amplified regions), AP-PCRs (Arbitrarily primed-PCRs), CAPSs (Cleaved amplified polymorphic sequences), SNPs (Single nucleotide polymorphisms), S-SAPs (Sequence-specific amplification polymorphisms), DArT (diversity arrays technology), EST-SSRs (Expression Sequence Tag based simple sequence repeats), TRAPs (target region amplification polymorphisms) etc., are some of the popular marker systems that are being used for various applications in variety of organisms in modern genetic analysis.

What are molecular markers and why?

The idea and concept of the markers begins from human evolution and domestication of plants and animals, which predominantly, selection based on desired phenotype. Continuous selection and conventional breeding efforts by farmers and breeders are responsible for sustained global ability for adequate food and fiber from domesticated plant species. The breeding efforts have been based primarily on the availability of genetic variation in germplasm collections and wild relatives. During the first half of the last century, studies on genetic variability were largely based on gross morphological, anatomical and behavioural features. Gregor Mendel was the one who systematically studied and proposed the Laws of Heredity based on pure phenotype-based studies in garden pea. Further in twentieth century, Morghan and group proposed and proved theory of genetic linkage based on studies of various mutant phenotypes (morphological marker) in Drosophila. Subsequently, after 1960s it was possible to look at more subtle variation in genes using a variety of molecular technologies. Since 1980s, it has become possible to explore and understand the extent of variation at the level of DNA itself. No matter at what level one is studying, variability-from subtle variations in DNA to an observable change in phenotype-distinct differences can be used as genetic markers to identify traits of interest and map to specific chromosomal regions. Since 1990's molecular markers involving macromolecules like DNA and proteins have been used as complimentary to phenotypic parameters.

Recent advances during last two decades has changed entire scenario of molecular markers through development in techniques, and their applications in cellular and molecular biology during the last two decades, have provided breeders with new tools such as i) recombinant DNA technology for genetic manipulation and moving genes across sexual barriers of plant genomes through tissue culture and ii) molecular marker technology, for enhancing speed and precision in selection. While genetic transformation is dependent on availability of candidate genes, tissue specific promoters and efficient transformation protocols, marker-assisted selection depends on availability of new markers and high-density linkage maps. Molecular marker techniques emerged as new reliable tools that are neither affected by the surrounding environment nor by growth stage of the plant (as in case of morphological characters). These can be applied for organizing and characterizing germplasm, identification of cultivars, assisting in the selection of parents for hybridization and reducing the number of accessions needed to ensure sampling a broad range of genetic variability.

Advances in marker technology have not only revolutionized the methods of genetic analysis but also helped greatly to accelerate breeding programmes for improvement of traits of complex inheritance. The most fundamental of these markers are DNA markers that detect differences in the genetic information carried by different individuals. The well-known application of these markers is in diversity analysis, construction of genetic linkage maps, which are used to identify and tag the QTLs that govern agronomically important traits. In order to be used effectively, a genetic or molecular marker associated with the trait of interest must be polymorphic between parents and must stably inherit.

Descriptions of Types of Molecular Markers

A molecular marker is defined as a particular segment of DNA that is representative of the differences at the genome level. Molecular markers may or may not correlate with phenotypic expression of a trait. Molecular markers offer numerous advantages over conventional phenotype based alternatives as they are stable and detectable in all tissues regardless of growth, differentiation, development, or defense status of the cell are not confounded by the environment, pleiotropic and epistatic effects.

An ideal molecular marker shares some of the following characters:

(1) The marker should be polymorphic between the two individuals of the choice

(2) Even distribution throughout the genome

(3) The marker should be simple to study, inexpensive and quick to assess

(4) The marker should require less quantity of the starting material (either tissue or DNA) for its analysis

(5) The marker should require no prior sequence information of the genome

(6) The marker should be highly reproducible and show strong heritability from generation to generation

(7) A marker should have close association with desired phenotype so as to follow itself rather than phenotype everytime

But, to date there is no perfect / ideal marker system that has all qualities and to be used for every purpose.

Various molecular marker systems differ from one another with respect to its total genome coverage, locus specificity, reproducibility across laboratories, stable inheritance, ease with which it can be detected and total cost effectiveness per sample analysis (Table 1). Depending on the need, there are enormous modifications in the techniques that are being used to study the genome, leading to a second generation of advanced molecular markers. In present paper we have tried to incorporate all technical advancements in the area of molecular markers, right from morphological markers that are being used to study inheritance to advanced functional markers as SNP's / EST-PCR markers that are being applied in area of plant sciences. Markers can be divided into morphological markers and molecular markers and further molecular markers can be divided into two categories: protein markers and DNA based markers.

Morphological Markers

These are first generation markers, which have been extensively used to develop maps in plants. The inheritance of these markers can be monitored visually without specialized biochemical or molecular techniques. The traits controlled by a single locus can be used as genetic markers provided their expression is reproducible over a range of environments (e.g., Internode colour and leaf sheath colour inheritance in foxtail millet) [**Fig. 2.1**]. But large effects on phenotype caused by most of morphological markers and the mask effect of minor gene(s) that are few or more in number make them undesirable in breeding programmes (Tanksley *et al.*, 1989).

The limitations of phenotype based genetic markers led to the development of more general and useful direct DNA based markers that became known as molecular markers.

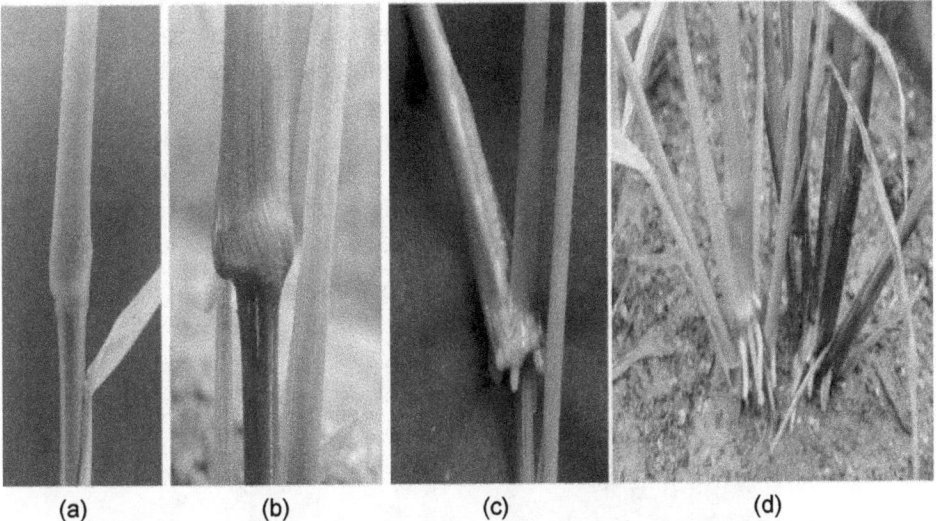

   (a)   (b)   (c)   (d)
Fig. 2.1: Foxtail Millet Internode Colour and Leaf Sheath Colour inheritance.
(a) Female Parent (b) Male parent (c) F1 Plant and (d) F2 Segregating population in field

Protein Markers

These are also known as isozyme markers. Until recently, isozyme or biochemical markers have been used very successfully in certain aspects of plant breeding and genetics as nearly-neutral genetic markers for variability analysis and construction of framework genome maps (Tanksley and Orton, 1983). Allozymes have been the most important type of genetic marker in forestry and are used in many species for many different applications (Conkle, 1981; Adams, 1992). The number of genetic markers provided by isozyme assays is insufficient. Further, stage dependant expression of isozyme loci proved a serious limitation to use of isozymes as genetic markers for assessment of genetic diversity in breeding material (**Fig. 2.2**).

Protein markers, including seed storage proteins, structural proteins, and isozymes were among the first group of molecular markers exploited for genetic diversity assessment and genetic linkage map development. They are the basis for a newly emerging research area called proteomics. They also provide some of the most cost-effective tools for data point generation, especially when iso-electric focusing equipment is used to precisely distinguish between very similar versions of proteins. The major limitations of these markers are

- that much of the genome (including much of the most polymorphic portions of it that are less subject to evolutionary' restrictions) does not code for proteins,

- different biochemical procedures are required to visualise allelic differences for enzymes having different functions, and

- many proteins are several post-transcnptional steps removed from underlying DNA sequence polymorphism and thus can mask variation present at that level (e.g., differences in tri-nucleotide sequences coding for the same amino acid, interon sequences that are post-transcriptionally removed from the mRNA and post-translational modification can all contribute to reduced polymorphism expression at the protein level compared to that at the DNA level).

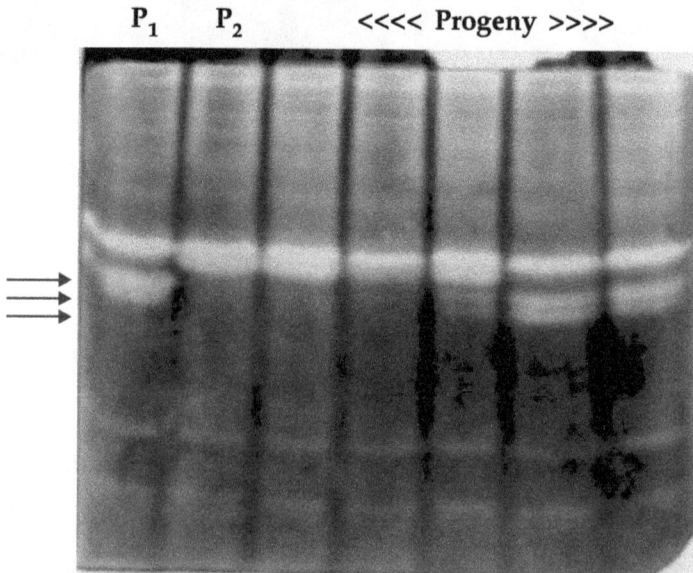

Fig. 2.2: Rice Ascorbate peroxidase isozyme inheritance. Lane 1- P_1 Female parent; Lane 2 - P_2 Male parent and Lanes 3 to 7 F_2 progeny (3 & 4 male parent type; 5 – heterozygote and 5 & 7 female parent type in segregating population)

DNA Markers

In 1980's, Botstein *et al.* (1980) reported that it could be possible to find numerous genetic markers by studying the DNA molecule itself. DNA markers are ubiquitous and innumerable, discrete and nondeleterious, inherited in Mendelian fashion, unaffected by environment, and are free of epistatic interactions and pleiotropic effects (Beckman and Soller, 1986; Tanksley, 1993; Tanksley *et al.*, 1989). Phenotypic neutrality has provided an easier and unbiased way to detect linkage between segregating markers and the polygenes (QTLs) and to estimate phenotypic effect of each polygene (QTL) without interference by the marker locus.

Most points on molecular marker-based genetic linkage maps are anonymous DNA polymorphisms (e.g., restriction fragment length polymorphism (RFLP), random amplified polymorphic DNA (RAPD), amplified fragment length polymorphism (AFLP), and micro-satellite markers) and do not correspond to any gene of known function. However, some molecular markers (including coding DNA (cDNA) and expressed sequence tag (EST) markers, as well as the protein markers described above) do pinpoint individual genes. Anonymous DNA markers are generated by a wide variety of techniques, differing greatly in their reliability (repeatability and robustness), difficulty, expense, and the nature of the polymorphism that they detect. Because of these differences, they also vary' greatly in their suitability for various uses. They may be hybridisation based (e.g., RFLP), or polymerase chain reaction (PCR) based (e.g., RAPD and AFLP); they may detect single locus, oligo- locus, or multiple locus differences; and the markers detected may be inherited in a presence/absence, dominant, or co-dominant manner. Brief descriptions of each of a number of the more widely used DNA marker groups are given below based on recent reviews of molecular markers useful in mappmg plant genomes (Karp *et al.*, 1997; Malyshev and Kartel, 1997; Mohan *et al.*, 1997).

Various types of molecular markers are utilized to evaluate DNA polymorphism and are generally classified as hybridization-based markers and polymerase chain reaction (PCR)-based markers. In the former, DNA profiles are visualized by hybridizing the restriction enzyme-digested DNA, to a labelled probe, which is a DNA fragment of known origin or sequence. PCR-based markers involve *in vitro* amplification of particular DNA sequences or loci, with the help of specifically or arbitrarily chosen oligonucleotide sequences (primers) and a thermostable DNA polymerase enzyme. The amplified fragments are separated electrophoretically and banding patterns are detected by different methods such as staining and autoradiography. These well-defined markers once mapped enable dissection of complex traits more precisely into component genetic units (Hayes *et al.*, 1993), thus providing breeders with new tools to manage complex traits more efficiently. Potential use of various types of DNA markers such as Restriction Fragment Length Polymorphism (RFLP), and PCR-based markers like microsatellites, minisatellites, RAPDs and their utility are described below in more detail.

Molecular markers based on Hybridization

Restriction Fragment Length Polymorphism (RFLP)

RFLPs are DNA markers, with differences in the nucleotide sequence (restriction sites) in the DNA of two different genotypes. Such differences may result from point mutations, deletions, insertions or sequence rearrangements that might have occurred during the course of evolution and are detected as variation (polymorphism) in the length of restriction fragments. Restriction enzymes cut DNA at restriction sites. Each different restriction enzyme recognizes a specific and characteristic nucleotide sequence within the genome. Because even a single nucleotide alteration can create or destroy a restriction site, mutations cause variation in the number of sites. Thus there is variation / polymorphism between individuals in the positions of cutting sites and the lengths of DNA between them, resulting in restriction fragments of different sizes (**Fig. 2.3**). Since the genome of most plants contains between 10^8 and 10^{10} nucleotides, changes in even a small proportion of these can yield a large number of potential DNA markers (Paterson *et al.*, 1988, Paterson *et al.*, 1991).

A particular restriction enzyme, say a four-base cutter, will generate a whole range of fragment sizes, and when the DNA digest is run out on an agarose gel it will form a smear with the larger pieces at the -ve end and the smaller at the + ve end. The range of fragment lengths will be different for different restriction enzymes: a six-base cutter will generate fewer, and on the average larger-sized, fragments than a four-base cutter. A small piece of cloned genomic DNA, from the same sample of DNA, will match the whole or part of *one* of the fragments in our smear, and if we label this cloned bit with a radioactive or chemical tag it will serve as a probe in a Southern hybridization and will detect the *single* fragment with which it has sequence homology. The procedures and principles of RFLP markers are summarized below:

(a) Digestion of the DNA with one or more restriction enzyme(s)

(b) Separation of the restriction fragments in agarose gel.

(c) Denaturation and transfer of separated fragments from agarose gel to a filter by Southern blotting.

(d) Detection of individual fragments by nucleic acid hybridization with a labelled probe(s) of known or unknown function

(e) Autoradiography (Perez de la Vega, 1993; Terachi, 1993; Landry, 1994) will give a band at the place where, the probe has sequence homology.

Mode of identification: A plant under study may show a single band in autoradiogram after hybridization and development with a specific DNA fragment, only if the two fragments from the diploid genome are homozygous, with restriction site identical places (so the probe detects both of them at the same place). On the contrary, the counter part, second plant might give a variant of the same fragment that differs in length because it is homozygous for a mutation which has either destroyed one of the restriction sites or else a new one within the original fragment. A case of third plant (lane16 of figure 2.3), where if it is a hybrid (heterozygous) between plant one and plant two, will show two bands corresponding to the fragment sizes of plants 1 and 2. The two different sized fragments are alleles of one locus. The rest of the plants are either homozygous for parent 1 or 2 respectively at that locus. The locus itself is generated by the probe used to detect it, and takes the name or number of that probe (**Fig. 2.3**).

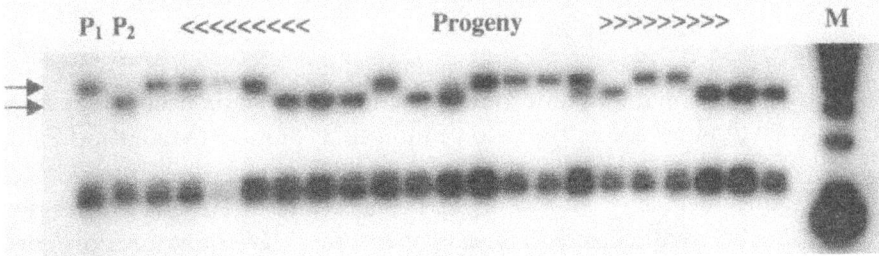

Fig. 2.3. Southern hybridization pattern in a F_2 segregating mapping population of rice, with a single probe at one locus using DNA from plants digested with one single restriction enzyme. Lane 1- parent one, Lane 2 – parent two and rest are progeny (lane 16 – is a hetetozygote). M – represent labelled lambda DNA double digested with EcoR1 and BamH1 restriction enzymes

RFLPs are co-dominant markers, highly polymorphic, reliable in linkage analysis and selection process. They can be used to study the nature of linked trait i.e., in hetero or homozygous state, where such information is highly desirable, especially for recessive traits.

Single copy genomic DNAs or cDNAs have been used as molecular probes to construct RFLP maps of model plant like *Arabidopsis* (Cheng *et al.,* 1988) and crop plants including rice (Mc Couch *et al.,*1988; Saito *et al.,* 1991), barley (Heun *et al.*, 1991), maize (Helentjaris, 1987), lettuce (Landry *et al.*, 1987), soybean (Keim *et al.*, 1990), tomato (Helentjaris *et al.*, 1986, Yang and Tanksley, 1989), potato (Gebhardt *et al.*, 1989) and wheat (Cho *et al.*, 1989). Genetic maps based on such markers have many applications in plant breeding. Dense genetic maps developed using these markers have been used to identify QTLs controlling water use efficiency (Martin *et al.*, 1989), salt tolerance (Breto *et al.*, 1994) in tomato; drought responses (Lebreton *et al.*, 1995), anthesis-silking interval (Ribaut *et al.*, 1996), grain yield (Austin and Lee, 1996) in maize; yield (Kjaer and Jensen, 1996), cold tolerance (Oziel *et al.*, 1996) in barley; and found useful in resolving of some of the long pending questions from classical genetic studies such as intergenomic relationships between maize and sorghum (Hulbert *et al.*, 1990) and between tomato and potato (Bonierbale *et al.*, 1988). In rice, RFLP maps have been used to identify QTLs for plant height and heading date (Li *et al.*, 1995; Zhuang *et al.*, 1997), grain yield (Xu *et al.*, 1995; Lin *et al.*, 1996; Wu *et al.*, 1996; Zhuang *et al.*, 1997), root traits related to drought avoidance and penetration ability (Champoux *et al.*, 1995; Ray *et al.*, 1996), osmotic adjustment (Lilley *et al.*, 1996), aluminium tolerane (Wu *et al.*, 2000), cell membrane stability (Tripathy *et al.*, 2000), leaf trait and ABA accumulation (Quarrie *et al.*, 1997), biomass and yield under field drought stress condition (Babu *et al.*, 2003; Lanceares *et al.*, 2004) have been identified. Inspite of its usefulness, RFLP marker technology is less used being time consuming and expensive.

PCR-based markers: The polymerase chain reaction (PCR), an *in vitro* method for enzymatic synthesis of specific DNA sequence, is yet another significant step forward in the field of molecular biology (Mullis and Faloona, 1987). It uses two oligonucleotide primers of about 10-20 nucleotides in length that specifically anneal to opposite strands flanking the region of interest to be synthesized. Several cycles of DNA denaturation, primer annealing and extension of annealed primers by DNA polymerase, produce an exponential amplification of a specific DNA sequence which can then directly detected on agarose or acrylamide gels by ethidium bromide or silver staining. There are different kinds of PCR based markers, which are useful in genome mapping and gene tagging:

Random Amplified Polymorphic DNA: Generation of RAPDs, which are dominant markers, involves use of single short but random oligonucleotide primers. The DNA amplification with random primers exposes polymorphisms that are distributed throughout the genome (Williams *et al.*, 1990). RAPD markers are based on PCR technique, and require neither cloning nor sequencing of DNA. Williams *et al.* (1990) discovered that, single primer of a short oligo of a 10 bases of arbitrary nucleotide bind at several loci with in genome based on sequence complimentarity. Total number of amplicons amplified depends on the ability of Taq to amplify the region between two complimentary binding sites on genome in opposite orientation with in given extension time. Primer will amplify different regions of a genome ranging between 200 bp to 2 kb long region, which lies between two inverted copies of itself, one copy binding to each strand of the DNA. In general, for the average-sized genome, between five and 10 fragments will be amplified to produce discrete DNA-banding patterns. Polymorphisms arise because sequence variation in the genome alters the primer binding sites. RAPDs are therefore dominant markers as a consequence of their presence/absence at particular loci, and they will segregate from a heterozygous diploid as Mendelian alleles (**Fig. 2.4**). RAPDs are much simpler and less expensive to work with than RFLPs because no prior knowledge of sequences is required and there is no need for radioactive probes. Many different primers can be designed, and there is virtually no limit to the numbers of RAPDs in a genome. RAPDs can be used for mapping, but because of the random nature of their generation, and short primer length, they cannot easily be transferred between species. Their main disadvantages are poor reliability and reproducibility, and their sensitivity to experimental conditions (Karp *et al.*, 1996).

Although suggested to be easy, inexpensive and fast, its reproducibility is sufficiently problematic (due to short primers being easily affected by annealing conditions) to make it inappropriate for any but phylogenetic studies unless great care is used to ensure stringent annealing conditions.

RAPDs have been mostly used in variety identification, genetic relationship and diversity studies in many crops species, such as rice (Yu and Nguyen, 1994), Sorghum (Vierling *et al.*, 1994), soybean (Mienie *et al.*, 1995) and in jatropha (Ram *et al.*, 2008). RAPD markers were exploited in genome mapping and gene tagging studies in plants like faba bean (Torres *et al.*, 1993), lettuce (Kesseli *et al.*, 1994), spring barley (Thomas *et al.*, 1995), tomato (Martin *et al.*, 1991), rice (Virk *et al.*, 1996), lettuce (Paran *et al.*, 1991), and *Arabidopsis* (Reiter *et al.*, 1992). Lack of reproducibility and stability, generation of spurious polymorphism and failure to detect heterozygotes are some of the limitations of RAPDs, which can however, be overcome by converting them into codominant RFLP markers for stable performance.

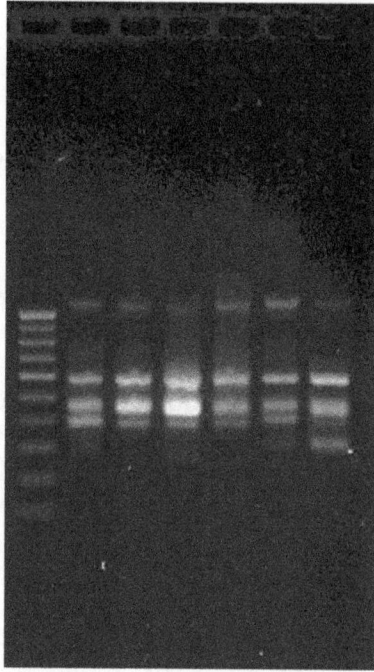

Fig. 2.4. Diversity analysis of rice genotypes using RADP marker. OPA-9 Lane 1-Marker; Lane 2 to 7 - RAPD profiling of six different genotypes of rice using single RAPD marker OPA-9

Amplified Fragment Length Polymorphism: AFLPs is a technique with the combined advantage of reliability of RFLP and convenience of PCR. It is based on Selective Restriction Fragment Amplification of DNA fragments (SRFA) (Zabeau and Vos, 1993). SRFA involves three major steps viz., i) restriction digestion of genomic DNA with restriction enzyme(s) ii) ligating double stranded oligo nucleotide adapters to the restriction fragments and iii) amplifying selective restriction fragments and gel analysis of amplified fragments. SFRA may be performed with a single enzyme, but the best results are achieved when two different enzymes, a rare cutter and a frequent cutter are used. AFLP is a powerful, reliable, stable and rapid assay with potential application in genome mapping, DNA fingerprinting and marker-assisted breeding (Vos *et al.*, 1995).

This PCR-based technique requires no sequencing or cloning. It is similar to RAPD (see above), but the primer consists of a longer fixed portion (of about 15 bases) and a short (2-4 base pairs) random portion. The fixed portion gives the primer stability (and hence repeatability) and the random portion allows it to detect many loci. Polymorphism is detected as band presence/absence (so it is usually interpreted as dominantly inherited, although claims for co-dominant inheritance are also made based on band intensity). AFLP markers are often inherited as tightly linked clusters in centromeric and telomeric regions of chromosomes, but randomly distributed AFLP markers also occur outside these clusters. The technique is difficult to master and is less appropriate than others for comparative mapping studies.

This PCR-based technique permits inspection of polymorphism at a large number of loci within a very short period of time, requires very small amounts of DNA and above all, is highly reproducible. Because of these advantages, AFLPs are currently considered as the molecular markers of choice for the genome mapping whose sequence is yet to be unravelled. AFLP markers were exploited in genetic diversity studies, genome mapping and gene tagging studies of rice (Mackill *et al.*, 1996; Nandi *et al.*, 1997), barley (Qi *et al.*, 1998), potato (De Jong *et al.*, 1997), soybean (Maughan *et al.*, 1996) and Sunflower (Quagliaro *et al.*, 2001).

Simple Sequence Repeats (SSRs): These are also known as Variable Number Tandem Repeats (VNTRs) which include microsatellites, minisatellities and hyper variable regions. Microsatellites are arrays of tandemly repeated DNA sequences, which are dispersed throughout the genome (Jeffreys *et al.*, 1985a) and are also referred to as "Sequence Tagged Microsatellite Sites (STMS)". Simple sequence repeat (SSR) markers were first developed for use in genetic mapping in humans (Litt and Luty, 1989; Weber and May, 1989). Microsatellites consist of around 10-50 copies of motifs ranging from 1 to 5 bp that occur in perfect tandem repetition, as in perfect repeats or in imperfect repeats (together with another repeat type). SSRs are highly variable and evenly distributed throughout the genome. This type of repeated DNA is common in eukaryotes, their number of repeated units varying widely among organisms to as high as 50 copies of the repeated unit. These polymorphisms are identified by constructing PCR primers for the DNA flanking the microsatellite region. The flanking regions tend to be conserved within the species, although sometimes they may also be conserved in higher taxonomic levels. The repeat number of microsatellites has been demonstrated to be highly variable in plants and animals and inherited in a co-dominant manner (Litt and Luty, 1989; Johansson *et al.*, 1992) (Fig. 2.5). Presence of microsatellites has been documented in many plants, which include *Arabidopsis* (Bell and Ecker, 1994), barley (Saghai *et al.*, 1994), *Brassica* (Langercrantz *et al.*, 1993), maize (Condit and Hubbel, 1991; Senior and Heun, 1993), rice (Wu and Tanksley, 1993), soybean (Akkaya *et al.*, 1992; Morgante and Olivieri, 1993), sorghum (Brown *et al.*, 1996; Taramino *et al.*, 1997; Dean *et al.*, 1999), peach (Wang *et al.*, 2002) and wheat (Roder *et al.*,1995). These markers have mostly been used in genetic relationship and diversity studies in different plants species, such as barley (Sanchez dela Hoz *et al.*, 1996), maize (Taramino and Tingey, 1996), rice (Wu and Tanksley, 1993; Panaud *et al.*, 1996; Thanh *et al.*, 1999).

Also, these markers were used along with other kinds of markers for genome mapping and QTL identification. Microsatellite molecular marker linkage maps developed in crops such as maize, barley, wheat, rice, soybean, common bean and *Arabidopsis* are gaining an increasing importance in recent years.

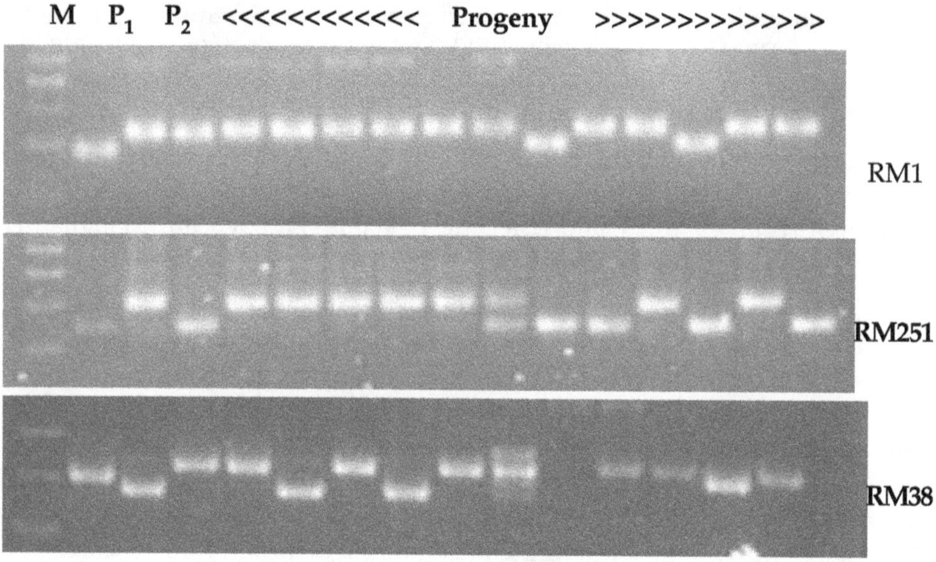

Fig. 2.5: Simple sequence repeat (SSR) inheritance pattern in a F_2 segregating mapping population of rice, Lane 1 (M) - 200 base pair marker; Lane 2- parent one, Lane 3 – parent two and rest are progeny (lane 10 – is a hetetozygote at locus RM251).

Minisatellites comprise a class of variable number of tandem repeat (VNTR) loci, in which the repeated sequences are short (<65 bp) and frequently GC rich (Jeffreys *et al.*, 1985a, b; Nakamura *et al.*, 1987). A 15bp repeat motif in the protein III gene of bacteriophage M13 appears to be an extremely useful in detecting polymorphism (Vassart *et al.*, 1987) and serves as a universal marker for DNA fingerprinting (Ryskova *et al.*, 1988). This was first used in plants to identify polymorphic regions in barely and fingerprinting Asian and African rice (Dallas, 1988), while a human minisatellite probe, pV 47 has been shown to be polymorphic in *indica*, *japonica* and wild rices (Ramakrishna *et al.*, 1995).

Inter Simple Sequence Repeats: Inter-simple sequence repeat analysis involves polymerase chain reaction (PCR) derived amplification of regions between adjacent, inversely oriented microsatellites using a single simple sequence repeat (SSR) containing primer (Zietkiewicz *et al.*, 1994). This technique can be applied to any species that contains sufficient number and distribution of SSR motifs. Another advantage of ISSR markers is that genomic sequence data of an organism is not required (Gupta *et al.,* 1994; Godwin *et al.,* 1997). The primer used in ISSR analysis is based on any of the SSR motifs (di-, tri-, tetra-, or penta-nucleotides) found at microsatellite loci, giving a wide array of possible amplification products, and can be anchored to genomic sequences flanking either side of the targeted simple sequence repeats. The potential use of ISSR markers depends on variety and frequency of microsatellites, which vary with species and SSR motifs, that are targeted (Morgante and Olivieri, 1993; Depeiges *et al.,* 1995). As the ISSR technique amplifies a large number of DNA fragments per reaction, representing multiple loci across the genome, it is an ideal method for fingerprinting rice varieties and a useful alternative to single-locus or hybridization-based methods (Zietkiewicz *et al.,* 1994; Godwin *et al.,* 1997). The ISSR method has proven its usefulness, especially in the Gramineae family for analysis of near isogenic lines (Akagi *et al.,* 1996) and varieties (Parsons *et al,* 1997; Swathi *et al,* 2000) of rice, inbred lines of corn (Kantety *et al.,* 1995), populations of finger millet (Salimath *et al.,* 1995) and accessions of sorghum (Yang *et al.,* 1996).

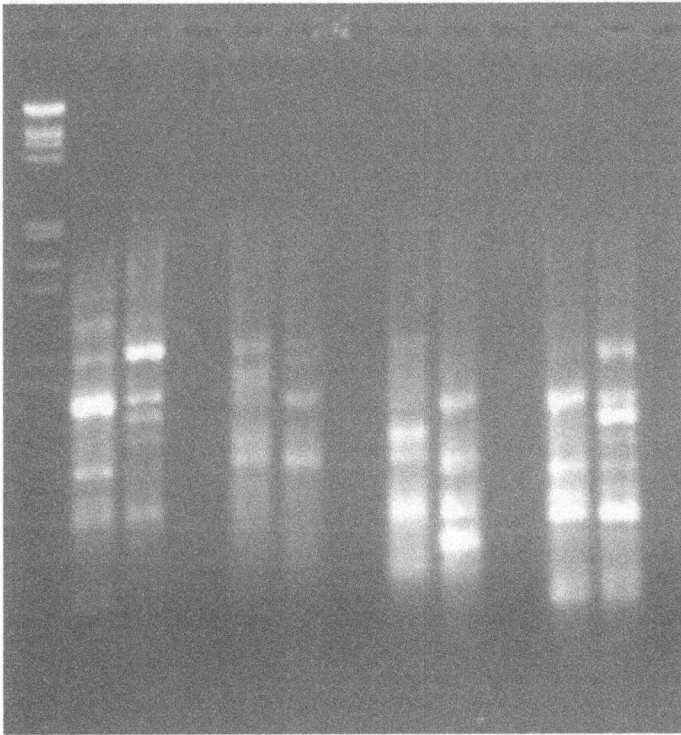

Fig. 2.6: Genotyping of two rice cultivars using ISR primers. Lane 1 – Marker; Lane 2 and 3 UBC primers

STS (sequence-tagged site)

The STS concept was introduced by Olson *et al.* (1989). In assessing the likely impact of the Polymerase Chain Reaction (PCR) on human genome research, they recognized that single-copy DNA sequences of known map location could serve as markers for genetic and physical mapping of genes along the chromosome. These PCR-based markers detect a single, unique, sequence-defined point in the genome. They are obtained by sequencing terminal regions of genomic fragments and cDNAs expressing RFLP. Primers of 18-20 base pairs are designed to amplify this short, unique fragment. Polymorphism is often reduced compared to the original RFLP marker, but can be increased at some additional cost by restricting the PCR products to increase the number of bands detected. Since they are longer than RAPD primers and based on a specific sequence, STS markers more reliably detect the same locus. They are good for both mapping studies and MAS, provided that polymorphism detected is adequate but requires knowledge on preexisting sequence data.

SCARs: Utility of RAPD marker(s) can be increased by cloning, sequencing its termini and designing longer primers for specific amplification of markers (Paran and Michelmore, 1993). These are more reproducible than RAPDs, some times co-dominant and used for mapping and QTL identification application (Nair *et al.*, 1995, 1996). Selective AFLP fragments are also processed the same way to be converted into SCAR markers which are co-dominant markrs. SCAR marker is a PCR-based secondary markers are detected with two (farward and reverse) 24-nucleotide primers homologous to sequenced ends of a selected RAPD / AFLP marker. They amplify a single fragment with high reproducibility and most of times a co-dominant type marker polymorphic between individuals of same species.

Steps of SCAR marker Development: The steps of SCAR marker development are as below (Fig. 2.7).

- Amplified with Pfu and gel elute the amplicon of choice
- Blunt end cloned into suitable Vector and transformed it into DH5a
- White colonies were selected, and clone was confirmed by colony PCR
- Plasmid was isolated from positive colonies and used for sequencing
- Blast analysis, multiple sequence alignment and primer design

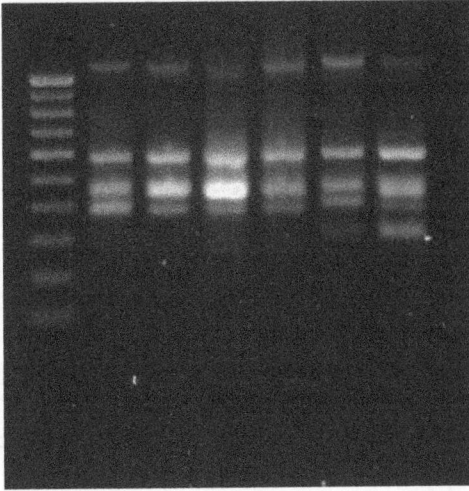
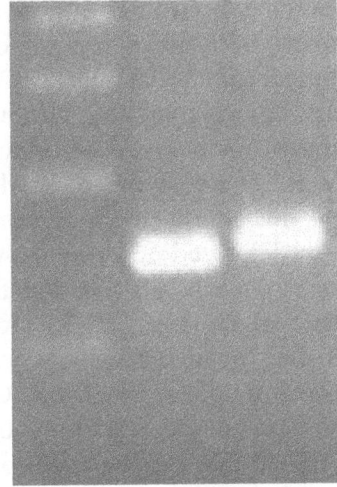

(a)   (b)

FP
→

```
Nipponbare  TACTGATGATCGCGAGTTGGAGCTAGCAGTTTTGAGCTCAACCAGCTTTGCTCCTCCTAT  60
CT9993      TACTGATGATCGCGAGTTGGAGCTAGCAGTTTTGAGCTCAACCAGCTTTGCTCCTCCTAT  60
BI306107    TACTGATGATCGCGAGTTGGAGCTAGCAGTTTTGAGCTCAACCAGCTTTGCTCCTCCTAT  60
            ************************************************************

Nipponbare  ACAGCTAAATACTGTAGGAGAAATTAATGGAGATTTTTTCCTTCTTTATTTTTTTTATAT  120
CT9993      ACAGCTAAATACTGTAGGAGAAATTAATGGAGATTTTTTCCTTCTTTATTTTTTTTATAT  120
BI306107    ACAGCTAAATACTGTAGGAGAAATTAATGGAGATTTTTTCCTTCTTTATTTTTTT-ATAT  119
            ******************************************************* ****

Nipponbare  TTTTTCCAAGATAAAATGGCTGAGCTGGTAGGGAGTTCTAGAAGGAGGAAATAAAGATTA  180
CT9993      TTTTTCCAAGATAAAATGGCTGAGCTGGTAGGGAGTTCTA---GGAGGAAATAAAGATTA  177
BI306107    TTTTTCCAAGATAAAATGGCTCAGCTGGTACGGACTTCTAGAAGGACGAAATAAAGATTA  179
            *********************  **   ** ** ** *** ******************

Nipponbare  CGAGATTTGTAATACTACTACGAGGGACAGCATGCAAAAAAGAAGTATAATTGCTATGCT  240
CT9993      CGAGATTTGTAATACTACTACGAGGGACAGCATGCAAAAAAGAAGTATAATTGCTATGCT  237
BI306107    CGAGATTTGTAATACTACTACGAGGGACAGCATGCAAAAAAGAAGTATAATTGCTATGCT  239
            ************************************************************

Nipponbare  CTCTCTCTCTCTCTCTCTCTCTTTTGTGTGTGGAGTGGAATAAAATGCCAGCTCTGCTTATGGT  300
CT9993      CTCTCTCTCTCTCTCTCTCT-TTNTGTGTGTGGAGTGGAATAAAATGCCAGCTCTGCTTATGG-  295
BI306107    CTCTCTCTCTCTCTCTCTCTCTTTTGGGTGTGGAGTGGAATAAAATGCCAGCTCTGCTTATGGT  299
            *****************  **  **  *****************************
```

← RP

(c)

Fig. 2.7: (a) Genotying of rice cultivars using RAPD primer OPA-1.

(b) Selecting and purification of amplicon

(c) Cloning, sequencing multiple sequence alignment and primer designing to convert into SCAR marker (FP-Farward Primer; RP-Reverse Primer)

Functional Markers: These are entirely new class of molecular markers that uses vast sequence information from the nucleotide data bases. They are derived from ESTs and genome sequences for development of markers from the transcribed regions of the genome (cDNA derived information), so called functional markers. ESTs are single pass partial sequences of cDNA clones available in public (www.ncbi.nlm.nih.gov) and private databases for a range of organisms including *Arabidopsis,* rice, wheat etc. They are of two types of markers that are developed from ESTs, they are single nucleotide polymorphisms (SNPs) as described by Rafalski, (2002) and EST-derived SSRs as described by Varshney *et al.* (2005) Therefore, molecular markers generated from expressed sequence data are known as 'functional markers' (FMs) (Andersen and Lubberstedt, 2003). FMs have been developed extensively for the plant species in which ESTs or gene sequence data are available (Gupta and Rustgi, 2004). Rudd *et al.* (2005) demonstrated the feasibility of predicting molecular markers (e.g., SSRs and SNPs) that can be used to develop FMs for several species. FMs have some advantages over random markers that are generated from anonymous region of the genome, because FMs are linked to the desired trait allele. Such markers are derived from the gene responsible for the trait of interest and target the functional polymorphism in the gene they allow selection in different genetic backgrounds without revalidating the marker–quantitative-trait-locus (QTL) allele relationship. An FM allows breeders to track specific alleles within pedigrees and populations, and to minimize linkage drag flanking the gene of interest. As markers become more abundant, breeders can develop strategies that are compatible with financial resources and breeding goals. Using ESTs in the place of anonymous molecular probes in mapping programmes is gaining importance. This approach has been incorporated into mapping and exploited in the candidate gene approach for the dissection of quantitative trait loci (Quarrie *et al.*, 1994). Molecular marker techniques based on PCR amplification of DNA in different cultivars, by using long primers based on known sequence motifs, like introns, promoter regions, 5' untranslated regions, 3' untranslated regions will help in knowing length variation in the targeted regions and can effectively use them as molecular markers in breeding and QTL identification programmes (Bhattramakki *et al.*, 2002). Use of EST-PCR markers to monitor intrapopulation variation in comparison with isoenzyme markers has been studied in *Picea* (Schubert *et al.*, 2001) and conversion of ESTs into PCR based markers to study expressed sequence tag polymorphism (ESTPS) and their application in genetic mapping has been demonstrated in *Pinus* (Harry *et al.*, 1998; Temesgen *et al.*, 2001; Cato *et al.*, 2001). These new generation sequence based markers, which can be used as simple genetic markers and also have a great potential in association genetics. They may be identified with in or in the vicinity of virtually every gene, and can be used for construction of haplotype-based genetic maps for various studies like linkage disequilibrium and association with phenotype or loci (QTL). Rafalski (2002) reviewed the use of SNPs in crop genetics and their application in improving quantitative and qualitative traits.

Expressed sequence tag - SSRs (EST-SSRs) are PCR-based genetic markers that are derived from expressed sequenced tags (ESTs). EST-SSRs are derived essentially from the sequence information derived from the nucleotide sequence with simple sequence repeat with in it. Development process of EST-SSR markers has four main steps. The steps are i) Collection of sequences from a target species ii) search and identification of sequence with simple sequence repeats iii) designing of primers to the flanking regions of simple sequence repeat present with in the sequence iv) screen the individuals of species under study for polymorphism. These can be genetically mapped by a variety of methods all of which rely on detecting polymorphism between individuals (**Fig. 2.8**).

M P₁ P₂ <<<<<<<<<<<<   Progeny   >>>>>>>>>>>>>>

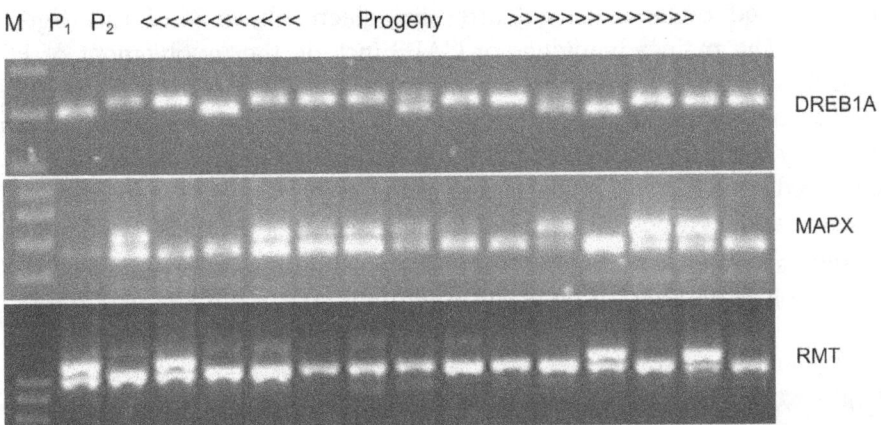

Fig. 2.8: Segregation analysis of EST-SSR (EST-PCR) markers in Doubled Haploid Line (DHL's) mapping population of rice. Lane 1 (M) - 200 base pair marker; Lane 2- parent one, Lane 3 – parent two and rest are progeny.

Single nucleotide polymorphisms (SNPs): SNPs (single nucleotide polymorphisms), which belong to the last-generation molecular markers, occur at high frequencies in both animal and plant genomes. The development of SNP markers allows to automatize and enhance ten folds the effectiveness of genotype analysis. By comparing sequences from a japonica rice cultivar to those from an indica cultivar, for example, Yu *et al.* (2002) identified, on average, one SNP every 170 bp and one InDel every 540 bp upon comparison of the sequence information of japonica and indica cultivars of rice. The abundance of these polymorphisms in plant genomes makes the SNP marker system an attractive tool for mapping, marker-assisted breeding and map-based cloning (Gupta *et al.*, 2001; Rafalski, 2002; Batley *et al.*, 2003). SNP marker is just a single base change in a DNA sequence, with a usual alternative of two possible nucleotides at a given position. Two allele-specific probes are designed, usually with the polymorphic base in a central position in the probe sequence. Under optimized assay conditions, only the perfectly matched probe-target hybrids are stable, and hybrids with one-base mismatch are unstable.

Cleaved Amplified Polymorphic Sequence (CAPS)

Cleaved Amplified Polymorphic Sequence markers are type of markers which use PCR and RFLP. For the development of these markers, prior information on the sequence / target is required from the same species / organism under study. Specific region of interest ranging from 300 to 800 base pairs is PCR amplified using specific primer pairs, followed by restriction digestion with an enzyme. Subsequently, length variation / polymorphisms resulting from variation in the occurrence / absence of restriction sites and are identified on agarose gel after gel electrophoresis of the digested products. The main advantages of CAPS include the involvement of PCR, hence the initial DNA requirement is very low when compared to RFLPs. Their codominance nature, high reproducibility and stable inheritance make them very ideal for molecular mapping and gene tagging studies (Fig. 2.9). Compared to RFLPs, CAPS require prior sequence data information for selection and synthesis of primers. They were extensively used in gene mapping studies across many crop plants including rice, tomato, sugarcane, etc., (Wang *et al.*, 2003; Zhang and Stommel 2001; Quint *et al.*, 2002)

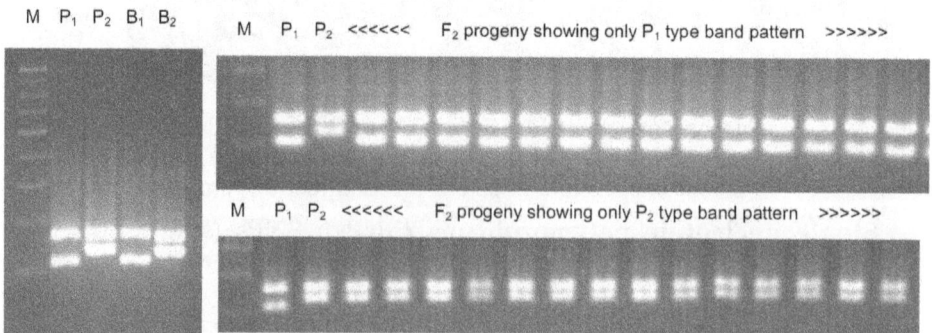

Fig. 2.9 . (a) Inheritance pattern of CAPS marker in parents and bulks for disease resistance and susceptible b) Segregation pattern of the same CAPS marker in debulks along with their parents.

DArT (diversity arrays technology): DArT is one of the recently developed molecular techniques and it has only been used in many crop and non-crop plant species including rice (Jaccoud *et al.*, 2001), barley (Wenzl *et al.*, 2004), eucalyptus (Lezar *et al.*, 2004), *Arabidopsis* (Wittenberg *et al.*, 2005), cassava (Xia *et al.*, 2005), wheat (Akbari *et al.*, 2006; Semagn *et al.*, 2006), and pigeon-pea (Yang *et al.*, 2006). The inventors promote it as an open source (non exclusive) technology with a great potential for genetic diversity and mapping studies in a number of 'orphan' crops relevant in developing countries (www.cambia.org or http://www.diversityarrays.com for information). DArT technique is a microarray hybridization-based technique that enables the simultaneous typing of several hundred polymorphic loci spread over the genome (Jaccoud *et al.*, 2001; Wenzl *et al.*, 2004). Details of the methodology for DArT was first described by Jaccoud *et al.* (2001).

For each individual under study, DNA sample are prepared by restriction enzymes digestion of genomic DNA followed by ligation of restriction fragments to specific adapters, followed by selective amplification of the fragments by PCR using adapter specific primers with selective overhangs. The fragments from representations are cloned, and cloned inserts are amplified using vector-specific primers, purified and arrayed onto a solid support (microarray) resulting in a "discovery array." Labelled genomic representations prepared from the individual genomes included in the pool are hybridized to the discovery array (Jaccoud *et al.*, 2001). Polymorphic clones (DArT markers) show variable hybridization signal intensities for different individuals. These clones are subsequently assembled into a "genotyping array" for routine genotyping of multiple individuals of same species.

Advantages of DArT markers

(a) No prior sequence information of the organism under study is required

(b) It is a highthroughput, quick, and highly reproducible method across labs.

(c) It is cost effective, with an estimated cost per data point ten fold lower than SSR markers (Xia *et al.*, 2005) or for that matter any other markers.

Common application of molecular markers

Molecular markers and marker mapping are part of the integrative approach of modern biology / biotechnology which covers quite wide areas of research that is thrusting its way into all areas of modern biology, from genomics to breeding, from transgenics to developmental biology, from systematics to ecology, and even, perhaps especially, into plant and crop physiology (Jones *et al.*, 1997). Molecular markers are rapidly being adopted for crop improvement by researchers globally as an effective and appropriate tool for basic and applied studies addressing biological components in agricultural production systems (Jones *et al.*, 1997; Mohan *et al.*, 1997; Prioul *et al.*, 1997). Now that we have the capacity to isolate and clone genes, and to map quantitative trait loci, geneticists and physiologists have passed through their courtship phase and gone into serious partnership. We have the technology, and we can glimpse the prize of making that vital connection between the gene and the character, but there are still many obstacles hindering consummation (Jones *et al.*, 1997). Molecular markers offer specific advantages in assessment of genetic diversity and in trait-specific crop improvement. Use of markers in applied breeding programmes can range from facilitating appropriate choice of parents for crosses, to mapping tagging of gene blocks associated with economically important traits (often termed "quantitative trait loci" (QTLs)). Gene tagging and QTL mapping in turn permit marker-assisted selection (MAS) in backcross, pedigree, and population improvement programmes (Mohan *et al.*, 1997). This is especially useful for crop traits that are otherwise difficult or impossible to deal with by conventional means. The near-isogenic products of marker-assisted backcrossing programmes provide genetic tools for crop physiologists and crop protection scientists to use in improving our understanding of the mechanisms of various abiotic stress tolerances (Jones *et al.*, 1997; Prioul *et al.*, 1997) and resistances to biotic production constraints such as diseases, insect pests, nematodes, and parasitic weeds like *Striga*. QTL mapping of yield and quality components, as well as components of other physiologically or biochemically complex pathways, can provide crop breeders with a better understanding of the basis for genetic correlations between economically important traits (linkage and/or pleiotropic relationships between gene blocks controlling associated traits; e.g., flowering time and biomass; inflorescence size and inflorescence number). This can facilitate more efficient incremental improvement of specific individual target traits. Further, specific genomic regions associated with QTLs of large effect for one target trait can be identified having minimal effects on otherwise normally correlated traits, permitting an improvement in the first trait that need not be accompanied by counterbalancing reductions in others. Finally, these molecular marker tools can also be used in ways that allow us to more effectively discover and efficiently exploit the evolutionary relationships between organisms, through comparative genomics.

Overall applications of the molecular markers in plant genetics and genomics studies is much diverse, of which following are few important

(a) Characterization and assessment of genetic variability present in crop plants
(b) Fingerprinting and identification of genotypes

Table 2.1: Comparison of the molecular markers with respect to various properties and their use.

Type of marker	Quality of DNA required	Quantity of DNA required	Abundance / Frequency	No. of alleles per loci	Reproducibility	Polymorphic type	Degree of polymorphism	Locus specificity	Cost per marker development	Cost per sample / array	Amicable to automation	Possibility for Association studies
Morphological	NR	NR	Low	Two	+*	Dominant	+	Low	Very less	Very Cheap	No	+
Isozyme	NR	NR	Low	Two to three	++#	Co-dominant	++	Low	Less	Cheap	No	++
RFLPs	High	More	High	Three to four	++++	Co-dominant	+++	High	Expensive	Expensive	No	++
RAPDs	Medium	Less	High	Two	++	Dominant	+++	Low	Moderate	Moderate	No	+
AFLPs	High	Less	High	Two	+++	Dominant	++++	High	Expensive	Moderate	Moderate	+
SCAR	Medium	Less	Low	Two	+++++	Co-dominant	+++	Very High	Expensive	Moderate	Moderate	++
SSRs	Medium	Less	Medium - High	three to five	+++++	Co-dominant	++++ +	Very High	Expensive	Moderate to Expensive	Yes	++
ISSRs	Medium	Less	Medium	Two	++	Dominant	+++	Low	Moderate	Moderate	Moderate	+
CAPS	Medium	Less	Low	Two to three	+++++	Co-dominant	++++	Very High	Moderate	Moderate	Moderate	++
SNP	Medium	Less	High	Two to Three	+++++	Co-dominant	++++	Very High	Expensive	Moderate to Expensive	Yes	+++
EST-PCR / EST-SSR	Medium	Less	Low - Medium	Two to Four	+++++	Co-dominant	+++	Very High	Moderate	Moderate	Yes	+++

NR- Not Required   * - Environment Dependent   # - Stage Specific

(c) Genetic distance estimation and population structure studies

(d) Tagging a quantitative trait loci (QTL) and marker assisted selection (MAS)

(e) Association studies

To conclude with we highlight the over all merits and demerits of different molecular marker systems, their genetic aspects and what they are used for (**Table 2.1**). These markers have become an increasingly helpful tool in plant genetic research and applications in areas of modern biotechnology and genomics. The basic premise behind molecular markers is that there is enormous natural genetic variation in individuals, and most of these genetic sequences are polymorphic, meaning they differ among to closely related individuals. Molecular markers seek to exploit this variation to identify and differentiate two individuals under study, in terms of phenotypic traits, or genes on the basis of genetic differences to create new sources of genetic variation by introducing new and favourable traits from various available sources like wild / landraces of plant species. These marker systems have thoroughly revolutionized and nevertheless it is not astonishing to claim that the molecular markers has changed entire scenario of modern plant breeding.

References

Adams, W.T. 1992. Gene dispersal within forest tree populations New Forests **6**: 217–240.

Akagi, H., Yokozeki, Y., Inagaki, A., Nakamura, A. and Fujimura, T. 1996. A co-dominant DNA marker closely linked to the rice nuclear restorer gene, Rf-1, identified with inter-SSR fingerprinting. Genome **39**: 1205-1209.

Akbari, M., Wenzl, P., Caig, V., Carlig, J., Xia, L., Yang, S., Uszynski, G., Mohler, V., Lehmensiek, A., Howes, N., Sharp, P., Huttner, E. and Kilian, A. 2006. Diversity arrays technology (DArT) for highthroughput profiling of the hexaploid wheat genome. Theor. Appl. Genet.**113**:1409-1420.

Akkaya, M.S., Bhagwat, A.A. and Cregan, P.B. 1992. Length polymorphisms of simple sequence repeat DNA in soybean. Genetics **132**: 1131-1139.

Anderson, J.R. and Lubberstedt, T. 2003. Functional markers in plants. Trends Plant Sci. **8**: 554–560.

Austin, D.E. and Lee, M. 1996. Comparative mapping in F2:3 and F6:7 generations of quantitative trait loci for grain yield and yield components in maize. Theor. Appl. Genet. **92**: 817-826.

Babu, R.C., Nguyen, B.D., Chamarerk, V., Shanmugasundaram, P., Chezhian, P., Jeyaprakash, P., Ganesh, S. K., Palchamy, A., Sadasivam, S., Sarkarung, S., Wade, L. J., and Nguyen, H.T. 2003. Genetic analysis of drought resistance in rice by molecular markers : association between secondary traits and field performance. Crop Sci. **43**:1457–1469.

Batley, J., Mogg, R., Edwards, D., O'Sullivan, H. and Edwards, K.J. 2003. A high-throughput SNuPE assay for genotyping SNPs in the flanking regions of *Zea mays* sequence tagged simple sequence repeats. Mol. Breed. 11:111–120.

Beckman, J.S and Soller, M. 1986. Restriction fragment length polymorphism and genetic improvement of agricultural species. Euphytica 35: 111-124.

Bell, C.J. and Ecker, J.R. 1994. Assignment of 30 microsatellite loci to the linkage map of *Arabidopsis*. Genomics 19: 137-144.

Bhattramakki, D., Dolan, M., Hanafey, M., Wineland, R., Vaskel, D., Register III, J.C., Tingey, S.V. and Rafalski, A. 2002. Insertion-deletion polymorphisms in 3_ regions of maize genes occur frequently and can be used as highly informative genetic markers. Plant Mol. Biol. 48: 539–547.

Bonierbale, M.W, Plaisted, R.L and Tanksley, S.D. 1988. RFLP maps based on a common set of clones reveal modes of chromosomal evolution in potato and tomato. Genetics 120: 1095-1103.

Botstein, D., White, R.L., Skolnick, M., and Davis, R.W. 1980. Construction of a genetic linkage map in man using restriction fragment length polymorphisms. Am. J. Hum. Genet. 32: 314-333.

Breto, M.P., Asins, M.J., and Carbonell, E.A. 1994. Salt tolerance in *Lycopersicon* species, III. Detection of quantitative trait loci by means of molecular markers. Theor. Appl. Genet. 88: 395-401.

Brown, S. M., Hopkins, M. S., Mitchell, S. E., Senior, M. L., Wang, T. Y., Duncan, R. R., Gonzalez-Candelas, F. and Kresovich, S. 1996. Multiple methods for the identification of polymorphic simple sequence repeats (SSRs) in sorghum [*Sorghum bicolor* (L.) Moench]. Theor. Appl. Genet. 93: 190 – 198.

Cato, S. A, Gardner. R.C, Kent. J., and Richardson. T. E. 2001. A rapid PCR-based method for genetically mapping ESTs. Theor. Appl. Genet. 102: 296-306.

Champoux, M.C., Wang, G., Sarkarung, S., MacKill, D.J., O'Toole, J.C., Huang, N., and McCouch, S.R. 1995. Locating genes associated with root morphology and drought avoidance in rice via linkage to molecular markers. Theor. Appl. Genet. 90: 969-981.

Cheng, C., Bowman, A.W., Lander, E.S. and Meyerowitz, E.W. 1988. Restriction length polymorphism linkage map of *Arabidopsis thaliana*. Proc. Natl. Acad. Sci. USA 88: 9828-9832.

Cho, S., Sharp, P.J., Worland, A.J., Warham, E.J., Koebner, R.M.D. and Gale, M.D. 1989. RFLP-based genetic maps of wheat homoeologous group 7 chromosomes. Theor. Appl. Genet. 78: 495-504.

Conkle, M.T. 1981. Isozyme variation and linkage in six conifer species. In: Proc. Symp. Isozymes of North American Forest Trees and Forest Insects. USDA For. Serv Gen Tech Rep, PSW 48: 11-17.

Condit, R. and Hubbel, S.P. 1991. Abundance and DNA sequence of 2-base repeat regions in tropical tree genomes. Genome 34: 66-71.

Dallas, J.F. 1988. Detection of DNA 'fingerprints' of cultivated rice by hybridization with a human minisatellite probe. Proc. Natl. Acad. Sci. USA. 85: 6831-6835.

De Jong, W., Forsyth, A., Leister, D., Gebhardt, C., and Baulcombe, D.C. 1997. A potato hypersensitive resistance gene against potato virus X maps to a resistance gene cluster on chromosome 5. Theor. Appl. Genet. **95**: 246-252.

Dean, R.E., Dahlberg, J.A., Hopkins, M.S., Mitchell, S.E., and Kresovich, S. 1999. Genetic redundancy and diversity among 'Orange' accessions in the U.S. national Sorghum collection as assessed with simple sequence repeat (SSR) markers. Crop Sci. **39**: 1215–1221.

Depeiges, A., Goubely, C., Lenior, A., Cocherel, S., Picard, G., Raynal, M., Grellet, F. and Delseny, M. 1995. Identification of the most represented repeated motifs in *Arabidopsis thaliana* microsatellite loci. Theor. Appl. Genet. **91**: 160-168.

Gebhardt, C., Ritter, E., Debene, T., Schachtschabel, U., Walkemeier, B., Uhrig, H., Salamini, F. 1989. RFLP analysis and linkage mapping in *Solanum tuberosum*. Theor. Appl. Genet. **78**: 65-75.

Godwin, I.D., Aitken, E.A.B. and Smith, L.W. 1997. Application of inter simple sequence repeat (ISSR) markers to plant genetics. Electrophoresis **18**: 1524-1528.

Gupta, P.K. and Rustgi, S. 2004. Molecular markers from the transcribed/expressed region of the genome in higher plants. Funct Integr Genomics **4**: 139–162.

Gupta, P.K., Roy, J.K. and Prasad, M. 2001. Single nucleotide polymorphisms: a new paradigm for molecular marker technology and DNA polymorphism detection with emphasis on their use in plants. Curr. Sci. **80**: 524–535.

Gupta, M., Chyi, Y.S., Romero-severson, J. and Owen, J.L. 1994. Amplification of DNA markers from evolutionarily diverse genomes using single primers of simple-sequence repeats. Theor. Appl. Genet. **89**: 998-1006.

Harry, D.E., Temesgen, B. and Neale, D.B. 1998. Codominant PCR-based markers for *Pinus taeda* developed from mapped cDNA clones. Theor. Appl. Genet. **97**: 327-336.

Hayes, P.M., Liu, B.H., Knapp, S.J., Chen, F. and Jones, B. 1993. Quantitative trait loci effects and environmental interaction in a sample of North American barley germplasm. Theor. Appl. Genet. **87**: 392–401.

Helentjaris, T. 1987. A genetic linkage map for maize based on RFLPs. Theor. Appl. Genet. **3**: 217-221.

Helentjaris, T., Solcum, M., Wright, S., Schaefer, A. and Nienhuis, J. 1986. Construction of genetic linkage maps in maize and tomato using restriction fragment length polymorphism. Theor. Appl. Genet. **72**: 761-769.

Heun., M., Kennedy, A.E., Anderson, J.A., Lapitan, N.L.V., Sorrells, M.E., and Tanksley, S.D. 1991. Construction of a restriction fragment length polymorphism map for barley (*Hordeum vulgare*). Genome **34**: 437-447.

Hulbert, S.H., Richter, T.E., Axtell, J.D. and Bennetzen, J.L. 1990. Genetic mapping and characterization of sorghum and related crops by means of maize DNA probes. Proc. Natl. Acad. Sci. USA. **87**: 4251-4255.

Jaccoud, D., Peng, K., Feinstein, D. and Kilian, A. 2001. Diversity arrays: a solid state technology for sequence information independent genotyping. Nucleic Acids Res. **29**: e25.

Jeffreys, A.J., Wilson, V. and Thein, S.L. 1985a. Hypervariable minisatellite regions in human DNA. Nature **314**: 67-73.

Jeffreys, A.J., Wilson, V. and Thein, S.L. 1985b. Individual specific 'fingerprints' of human DNA. Nature **316**: 76-79.

Johansson, M., Ellegren, H. and Andersson, L. 1992. Cloning and chracterization of highly polymorphic porcine microsatellites. J. Heredity **83**: 196-198.

Jones, N., Ougham, H. and Thomas, H. 1997. Markers and mapping: we are all geneticists now. New Phytologist **137**: 165-177.

Kantety, R., Zeng, X., Bennetzen, J. and Zehr, B.E. 1995. Assesment of genetic diversity in dent and popcorn (*Zea mays* L.) inbred lines using inter-simple sequence repeat (ISSR) amplification. Mol. Breed. **1**: 365-373.

Karp, A., Seberg, O., Buiatti, M. 1996. Molecular techniques in the assessment of botanical diversity. Ann. Bot. **78**: 143–149.

Karp, A., Kresovich, S., Bhat,.V., Ayad, W.G. and Hodgkin, T. 1997. Molecular tools in plant genetic resources conservation, a guide to the technologies. IPGRI Technical Bulletin No. 2. International Plant Genetic Resources Institute: Rome, Italy.

Keim, P., Diers, B.W., Olson, T.C. and Shoemaker, R.C. 1990. RFLP mapping in Soybean: association between marker loci and variation in quantitative traits. Genetics **126**: 735-742.

Kesseli, R.V., Paran, I., and Michelmore, R. W. 1994. Analysis of detailed genetic linkage map of *Lactuca sativa* (Lettuce) constructed from RFLP and RAPD markers. Genetics **136**: 1435-1446.

Kjaer, B. and Jensen, J. 1996. Quantitative trait loci for grain yield and yield components in a cross between six rowed barley. Euphytica **90**: 39-48.

Lanceras, J.C., Pantuwan, G., Jongdee, B., and Toojinda, T. 2004. Quantitative trait loci associated with drought tolerance at reproductive stage in rice. Plant Physiology **135**: 384–399.

Landry, B.S. 1994. DNA mapping in plants. In: Glick, B.R., Thompson, J.E. (eds) Methods in plant molecular biology and biotechnology. CRC Press, Boca Raton. pp 269-285.

Landry, B.S., Kesseli, R.V., Farrara, B. and Michelmore, R.W. 1987. A genetic map of lettuce (*Lactuca sativa* L.) with restriction fragment length polymorphism, isozyme, disease resistance, and morphological markers. Genetics **116**: 331-337.

Langercrantz, U., Ellegren, H. and Anderson, L. 1993. The abundance of various polymorphism microsatellite motifs differs between plants and vertebrates. Nucl. Acids. Res. **21**: 1455-1460.

Lebreton, C., Lazic-Jancic, V., Steed, A., Pekic, S. and Quarrie, S.A. 1995. Identification of QTLs for drought response in testing causal relationships between traits. J. Exptl. Bot. **46**: 853-865.

Lezar, S., Myburg, A.A., Berger, D.K., Wingfield, M.J. and Wingfield, B.D. 2004. Development and assessment of microarray-based DNA fingerprinting in *Eucalyptus grandis*. Theor. Appl. Genet. **109**: 1329– 1336.

Li, Z., Pinson, S.R.M., Stansel, J.W. and Park, W.B. 1995. Identification of quantitative trait loci (QTLs) for heading date and plant height in cultivated rice (*Oryza sativa* L.). Theor Appl. Genet. **91**: 374-381.

Lilley, J.M., Ludlow, M.M., McCouch, S.R., and O'Toole. J.C. 1996. Locating QTLs for osmotic adjustment and dehydration tolerance in rice. J. Exp. Bot. **47**: 1427-1436.

Lin, H.X., Qian, H.R., Zhuang, J.Y., Lu, J., Min, S.K., Xiaong, Z.M., Huang, N. and Zheng, K.L. 1996. RFLP mapping of QTLs for yield and related characters in rice (*Oryza sativa* L.). Theor. Appl. Genet. **92**: 920-927.

Litt, M. and Luty, J. A. 1989. A hypervariable microsatellite revealed by *in vitro* amplification of a dinucleotide repeat within the cardiac muscle actin gene. Amer. J. Human Genet. **4**: 397-401.

Mackill, D.J., Zhang, Z., Redona, E.D., Colowit, P.M. 1996. Level of polymorphism and genetic mapping of AFLP markers in rice. Genome **39**: 969-977.

Malyshev, S.V. and N.A. Kartel. 1997. Molecular markers in mapping of plant genomes. Belarus Mol. Biol. **31**: 163-171.

Martin, G.B., Nienhuis, J., King, G., and Schaefer, A. 1989. Restriction fragment length polymorphisms associated with water use efficiency in tomato. Science **243**: 1725-1728.

Martin, G.B., Williams, J.G.K. and Tanksley, S.D. 1991. Rapid identification of markers linked to a *Pseudomonas* resistance gene in tomato by using random primers and near-isogenic lines. Proc. Natl. Acad. Sci. USA. **88**: 2336-2340.

Maughan, P.J., Maroof, M.A.S., Buss, G.R., and Huestis, G.M. 1996. Amplified fragment lengh polymorphism (AFLP) in soybean: species diversity, inheritance and near isogenic line analysis. Theor. Appl. Genet. **93**: 392-401.

McCouch, S.R, Kochert, G, Yu, Z.H., Wang, Z.Y., Khush, G.S., Coffman, W.R and Tanksley, S.D. 1988. Molecular mapping of rice chromosome. Theor. Appl. Genet. **76**: 815-829.

Mienie, C.M.S., Smit, M.A., Cobos, S., and Torres, A.M. 1995. Using RAPDs to study phylogenitic relationships in *Rosa*. Theor. Appl. Genet. **92**: 273-277.

Mohan, M., Sathyanarayanan. P.V., Kumar, A., Srivastava, M.N, and Nair, S. 1997. Molecular mapping of a resistance-specific PCR-based marker linked to a gall midge resistance gene (*Gm4t*) in rice. Theor. Appl. Genet. **95**:777-782.

Morgante, M. and Olivieri, A.M. 1993. Abundance, variability and chromosomal location of microsatellites in wheat. Mol. Gen. Genet. **246**: 327-333.

Mullis, K.B and Faloona, F.A. 1987. Specific synthesis of DNA *in vitro*: A polymerase catalyzed chain reaction. Methods Enzymol. **155**: 335-351.

Nair, S., Bentur, J.S., Prasad Rao, U. and Mohan, M. 1995. DNA markers tightly linked to a gall midge resisance gene (Gm2) are potentially useful for marker-aided selection in rice breeding. Theor. Appl. Genet. **91**: 68-73.

Nair, S., Kumar, A., Srivastava, M.N., and Mohan, M. 1996. PCR-based markers linked to a gall midge resistance gene, Gm4, has potential for marker aided selection in rice. Theor. Appl. Genet. **92**: 660-665.

Nakamura, Y., Leppert, M., Connell, P., Wolf, R., Holm, T., Culver, M., Martin, C., Fujimoti, E., Hoff, M., Kumalin, E. and White, R. 1987. Variable number of tandem repeats (VNTR) markers for human gene mapping. Science (Washington, D.C.) **235**: 1616-1622.

Nandi, S., Subudhi, P.K., Senadhera, D., Manigbas, N.L., Sen-Mandi, S., and Huang, N. 1997. Mapping QTLs for submergence tolerance in rice by AFLP analysis and selective genotyping. Mol. Gen. Genet. **255**: 1-8.

Olson, M., Hood, L., Cantor, C., and Botstein, D. 1989. A common language for physical mapping of the human genome. Science. **29;245(4925):1434-5**.

Oziel, A., Hayes, P.M., Chen, F.Q. and Jones, B. 1996. Application of quantitative trait locus mapping to the development of winter habit malting barley. Plant Breeding **115**: 43-51.

Panaud, O., Chen, X. and McCouch, S. R. 1996. Development of microsatellite markers and characterization of simple sequence length polymorphism (SSLP) in rice (*Oryza sativa* L.). Mol. Gen. Genet. **252**: 597-607.

Paran, I. and Michelmore, R,W. 1993. Development of reliable PCR-based markers linked to downy mildew resistance genes in lettuce. Theor. Appl. Genet. **85**: 985-993.

Paran, I., Kesseli, R. and Michelmore, R. 1991. Identification of RFLP and RAPD markers linked to downy mildew resistance genes in lettuce using near-isogenic lines. Genome. **34**: 1021-1027.

Parsons, B.J, Newbury, H.J, Jackson, M.T and Ford-Lloyd, M.V. 1997. Contrasting genetic diversity relationship are revealed in rice (*Oryza sativa* L) using different marker types. Molecular Breeding **3**: 115-125.

Paterson, A.H, Lander, E.S, Hewitt, J.D, Paterson, S., Lincon, S.E. and Tanksley, S.D. 1988. Resolution of quantitative traits into Mendelian factors by using a complete linkage map of restriction fragment length polymorphism. Nature. **335**: 721-726.

Paterson, A.H., Tanksley S.D., and Sorrells M.E. 1991. DNA markers in plant improvement. Advances in Agronomy **46**: 39-90.

Perez de la Vega, M.P. 1993. Biochemical characterization of populations. In: Hayward, M.D., Bosemark, N.O., Romagosa, I. (eds) Plant breeding: Principles and prospects. Chapman and Hall, London, pp 182-201.

Prioul, J.-L., S. Quarrie, M. Causse, and D. de Vienne. 1997. Dissecting complex physiological functions through the use of molecular quantitative genetics. J. Exper. Bot. **48**:1151-1163.

Qi, X., Nikes, R.E., Stam, P., and Lindout, P. 1998. Identification of QTLs for partial resistance to leaf rust (*Puccinia horsei*) in barley. Theor. Appl. Genet. **96**: 1205-1215.

Quagliaro, G.M., Vischr, M., Tyrka, M. and Olivieri, A.M 2001. Identification of wild and cultivated sunflower for breeding purposes by AFLP markers. J. Hered. **92**: 38–42.

Quarrie, S. A., Laurie, D.A., Zhu, J., Lebreton, C., Semikhodskii, A., Steed, A., Witsenboer, H., and Calestani. C. 1997. QTL analysis to study the association between leaf size and abscisic acid accumulation in droughted rice leaves and comparisons across cereals. Plant Mol. Biol. **35**: 155-165.

Quarrie, S., Lebreton, C., Gulli, M., Calestani, C., and Marmiroli, N 1994. QTL analysis of ABA production in wheat and maize and associated physiological traits. Russ J. Plant Phys **41**: 565-571.

Quint, M., Mihaljevic. , R., Dussle. C., Xu. M., Melchinger, A. and Lübberstedt, T. 2002. Development of RGA-CAPS markers and genetic mapping of candidate genes for sugarcane mosaic virus resistance in maize. Theor. Appl. Genet. **105 (2-3)**: 355-363.

Rafalski, A. 2002. Applications of single nucleotide polymorphisms in crop genetics. Curr. Opin. Plant Biol. **5**: 94–100.

Ram, S. G.,. Parthiban, K. T., Kumar, S R., Thiruvengadam, V., and Paramathma, M. 2008. Genetic diversity among *Jatropha* species as revealed by RAPD markers. Genet Resour Crop Evol **55**: 803–809.

Ramakrishna, W., Chowdari, K.V., Lagu, M.D., Gupta, V.S. and Ranjekar, P.K. 1995. DNA fingerprinting to detect genetic variation in rice using hypervariable DNA sequences. Theor. Appl. Genet. **88**: 402-406.

Ray, J. D., Yu, L., McCouch, S. R., Champoux, M. C., Wang, G., Nguyen. H. T. 1996. Mapping quantitative trait loci associated with root penetration ability in rice (*Oryza sativa* L.). Theor. Appl. Genet. **92**: 627 – 636.

Reiter, R.S., Williams, J., Feldmann, K.A., Rafalski, J.A., Tingey, S.V. and Scolink, P.A. 1992. Global and local genome mapping in *Arabidopsis thaliana* by using recombinant inbred lines and random amplified polymorphic DNAs. Proc. Natl. Acad. Sci. **89**: 1477-1481.

Ribaut, J.M., Hoisington, D.A., Deutsch, J.A., Jiang, C., and Gonzalez-de-Leon, D. 1996. Identification of quantitative trait loci under drought conditions in tropical maize.1. Flowering parameters and the anthesis-silking interval. Theor. Appl. Genet. **92**: 905-914.

Roder, M.S., Plaschke, J., Konig, S.U, Borner, A., Sorrells, M.E., Tanksley, S.D. and Ganal, M.W. 1995. PCR-amplified microsatellite markers in plant genetics. Plant J. **3**: 175-182.

Rudd, S., Schoof, H. and Klaus, M. 2005. Plant Markers-A database of predicted molecular markers from plants. Nucleic Acids Res **33**: D628–D632.

Ryskova, A.P., Jincharadze, A.G., Prosnyak, M.I., Ivanov, P.L. and Limboraska, S.A. 1988. M 13 phage DNA as a universal marker for DNA fingerprint of animals, plants and microorganisms, FEBS Lett **233**: 388-392.

Saghai, M.A., Biyashev, R.M., Yang, G.P., Zhang, Q. and Allard, R.W. 1994. Extraordinary polymorphism microsatellite DNA in barely: Species diversity, chromosomal locations, and population dynamics. Proc.Natl. Acad. Sci. USA. **91**: 5466-5470.

Saito, A., Yano, M., Kishimoto, N., Nakahara, M., Yoshimura, A., Saito, K., Kuhara, Y., Ukai, M., Kawase, M., Nagamine, T., Yoshimura, S., Ideta, O., Ohsawa, R., Hayano, Y., Iwata, N., Sugiura, M. 1991. Linkage map of restriction fragment length polymorphism loci in rice. Jpn J. Breed. **41**: 665-670.

Salimath, S.S., De-Olivier, A.C., Godwin, I.D. and Bennetzen, J.L. 1995. Assessment of genome origins and genetic diversity in the genus *Eleusine* with DNA markers. Genome. **38**: 757-763.

Sanchez dela Hoz, M.P., Davila, J.A., Loarce, Y., and Ferrer, E. 1996. Simple sequence repeat primers used in polymerase chain reaction amplifications to study genetic diversity in barley. Genome **39**: 112-117.

Schubert, R., Mueller-Starck, G., Riegel, R. 2001. Development of EST-PCR markers and monitoring their intrapopulational genetic variation in *Picea abies* (L.) Karst. Theor. Appl. Genet. **103**: 1223-1231.

Semagn, K., Bjørnstad, Å., Skinnes, H., Marøy, A.G., Tarkegne, T. and William, M. 2006. Distribution of DArT, AFLP, and SSR markers in a genetic linkage map of a doubled haploid hexaploid wheat population. Genome **49**: 545–555.

Senior, M.L. and Heun, M. 1993. Mapping maize microsatellites and polymerase chain reaction confirmation of the targeted repeats using a CT primer. Genome. **36**: 884-889.

Swathi, S.P., Gupta, V.S., Aggarwal, R.K., Ranjekar, P.K. and Brar, D.S. 2000. Genetic diversity and phylogenetic relationship as revealed by inter simple sequence repeat (ISSR) polymorphism in the genus *Oryza*. Theor. Appl. Genet. **100**: 1311-1320.

Tanksley, S.D. 1993. Mapping polygenes. Annu. Rev. Genet. **27**: 205-233.

Tanksley, S.D., Young, N.D., Paterson, A.H. and Bonierbale, M.W. 1989. RFLP mapping in plant breeding: New tool for an old science. Bio Technology. **7**: 257-264.

Tanksley. S.D. and Orton, T. J. 1983. Isozymes in plant genetics and breeding, Parts 1A and 1B. Elsevier, Amsterdam.

Taramino, G., Tarchini, R., Ferrario, S., Lee, M., Pe, M.E. 1997. Characterization and mapping of simple sequence repeats (SSRs) in *Sorghum bicolor*. Theor. Appl. Genet. **95**: 66-72.

Taramino,G. and Tingey, S. 1996. Simple sequence repeats for germplasm analysis and mapping in maize. Genome **39**: 277-287.

Temesgen, B., Brown, G. R., Harry, D. E., Kinlaw ,C. S., Sewell, M. M. and Neale, D. B.. 2001. Genetic mapping of expressed sequence tag polymorphism (ESTP) markers in loblolly pine (*Pinus taeda* L.). Theor. Appl. Genet. **102**: 664-675.

Terachi T 1993. The progress of DNA analyzing techniques and its impact on plant molecular systematics. J. Plant Res. **106**: 75-79.

Thanh, N. D., Zheng, H. G., Dong, N. V., Trinh, L. N., Ali, M. L., and Nguyen, H. T., 1999. Genetic variation in root morphology and microsatellite DNA loci in upland rice (*Oryza sativa* L.) from Vietnam. Euphytica. **105**: 43-51.

Thomas, C. M., Vos, P., Zabeau, M., Jones, D. A., Norcott, K. A., Chadwick, B. P., Jones, J. D. G. 1995. Identification of amplified restriction fragment polymorphism (AFLP) markers tightly linkled to the tomato Cf-9 gene for resistance to *Cladosporium fulvum*. The Plant Journal **8**: 785-794.

Torres, A.M., Weeden, N.F., and Martin, A. 1993. Linkage among isozyme, RFLP and RAPD markers in *Vicia faba*. Theor. Appl. Genet. **85**: 937-945.

Tripathy, J. N., Zhang, J., Robin, S., Nguyen, T.T. and Nguyen, H. T. 2000. QTLs for cell-membrane stability mapped in rice (*Oryza sativa* L.) under drought stress. Theor. Appl. Genet. **100**: 1197-1202.

Varshney, R.K. Sigmund R, Börner A, Korzun V, Stein N, Sorrells M, Langridge P, Graner A. 2005. Interspecific transferability and comparative mapping of barley EST-SSR markers in wheat, rye and rice. Plant Sci. **168**: 195–202.

Vassart, G., Georges, M., Monsieuer, R., Brocas, H., Lequarre, A.S. and Christophe, D. 1987. A sequence of M 13 phage detects hypervariable minisatellites in human and animal DNA. Science. **235**: 683-684.

Vierling, R.A., Xiang, Z., Joshi, C.P., Gilbert, M.L., Nguyen, H.T. 1994. Genetic diversity among elite sorghum lines revealed by restriction fragment length polymorphisms and random amplified polymorphic DNAs. Theor. Appl. Genet. **87**: 816-820.

Virk, S. P., Ford-lloyd, V.B., Jackson, M.T., Pooni, S. H., Clemeno, T.P., and Newbury, H.J. 1996. Predicting quantitative variation within rice germplasm using molecular markers. Heredity **76**: 296-304.

Vos, P., Hogers, R., Bleeker, M., Reijans, M., Van de Lee, T., Horens, M., Frijters, A., Pot, J., Peleman, J., Kuiper, M. and Zabeau, M. 1995. AFLP: a new technique for DNA fingerprinting. Nucleic Acids Res. **23**: 4407-4414.

Wang, Y., Georgi, L.L., Zhebentyayeva, T. N., Reighard, G.L., Scorza, R, and Abbott. A. G. 2002. High-throughput targeted SSR marker development in peach (*Prunus persica*). Genome **45**: 319-328.

Wang, Y. G., Xing. Q. H., Deng, Q. Y., Liang, F. S., Yuan, L. P., Weng, M. L. and Wang, B. 2003. Fine mapping of the rice thermo-sensitive genic male-sterile gene *tms5*. Theor. Appl. Genet.**107 (5):** 917-921.

Weber, J. L. and May, P. E. 1989. Abundant class of human DNA polymorhisms which can be typed using the polymerare chain reaction. Am. J. Hum. Genet. **44**: 388-396.

Wenzl P, Carling J, Kudrna D, Jaccoud D, Huttner E, Kleinhofs A, Kilian A 2004. Diversity Arrays Technology (DArT) for whole-genome profiling of barley. Proc. Natl. Acad. Sci. USA **101**: 9915-9920.

Williams, J.G.K., Kubelik, A.R., Livak, K.J., Rafalski, J.A. and Tingey, S.V. 1990. DNA polymorphisms amplified by arbitrary primers are useful as genetic markers. Nucl. Acid. Res. **18**: 6531-6535.

Wittenberg AHJ, van der Lee T, Cayla C, Kilian A, Visser RGF, Schouten HJ 2005. Validation of the high-throughput marker technology DArT using the model plant *Arabidopsis thaliana*. Mol. Genet. Genomics **274**: 30-39.

Wu, K. S. and S. D. Tanksley, 1993. Abundance, polymorphism and genetic mapping of microsatellites in rice. Mol. Gen. Genet. **241**: 225-35.

Wu, P., Hu, B., Yi, K.K. and Liao, C.Y. 2000. QTLs and epistasis for tolerance for Al toxicity in rice at different seedling stages. Theor. Appl Genet **100**:1295-1303.

Wu, P., Zhang, G. and Huang, N. 1996. Identification of QTLs controlling quantitative characters in rice using RFLP markers. Euphytica **89**: 349-354.

Xia, L., Peng, K., Yang, S., Wenzl. P., de Vicente, M.C., Fregene, M. and Kilian, A. 2005. DArT for high-throughput genotyping of Cassava (*Manihot esculenta*) and its wild relatives. Theor. Appl. Genet. **110:** 1092-1098.

Xu, Y., Shen, Z., Chen, Y. and Zhu, L. 1995. Molecular mapping of quantitative trait loci controlling yield component characters using a maximum likelihood method in rice (*Oryza sativa* L.). Chines J. Genet. **22:** 37-42.

Yang, S., Pang, W., Ash, G., Harper, J., Carling, J., Wenzl, P., Huttner, E., Zong, X., Kilian, A 2006. Low level of genetic diversity in cultivated Pigeon pea compared to its wild relatives is revealed by diversity arrays technology. Theor. Appl. Genet. *113:* 585-595.

Yang, N. and Tanksley, S.D. 1989. RFLP analysis of the size of chromosomal segments retained around the *Tm-2* loucs of tomato during back cross breeding. Theor. Appl. Genet. **77:** 353-359.

Yang, W., De-Oliveira, A.C., Godwin, I., Schertz, K. and Bennetzen, J.L. 1996. Comparison of DNA marker technologies in characterizing plant genome diversity: variability in Chinese sorghums. Crop Sci. **36** : 1669-1676.

Yu, J., Hu, S., Wang, J., Wong, G.K.S., Li, S., Liu, B., Deng, Y., Dai, L., Zhou, Y. and Zhang, X. 2002. A draft sequence of the rice genome (*Oryza sativa* L. ssp. indica). Science, **296:** 79–92.

Yu, L.X. and Nguyen, H.T. 1994. Genetic variation detected with RAPD markers among upland and lowland rice cultivars (*Oryza sativa* L.). Theor. Appl. Genet. **87:** 668-672.

Zabeau, M. and Vos, P. 1993. Selective restriction fragment amplification, a general method for DNA fingerprinting. European patent Application Number 0534858 A1.

Zhang. Y and Stommel, J. R. 2001. Development of SCAR and CAPS Markers Linked to the *Beta* Gene in Tomato. Crop Sci. **41:** 1602–1608.

Zhuang, J.Y., Lin, H.X., Lu, J., Qian, H.R., Hittalmani, S., Huang, N. and Zheng, K.L. 1997. Analysis of QTL X environment interaction for yield components and plant height in rice. Theor. Appl. Genet. **95:** 799-808.

Zietkiewicz, E., Rafalski, A. and Labuda, D. 1994. Genome fingerprinting by simple sequence repeat (SSR) anchored polymerase chain reaction amplification. Genomics **20:** 176-183.

Chapter 3

In-silico re-evaluation of DNA Sequences for Drawing Phylogeny of Orchids with Special Emphasis on *Dendrobium* Sw.

Pritam Chattopadhyay and Nirmalya Banerjee

Introduction

Knowledge of DNA sequences has become indispensable for basic biological research and other numerous applied fields like diagnostic, biotechnology, forensic biology and biological systematics. Orchids are considered to be among one of the premier groups of flowering plants for evolutionary studies. For management of huge DNA data set and their efficient application, different databases have emerged about these wonderful plants. *In-silico* analysis of these DNA sequences required application of phylogenetic software employing tools used in comparative genomics, cladistics, and bioinformatics include neighbor-joining, maximum parsimony, UPGMA, Bayesian phylogenetic inference, maximum likelihood and distance matrix methods. In this review we have attempted to accumulate different DNA sequence based technologies, different locus of phylogenetic interests, different databases managing DNA sequences, different *in-silico* methods of DNA sequence analysis which all together contribute to develop new understanding about orchid systematic so that the new researchers find these useful information in single place and may make use of them.

Orchids (members of the family Orchidaceae) constitute an order of royalty in the world of ornamental plants and they are of immense horticultural importance and play a very useful role to balance the forest ecosystem (Kaushik, 1983). Orchids grow over a very wide range of conditions, from the icy wastes of Greenland to the hot and moist equatorial regions, and from sea level to about 14, 000 feet altitude. They have been found in swamps, deserts, forests and grasslands. The numbers of different kinds of orchids is truly amazing. Orchidaceae is one of the largest families of flowering plants, with between 21,950 and 26,049 currently accepted species, distributed in 880 genera (Stevens, 2001; Govaerts, 2010). The number of orchid species equals more than twice the number of bird species, and about four times the number of mammal species. It also encompasses about 6–11% of all seed plants. The largest genera are *Bulbophyllum* (2,000 species), *Epidendrum* (1,500 species), *Dendrobium* (1,400 species) and *Pleurothallis* (1,000 species). The genus *Dendrobium* is one of the most important genera in the orchid family with 1190 species as listed by the Royal Botanic Gardens, Kew, UK (http://en.wikipedia.org/wiki/List_of_Dendrobium_species). Moreover, since the introduction of tropical species in the 19th century, horticulturists have produced more than 100,000 hybrids and cultivars. India's biogeographical location at the junction of the Agrotropical, Indo-Malayan and Paleo-Arctic realms has contributed to the biological richness of the country. Like other important orchid growing countries, India is blessed with a wealth of orchid flora, and about 1300 species representing about 161 genera are estimated to occur in this country (Singh *et al.*, 2001). Characterization of the genetic diversity and examination of the genetic relationship among different orchid genera and species are important for the sustainable conservation and increased use of plant genetic resources.

The extensive development of molecular techniques for genetic analysis in the past decade has led to the increase of knowledge of orchid genetic diversity. Molecular techniques, in particular the use of molecular markers, have been used to monitor DNA sequence variation in and among species and cultivars of orchid (Wang *et al.*, 2009). As for example, cladistic parsimony analyses of *rbcL* nucleotide sequence data from 171 taxa representing nearly all tribes and subtribes of Orchidaceae divide the family into five primary monophyletic clades: apostasioid, cypripedioid, vanilloid, orchidoid, and epidendroid orchids, arranged in that order. These clades, with the exception of the vanilloids, essentially correspond to currently recognized subfamilies. A distinct subfamily, based upon tribe Vanilleae, is supported for *Vanilla* and its allies (Cameron *et al.*, 1999).

When public databases are flooded with different sequences for different loci of many species, it is important to undertake comparative investigation study for all possible loci, for deducing most consensus phylogenetic tree with all available species which may be used as a reference phylogenetic tree for further analysis. With the advancement in biological database management systems, analytical software and international collaborations, it is now possible to undertake comparative investigation study for all possible loci to re-evaluate phylogenetic relationship within this extraordinary group through *in-silico* experimentation

DNA based molecular techniques

Benefiting from molecular cloning and PCR techniques, DNA markers have now become a popular means for identification and authentication of plant and animal species. DNA-based markers are less affected by age, physiological condition of samples and environmental factors (Levine, 2000; Sanchez and Kron, 2008). They are not tissue-specific and thus can be detected at any phase of organism development. Only a small amount of sample is sufficient for analysis and the physical form of the sample does not restrict detection. These non-stringent requirements are particularly relevant for Orchid materials that are expensive or in limited supply. The power of discrimination of DNA-based markers is so high that very closely related varieties can be differentiated.

A diverse array of DNA-based marker technologies has been established to explore various DNA polymorphism and they can be classified into three broad types:

1. hybridization-based markers (First generation markers)
2. polymerase chain reaction (PCR)-based markers (Second and Third generation markers)
3. sequencing-based markers (Fourth generation markers).

1. *Hybridization-based Markers*

Hybridization-based marker technologies are considered to be the first generation markers which use cDNA, cloned DNA elements, or synthetic oligonucleotides as probes, which are labelled with radioisotopes or with conjugated enzymes that catalyse a coloured reaction, to hybridise DNA. The concept of using variations at DNA level as genetic markers started with the Restriction Fragment Length Polymorphism (RFLP). Subsequent to RFLP, several other methods such as Variable Number Tandem Repeats

(VNTR), Allele Specific Oligonucleotide (ASO), Allele specific Polymerase Chain Reaction (AS-PCR), Oligonucleotide Polymorphism (OP), Single Stranded Conformational Polymorphism (SSCP) and Sequence Tagged Sites (STS) have been developed (Kochert, 1994; Heslop-Harrison and Schwarzacher, 1996; Paterson, 1996). The conventional hybridization based assay of detecting DNA level variations was replaced by the Polymerase Chain Reaction based assay and it has been evolved to detect variations at DNA level.

DNA probes for detecting repeated DNA sequences can be divided into two types: (1) tandem repeats which are present as clusters along chromosomes and (2) dispersed repeats which are distributed over all the chromosomes. Shorter repeating sequences, that are generally less than 6 bp in length and are repeated from a few to many thousand times, are abundant in eukaryotes. They are designated as 'microsatellites' (Litt and Luty, 1989) or SSR. SSR sequence may act as a probe in DNA fingerprinting technologies (Leung and Ho, 1998). The probe, low C_0t DNA, is produced by shearing the isolated DNA from a reference DNA and the repetitive sequence is renatured and labelled. It is then hybridised to the membrane that has been fixed with restriction enzyme-digested DNA from the examined subject, which has been separated by electrophoresis. As a result, a specific DNA fingerprint is displayed on the autoradiogram. The abundance and ubiquitous distribution of microsatellites make them very valuable in linkage mapping, in identification of quantitative traits loci and in forensic cases. Additionally, a remarkable feature of microsatellites, not shared by minisatellites is that, the primers developed in one species can be used in related taxa (Coote and Bruford, 1996).

2. PCR-based Markers

PCR enzymatically multiplies a defined region of the template DNA. In 1990, two teams simultaneously reported the development of PCR-based, novel, genetic screening techniques: random amplified polymorphic DNA (RAPD) and arbitrarily-primed PCR (AP-PCR) (Williams *et al.*, 1990; Welsh and McClelland, 1990). These discoveries eventually led to the second generation of molecular markers. The second generation of molecular markers responsible for various revolutions in the field of molecular genetics, are micro satellites- arrays of tandemly repeated di-, tri-, tetra- and penta-nucleotide DNA sequences which occur dispersed throughout the genome of all eukaryotic organisms investigated to date. The micro satellites are otherwise called as Sequence Tagged Micro satellite Sites (STMS) or Simple Sequence Repeats (SSR). SSRs are currently considered the molecular markers of choice within the genome mapping community and are rapidly being adopted by plant researchers as well. SSRs consist of around 10-50 copies of motifs from 1 to 5 base pairs that can occur in perfect tandem repetition, as imperfect (interrupted) repeats or together with another repeat type. These repeated motifs are flanked by unique or single copy sequences, which provide a foot hold for specific amplification *via* PCR. Primers complimentary to the unique sequences in those flanking regions can be designed to amplify single copy products. The other marker systems developed during this period include Restriction Landmark Genome Scanning (RLGS), Cleaved Amplified Polymorphic Sequence (CAPS), Degenerate Oligonucleotide Primer PCR (DOP-PCR), Single Strand Conformation Polymorphism (SSCP), Multiple Arbitrary Amplicon Profiling (MAAP) and Sequence Characterized Amplified Region (SCAR). For example, species-specific fingerprints were identified in bacteria, fungi, mussels, ginseng and *Amaranthus* (Leung, 1999). Primers for SSR analysis can also be constructed by searching the GenBank for SSR loci of related species or by screening genomic libraries. Searching databases for existing information is a quick way to develop a probe for DNA fingerprinting and no prior knowledge of the sequence is needed. In addition, primers can be synthesized based on a repeat sequence of (CA)n, for example, primers (CA)8RG or (AGC)6TY have a degenerate 3'-anchor (Godwin *et al.*, 1997). SSR has been successfully used to construct detailed genetic maps of several organisms and to study genetic variation within populations of the same species, such as grapes, honeybees and tropical trees (Brown *et al.*, 1996).

The third generations of markers are the result of rapid developments in molecular biology that have opened the possibility of employing various types of molecular tools to identify and use genomic variation improvement of various organisms. Information concerning the basis of these techniques and their applications are from the technology spill over of several genome projects. The last decade have witnessed the birth of an array of molecular markers with high-throughput performance coupled with shift from manual mode of detection to complete automation. The examples are: Inter Simple Sequence Repeats (ISSR), Selective Amplification of Micro Satellite Polymorphic Loci (SAMPL), Single Nucleotide Polymorphisms (SNP), Amplified Fragment Length Polymorphism or selective Restriction Fragment Amplification (AFLP/SRFA), Allele Specific Associated Primers (ASAP), Cleavage Fragment Length Polymorphism (CFLP), Inverse Sequence-tagged Repeats (ISTR), Directed Amplification of Mini Satellite DNA-PCR (DAMD-PCR), Sequence-specific Amplified Polymorphism (S-SAP), Retrotransposon Based Insertional Polymorphism (RBIP), Inter-retrotransposon Amplified Polymorphism (IRAP), Retrotransposon-Microsatellite Amplified Polymorphism (MSAP), Methylation Sensitive Amplification Polymorphism

(REMAP), Miniature Inverted-repeat Transposable Element (MITE), Three Endonuclease AFLP (TE-AFLP), Inter-MITE Polymorphisms (IMP) and Sequence-related Amplified Polymorphism (SRAP) (Zeitkiewicz *et al.*, 1994; Morgante and Vogel, 1994; Jordan and Humperies, 1994; Voes *et al.*, 1995; Gu *et al.*, 1995; Brow, 1996; Bebeli *et al.*, 1997; Waugh *et al.*, 1997; Flavell *et al.*, 1988; Kalender *et al.*, 1999; Casa *et al.*, 2000; van der Wurff *et al.*, 2000; Chang *et al.*, 2001; Li and Quiros, 2001).

3. *Sequencing-based Markers*

DNA sequencing is a definitive means for identifying plant material. Further, variations due to transversion, transition, insertion or deletion can be accessed directly and information on a defined locus can be obtained (Franca *et al.*, 2002). The DNA sequencing techniques may be categorised as:

(i) First generation sequencing (e.g., the Sanger method and its most important variants and the Maxum and Gilbert method and other chemical methods);

(ii) Second generation sequencing (e.g., the Pyrosequencing method of DNA sequencing in real time by the detection of released pyrophosphate, PPi);

(iii) Third generation sequencing (e.g., single molecule sequencing with exonuclease);

(iv) Fourth generation sequencing (e.g., optical detection to decode nucleotide sequences).

(i) First generation sequencing: The classical chain-termination method was first developed by Sanger and Coulson (1975) and thereafter modified by several workers. It requires a single-stranded DNA template, a DNA primer, a DNA polymerase, normal deoxynucleotidetriphosphates (dNTPs), and modified nucleotides (dideoxyNTPs) that terminate DNA strand elongation. These ddNTPs will also be radioactively or fluorescently labelled for detection in automated sequencing machines. The DNA sample is divided into four separate sequencing reactions, containing all four of the standard deoxynucleotides (dATP, dGTP, dCTP and dTTP) and the DNA polymerase. To each reaction is added only one of the four dideoxynucleotides (ddATP, ddGTP, ddCTP, or ddTTP) which are the chain-terminating nucleotides, lacking a 3'-OH group required for the formation of a phosphodiester bond between two nucleotides, thus terminating DNA strand extension and resulting in DNA fragments of varying length. The newly synthesized and labelled DNA fragments are heat denatured, and separated by size (with a resolution of just one nucleotide) by gel electrophoresis on a denaturing polyacrylamide-urea gel with each of the four reactions run in one of four individual lanes (lanes A, T, G, C); the DNA bands are then visualized by autoradiography or UV light, and the DNA sequence can be directly read on the X-ray film or gel image. In the image on the right, X-ray film was exposed to the gel, and the dark bands correspond to DNA fragments of different lengths. A dark band in a lane indicates a DNA fragment that is the result of chain termination after incorporation of a dideoxynucleotide (ddATP, ddGTP, ddCTP, or ddTTP). The relative positions of the different bands among the four lanes are then used to read (from bottom to top) the DNA sequence.

The chemical sequencing method was first developed by Gilbert and Maxam (1973) and modified thereafter. The method requires radioactive labeling at one 5' end of the DNA (typically by a kinase reaction using gamma-^{32}P ATP) and purification of the DNA fragment to be sequenced. Chemical treatment generates breaks at a small proportion of one or two of the four nucleotide bases in each of four reactions (G, A+G, C, C+T). For example, the purines (A+G) are depurinated using formic acid, the guanines (and to some extent the adenines) are methylated by dimethyl sulfate, and the pyrimidines (C+T) are methylated using hydrazine. The addition of salt (sodium chloride) to the hydrazine reaction inhibits the methylation of thymine for the C-only reaction. The modified DNAs are then cleaved by hot piperidine at the position of the modified base. The concentration of the modifying chemicals is controlled to introduce on average one modification per DNA molecule. Thus a series of labelled fragments is generated from the radio labelled end to the first "cut" site in each molecule. The fragments in the four reactions are electrophoresed side by side in denaturing acrylamide gels for size separation. To visualize the fragments, the gel is exposed to X-ray film for autoradiography, yielding a series of dark bands each corresponding to a radiolabeled DNA fragment, from which the sequence may be inferred.

(ii) Second generation sequencing: The high demand for low-cost sequencing has driven the development of highthroughput sequencing technologies that could produce thousands or millions of sequences at once (Hall, 2007; Church, 2006; Schuster, 2008). The highthroughput sequencing was started with Lynx Therapeutics' Massively Parallel Signature Sequencing (MPSS) which was developed in the 1990s at Lynx Therapeutics. MPSS was a bead-based method that used a complex approach of adapter ligation followed by adapter decoding, reading the sequence in increments of four nucleotides. Subsequently many technologies were developed like Polony sequencing (Church, 2006), 454 pyrosequencing (Margulies *et al.*, 2005; Schuster, 2008), Illumina (Solexa) sequencing (Mardis, 2008); SOLiD sequencing (Schuster, 2008) and DNA nanoball sequencing (Drmanac *et al.*, 2010; Porreca *et al.*, 2008). A parallelized version of pyrosequencing was developed by 454 Life Sciences, which has since been acquired by Roche Diagnostics. The method amplifies DNA inside water droplets in an oil solution (emulsion PCR), with each droplet containing a single DNA template attached to a single primer-coated bead that then forms a clonal colony. The sequencing machine contains many picolitre-volume wells each containing a single bead and sequencing enzymes. Pyrosequencing uses luciferase to generate light for detection of the individual nucleotides added to the nascent DNA, and the combined data are used to generate sequence read-outs. This technology provides intermediate read length and price per base compared to Sanger sequencing on one end and Solexa and SOLiD on the other.

(iii) **Third generation sequencing:** In ultra-high-throughput sequencing as many as 500,000 sequencing-by-synthesis operations may be run simultaneously (Gilbert and Robert, 1992; Ten Bosch and Grody, 2008; Tucker *et al.*, 2009). It includes Helioscope™ single molecule sequencing, single molecule SMRT™ sequencing, single molecule real time (RNAP) sequencing etc. SMRT sequencing is based on the sequencing by synthesis approach. The DNA is synthesized in zero-mode wave-guides (ZMWs) - small well-like containers with the capturing tools located at the bottom of the well. The sequencing is performed with use of unmodified polymerase (attached to the ZMW bottom) and fluorescently labelled nucleotides flowing freely in the solution. The wells are constructed in a way that only the fluorescence occurring by the bottom of the well is detected. The fluorescent label is detached from the nucleotide at its incorporation into the DNA strand, leaving an unmodified DNA strand. This methodology even allows detection of nucleotide modifications (such as cytosine methylation). This happens through the observation of polymerase kinetics. This approach allows reads of up to 15,000 nucleotides, with mean read lengths of 2.5 to 2.9 kilobases (http://www.genomeweb.com/sequencing).

(iv) **Fourth generation sequencing:** In 2010, Life Technologies officially launched the first commercial sequencing instrument to avoid optical detection. The Personal Genome Machine (PGM), the product of Life Technologies' recent $375 million acquisition of Ion Torrent, monitors nucleotide incorporation electrochemically. Indeed, waiting in the wings, other companies are pursuing detection strategies based on nanopores and electron microscopy. The DNA passing through the nanopore changes its ion current. This change is dependent on the shape, size and length of the DNA sequence. Each type of the nucleotide blocks the ion flow through the pore for a different period of time (Branton *et al.*, 2008).

Importance of DNA Sequence based Technologies

A study appearing in the *Proceedings of the National Academy of Sciences* unravels 100 million years of evolution through an extensive analysis of plant genomes. It targets one of the major moments in plant evolution, when the ancestors of most of the world's flowering plants split into two major groups. Earlier studies were limited by technology and involved only four or five genes. The new study at Florida Museum of Natural History analyzed 86 complete plastid genome sequences from a wide range of plant species which proved that the taxa Pentapetalae failed to untangle the relationships among living species, suggesting that the plants diverged rapidly over 5 million years. Researchers selected genomes to sequence based on their best guess of genetic relationships from the previous sequencing work. The study provides an important framework for further investigating evolutionary relationships by providing a much clearer picture of the deep divergence that led to the split within flowering plants, which then led to speciation in the two separate branches (http://news.ufl.edu/2010/02/23/flower-evolution-2/).

There are several examples in the literature where DNA analysis has led to the resolution of longstanding controversies in plant classification. Brunsfeld *et al.* (1994) showed that species in the Taxodiaceae (redwood) family had only minor sequence differences in their chloroplast *rbc*L genes compared to species in the Cupressaceae. The buckwheat family (Polygonaceae) is widely recognized as a monophyletic group, but the membership of genera within its two subfamilies (Erigonoideae and Polygonoideae) has been controversial due to incongruencies in morphological character and phylogeny. Sequence data from the chloroplast genes *rbc*L, *mat*K and *ndh*F helped to redefine the tribes and genera within these subfamilies (Sanchez and Kron, 2008). By far, the nuclear ITS regions have proven most useful in deducing the correct phylogenies of plant groups. An ITS-derived phylogeny resolved the evolutionary relationships between genera in the Nyctagineae tribe of Nyctaginaceae while the *rbc*L-*acc*D IGS sequence provided relative few informative sites (Levin, 2000). A highly resolved phylogeny of 115 species of the widespread genus *Astragalus* was constructed from ITS sequence data and showed a distinct monophyly of New World *Astragalus* species (Wojchiechowski *et al.*, 1999). The ribosomal ITS-2 sequences were recomended as suitable tool for constructing phylogeny among very closely related taxa, such as members under a single genus or varieties of a particular species (Zhang *et al.*, 2005).

Identification of Proper Locus for Determination of Genetic Diversity

DNA sequencing and fingerprinting methods for the analysis of plant evolution and diversity have been in use for over two decades. Thousands of research reports have been published which have provided insight not only into the evolutionary pathways of plant species but also into the evolution of several genetic loci. Surprisingly, much of the research in this area has been accomplished by studying only a few regions of the plant genome. The smallest known haploid genome of a flowering plant still contains over 63 million base pairs of DNA while others contain over 100 billion base pairs. Therefore, it is crucial to locate which part of the genome (i.e., which *locus*) should be examined while searching for genetic variation.

Regulatory and coding sequences tend to be highly conserved and will generally only show variability when comparing plants belonging to different genera, families, orders, classes and divisions – with more sequence variation apparent at the higher taxonomic levels. Examples of these loci include the *rbcL* chloroplast gene and the nuclear 28S rDNA gene. Cladistic parsimony analyses of *rbcL* nucleotide sequence data from 171 taxa representing nearly all tribes and subtribes of Orchidaceae showed that the family may be divided into five primary monophyletic clades: apostasioid, cypripedioid, vanilloid, orchidoid, and epidendroid orchids, arranged in that order (Cameron *et al.*, 1999). These clades, with the exception of the vanilloids, essentially correspond to currently recognized subfamilies (Cameron *et al.*, 1999). Although powerful in assessing monophyly of clades within the family, in this case *rbcL* fails to provide strong support for the interrelationships of the subfamilies (Cameron *et al.*, 1999).

The greatest controversies in plant systematics arise at the lower taxonomic levels (genus and species) where morphological similarities can be quite high that create confusion in proper classification. On the scale of geological time, these hierarchial units have only recently diverged from common ancestors and their DNA has had much less time to accumulate mutations. Therefore, when comparing plants at these levels it is absolutely necessary to study loci that can pass mutations very rapidly without having a deleterious effect on the organism. Formerly known as "junk" DNA, these regions are interspersed between the conserved structural genes. A phylogenetic analysis of the major lineages of the sexually deceptive orchid genus *Ophrys* based on nuclear ribosomal (nr) DNA (internal transcribed spacer region) and noncoding chloroplast (cp) DNA (*trnL–trnF* region) sequences resulted in poorly resolved trees (Soliva *et al.*, 2001). A representative sample is the internal transcribed spacers (ITS) from ribosomal DNA (rDNA). The nrDNA ITS has proved to be a useful sequence for phylogenetic studies in many angiosperm families (Baldwin *et al.*, 1995;

Mitchell and Wagstaff, 1997). The level of ITS sequence variation suitable for phylogenetic analysis is found at various taxonomic levels within families, depending on the lineage (Fu *et al.*, 2001; Ngan *et al.*, 1999; Lau *et al.*, 2001). In general, sequence homologies within species were found to be high and those between species or families low, indicating that the ITS region can be used as a marker from family to intra-specific level. More extensive taxonomic sampling of both the nrDNA ITS and as well as a portion of 26S nrDNA were also recommended for more satisfactory estimate of relationships among orchids (Chase *et al.*, 2005).

In recent years there has been the realization that DNA analysis, i.e., molecular systematics, does not always provide clear cut answers to every problem. The *gene tree* (the phylogeny estimated by DNA analysis) does not always match the *species tree* (the true phylogeny of organisms) (Lyons-Weiler and Milinkovitch, 1997). The potential reasons for this incongruence are many, but the most common ones are caused by the evolutionary processes of rapid diversification, hybridization, introgression, incomplete lineage sorting, and inter-locus concerted evolution (Comes and Abbot, 2001). In other words, not all genetic changes reflect the same type of evolutionary divergence that is measured by morphological characters. Therefore, it is important to determine the appropriate locus for phylogenetic analysis.

The Consortium for the Barcode of Life (CBOL) has recommended that only *rbcL* and *matK* are approved and required barcode regions for land plants (CBOL, 2009). One of the major issues concerning the inclusion of molecular information into taxonomic aspects of biology is DNA barcodes. There are two separate tasks to which DNA barcodes are presently applied. The first one is to distinguish between species and the second one is to discover new species. Strong phylogenetic signal from *matK* has rendered it an invaluable gene in plant systematic and evolutionary studies at various evolutionary depths. Further, *matK* is proposed as the only chloroplast-encoded group II intron maturase, thus implicating *matK* in chloroplast post transcriptional processing. For a protein-coding gene, *matK* has an unusual evolutionary mode and tempo, including relatively high substitution rates at both the nucleotide and amino acids levels (Cullings and Vogler, 1998).

Microsatellite Analysis of Species and Populations

When analyzing intra-specific genetic variation, i.e., within a population or between populations of a species, another type of variable region, called a microsatellite, is often studied. Microsatellites, also referred to as Short Tandem Repeats (STRs), provide the highest resolution since their alleles can mutate from one generation to the next. Because of their abundance in nuclear DNA and the fact that they are inherited in a Mendelian fashion, microsatellite analysis (a type of DNA fingerprinting) is useful in determining kinship and gene flow and is most powerful when applied at the intra-specific level.

Nuclear microsatellites, being biparentally inherited and codominant, are ideal single-locus markers for population genetics, gene flow, hybridization, introgression, breeding and cultivar identification studies. Bredemeijer *et al.* (1998) developed a semi-automated detection method for tomato cultivar identification using microsatellite markers. Comparison of microsatellite loci of the sunflower species *Helianthus deserticola* with its presumptive parents provided strong evidence of its hybrid origin (Gross *et al.*, 2003). Genetic maps of nuclear microsatellite loci in some of the world's most important crops, such as wheat (Roder *et al.*, 1998), have proved to be extremely useful in breeding programmes.

Chloroplast microsatellites have found more utility in the assessment of genetic diversity within and among populations. Seven populations covering the entire range of the endangered Brazilian plant *Caesalpinia echinata* were analyzed at seven chloroplast microsatellite loci (Lira *et al.*, 2003). Isolation and genetic drift resulted in the fixation of haplotypes in five of those populations. A chloroplast microsatellite locus was used to compare haplotype frequencies between redwood (*Sequoia sempervirens*) populations in the southern and northern parts of their range (Brinegar *et al.*, 2007).

Twelve sets of nuclear SSR markers (Gu *et al.*, 2007) and nine sets of chloroplast SSR markers (Xu *et al.*, 2011) have been proposed in *Dendrobium officinale*. The set is based on six highly reproducible microsatellites with di-nucleotide repeats. In order to analyse the genetic diversity and structure of *D. fimbriatum*, 10 sets of SSR markers were developed (Fan *et al.*, 2009). In another study, 42 *Dendrobium* hybrids were identified with the help of 14 sets of SSR markers (Yu *et al.*, 2011) while 49 *Dendrobium* hybrids were investigated using nine sets of SSR markers (Boonsrangsom *et al.*, 2008). In a comparative study of RAPD and SSR with five selected *Dendrobium* species by our group, it was found that SSR markers are superior in the detection of polymorphism among *Dendrobium* species segregating them into their respective pre-defined morphological sections, than RAPD markers (Chattopadhyay *et al.*, 2012).

Discordance between nuclear and plastid phylogenies

In some of the previous examples, chloroplast DNA showed very little variation between taxa – not enough to provide a well resolved phylogeny.

Low genetic differentiation of organelle DNAs between closely related species is not unusual (Wolf *et al.*, 1997; Tsumura and Suyama, 1998; Domulin-Lapegue *et al.*, 1999). However, even when a chloroplast locus shows significant sequence variation, it may not provide a similar gene tree as a nuclear locus.

In fact, such discordance between nuclear and plastid phylogenies is fairly common. One of the most common reasons for this discordance is the phenomenon of incomplete lineage sorting. Ancestral chloroplast sequence haplotypes, sometimes more than a million years old, can survive in descendents through speciation events so that different species of a genus may carry the same haplotype. In a section of the genus *Senecio* where 502 individuals of 18 species were analyzed, only a few *trn*K haplotypes were found to be species specific (Comes and Abbott, 2001). In their *trn*L-F analysis of 875 individuals in 31 species of *Hordeum*, Jakob and Blattner (2006) found up to 18 haplotypes per single species with some haplotypes estimated to have survived for 4 million years. In both cases, chloroplast DNA phylogenies were determined to be unreliable compared to species trees estimated by ITS sequence data. An incongruence length difference test revealed that nrDNA and cpDNA data sets were not incongruent while phylogenetic analysis was carried of the major lineages of the sexually deceptive orchid genus *Ophrys* based on nuclear ribosomal (nr) DNA (internal transcribed spacer region) and noncoding chloroplast (cp) DNA (*trn*L–*trn*F region) sequences (Soliva *et al.*, 2001).

DNA sequencing of nrDNA ITS as the distinguishing markers for Orchid

The greatest controversies in plant systematics occur at the lower taxonomic levels (genus and species) where morphological similarities can be so great as to confound proper classification. On the scale of geological time, these hierarchial units have only recently diverged from common ancestors and their DNA has had much less time to accumulate mutations. Therefore, when comparing plants at these levels, it is necessary to study loci that can amass mutations very rapidly without having a deleterious effect on the organism. Formerly known as "junk" DNA, these regions are interspersed between the conserved structural genes.

The internal transcribed spacers (ITS1 and ITS2) of nuclear ribosomal DNA (rDNA), are nonstructural in nature and have much higher mutation rates relative to the 18S, 5.8S and 28S rDNA genes which they separate. These ITS regions are generally conserved within a species but show enough variation between species and genera to be useful in the construction of phylogenetic trees. These two loci, usually amplified and sequenced together with the small 5.8S rDNA region, have been tremendously helpful in delimiting plant taxa. An ITS-derived phylogeny resolved the evolutionary relationships between genera in the Nyctagineae tribe of Nyctaginaceae while the *rbcL-accD* IGS sequence provided relatively few informative sites (Levin, 2000). A highly resolved phylogeny of 115 species of the widespread genus *Astragalus* was constructed from ITS sequence data and showed a distinct monophyly of new world *Astragalus* species (Wojchiechowski *et al.*, 1999). Not only that, ITS sequences were reported to resolve taxa even at the level of subspecies and varieties of *Erigeron thunbergii* in Japan (Kawase *et al.*, 2007). Therefore, ITS sequences may be the best choice for studying genetic diversity in Orchids where diversity even at intra species level is highly vivid and prominent.

In orchids, in particular in Orchidinae, such an attempt is still lacking, but in recent years several independent and largely congruent studies (Bateman *et al.*, 1997; Pridgeon *et al.*, 1997; Aceto *et al.*, 1999; Cozzolino *et al.*, 2000) defined the patterns of phylogenetic relationships of the subtribe Orchidinae (Orchidoideae) based on nrDNA ITS sequences. In particular, most members of the old genus *Orchis* have been split into three related genera: *Anacamptis, Neotinea* and *Orchis* (s.s.) (Bateman 2009). These clades found support from karyological data and root tuber characteristics, though, in general, additional morphological synapomorphies defining these clades are still wanted (Bateman 2009). Douzery *et al.* (1999) sequenced the ITS genes of 13 members of *Disa*, thereby inferring some relationships within the Disinae. Separate maximum parsimomy analyses of RFLP, *matK* and nrDNA ITS sequences, macromorphological and anatomical data collected for 27 *Coelogyne* species and 13 representatives of related genera produce largely congruent results incating that Coelogyninae are monophyletic and diverged early into three major clades (Gravendeel, 2000). Epidendreae were studied intensively by van den Berg (2000), and his results relying upon *rbcL*, *matK* , *trnL-F* and ITS provide a definitive circumscription. Molecular systematic research using ITS on representatives of most taxonomic units within the Dendrobiinae has provided independent support, in addition to morphological and biological data, for the phylogenetic reassessment of the taxon (Clements, 2003). At a broad level, the Dendrobiinae is polyphyletic with *Epigeneium* forming an independent clade; *Dendrobium* section *Oxystophyllum* is deeply embedded within one of outgroups, subtribe Eriinae: Podochileae; and the remaining taxa isolated into two major groups, viz. the Asian and Australasian clades. A detailed study of part of the Asian clade, with emphasis on representatives of the morphologically based *Dendrobium* section *Pedilonum*, groups species into seven major clades (Clements, 2003).

In an attempt of rDNA analysis with three *Dendrobium* species (*D. tosaense, D. officinale,* and *D. moniliforme*) it has been shown that the 18S, 5.8S, and 28S are highly conserved sequences, making the plants difficult for inclusion in the same family (Blattner, 1999). The ITS1 and ITS2 regions were found to be heterologous in sequence and copy number, but low percentage differences were found between the samples at inter and intraspecific levels. In another attempt the genetic relationship of 36 *Dendrobium* species in China was determined on the basis of ITS sequences (Yuan *et al.,* 2009). The nrDNA ITS1 of *Dendrobium* was 225-234 bp and ITS2 was 239-248 bp and the phylogenetic relationship revealed by these sequences was partially supported by previously published monographic data (Yuan *et al.,* 2009). ITS-based analysis was successfully employed to differentiate and to ascertain the phylogenetic relationship among the 11 medicinal *Dendrobium* and two adulterant species *Pholidota articulata* and *Flickingeria comate. In-silico* analysis of nrDNA ITS2 by our group showed congruency with recent classification by Wood (2006) (Chattopadhyay *et al.,* 2010; Chaudhary *et al.,* 2012).

ITS microarray for rapid evaluation

Armed with the ITS sequences ITS microarray was established for high throughput authentication of medicinal *Dendrobium* species. Microarray of the ITS1-5.8S-ITS2 regions from 24 *Dendrobium* species, two other orchids and two non-orchids were generated. Distinctive hybridization profiles showed that 24 *Dendrobium* species could be differentiated from one another. The differentiation of *D. officinale* and *D. hercoglossum, D. nobile* and *D. moniliforme* was achieved by 5S rRNA array (Zhang *et al.,* 2005). This work has shown that ITS microarray not only could be used to establish the identities of the various *Dendrobium* species, but also to authenticate the medicinal *Dendrobium* from adulterant orchids. Orchid biochip was developed for *Phalaenopsis* microarray and is available for analysis in Orchidstra (http://orchidstra.abrc.sinica.edu.tw/chips).

In-silico analysis of DNA sequences

DNA sequencing and fingerprinting methods for the analysis of plant evolution and diversity have been in use for over two decades. Thousands of research reports have been published which have provided insight not only into the evolutionary pathways of plant species but also into the evolution of several genetic loci.

In bioinformatics, a sequence alignment is a way of arranging the sequences of DNA (even RNA or protein) to identify regions of similarity that may be due to functional, structural, or evolutionary relationships between the sequences (Mount, 2004). If two sequences in an alignment share a common ancestor, mismatches can be interpreted as point mutations and gaps as insertion or deletion mutations (indels) introduced in one or both lineages in the time since they diverged from one another. In sequence alignments of proteins, the degree of similarity between amino acids occupying a particular position in the sequence can be interpreted as a rough measure of how is a particular region or sequence motif is conserved among lineages. The absence of substitutions, or the presence of only very conservative substitutions (that is, the substitution of amino acids whose side chains have similar biochemical properties) in a particular region of the sequence, suggest that this region has structural or functional importance (Ng *et al.*, 2001). Although DNA and RNA nucleotide bases are more similar to each other than are amino acids, the conservation of base pairs can indicate a similar functional or structural role.

Computational phylogenetics make extensive use of sequence alignments in the construction and interpretation of phylogenetic trees, which are used to classify the evolutionary relationships between homologous genes represented in the genomes of divergent species. The degree to which sequences in a query set differ is qualitatively related to the sequences' evolutionary distance from one another. Roughly speaking, high sequence identity suggests that the sequences in question have a comparatively young most recent common ancestor, while low identity suggests that the divergence is more ancient. This approximation, which reflects the "molecular clock" hypothesis that a roughly constant rate of evolutionary change can be used to extrapolate the elapsed time since two genes first diverged (that is, the coalescence time), assumes that the effects of mutation and selection are constant across sequence lineages. Therefore, it does not account for possible difference among organisms or species in the rates of DNA repair or the possible functional conservation of specific regions in a sequence. (In the case of nucleotide sequences, the molecular clock hypothesis in its most basic form also discounts the difference in acceptance rates between silent mutations that do not alter the meaning of a given codon and other mutations that result in a different amino acid being incorporated into the protein.) More statistically accurate methods allow the evolutionary rate on each branch of the phylogenetic tree to vary, thus producing better estimates of coalescence times for genes.

Phylogenetic comparative methods

Phylogenetic comparative methods (PCMs) use information on the evolutionary relationships of organisms (phylogenetic trees) to compare species (Harvey and Pagel, 1991). The most common applications are to test for correlated evolutionary changes in two or more traits, or to determine whether a trait contains a phylogenetic signal (the tendency for related species to resemble each other). Several methods are available to relate particular phenotypic traits to variation in rates of speciation and/or extinction, including attempts to identify evolutionary key innovations. Most studies that employ PCMs focus on extant organisms. However, the methods can also be applied to extinct taxa and can incorporate information from the fossil record. Owing to their computational requirements, they are usually implemented by computer programmes (hence *in-silico*). PCMs can be viewed as part of evolutionary biology, systematics, phylogenetics, bioinformatics or even statistics, as most methods involve statistical procedures and principles for estimation of various parameters and drawing inferences about evolutionary processes. Methods for estimating phylogenies include neighbor-joining, maximum parsimony (also simply referred to as parsimony), UPGMA, Bayesian phylogenetic inference, maximum likelihood and distance matrix methods.

What distinguishes PCMs from most traditional approaches in systematics and phylogenetics is that they typically do not attempt to infer the phylogenetic relationships of the species under study. Rather, they use an independent estimate of the phylogenetic tree (topology plus branch lengths) that is derived from a separate phylogenetic analysis, such as comparative DNA sequences that have been analyzed by maximum parsimony or maximum likelihood methods. PCMs are consumers of phylogenetic trees, not primary producers of them. Different phylogenetics softwares show little overlap with the programmes for PCMs and a list of the softwares is available at http://en.wikipedia.org/wiki/List_of_phylogenetics_software.

Databases of universal primers for plants

Primers, that are designed to amplify important locus for a broad range of species distributed among phylogenetically distant families, are known as universal primers. In plants universal primers for *rbc*L, *mat*K, 5.8S rDNA and ITS were designed and under contentious development to maximize universality and resolvability (Chiang *et al.*, 1998; Cullings and Vogler, 1998; Linder *et al.*, 2000; Wicke and Quandt, 2009). To manage the continuously developing information regarding universal primers for DNA sequencing, different databases have emerged.

A primer database is maintained in Crandall lab, Brigham Young University, United States where all primers for 18S, 28S, CytB and ITS were maintained in a systematic order (http://crandalllab.byu.edu/ PrimerDatabase.aspx). Another database (available at http://bfw.ac.at/200/ 1859.html) was designed to serve as a resource for researchers who are venturing into the study of poorly described chloroplast genomes, whether for large- or small scale DNA sequencing projects, to study molecular variation or to investigate chloroplast evolution (Heinze, 2007).

Database of probes for orchid species

Though there are few databases featuring probes for DNA hybridization based technologies like 'plantGBD' (http://www.plantgdb.org/search/misc), 'probebase' (http://www.microbial-ecology.net/probebase/) and 'probe' (http://www.ncbi.nlm.nih.gov/probe/) most of them lacking information related to orchid. 'Probe' database is a public registry of nucleic acid reagents designed for use in a wide variety of biomedical research applications, together with information on reagent distributors, probe effectiveness, and computed sequence similarities maintained by National centre for biotechnology information, USA. Recently there was 22 SSR probe designed for different orchid species available for analysis (http://www.ncbi.nlm.nih.gov/probe?term=orchidaceae).

Databases of different DNA sequences for orchid species

In NCBI there is 285806 DNA sequence information of Orchid available for analysis in its Neucleotide database till 25th June, 2012 and counting is running (http://www.ncbi.nlm.nih.gov/sites/entrez?db=nucleotide&term=txid4747[Organism]&cmd=search). Among them 2791 sequences are from the Genus *Dendrobium* (http://www.ncbi.nlm.nih.gov/sites/entrez?db=nucleotide&term=txid37818[Organism]&cmd=search). Kew's DNA bank contains more than 1000 DNA sequences of orchid species including 20 species of the genus *Dendrobium* (http://apps.kew.org/dnabank/homepage.html). The plantgdb.org web resource is currently being managed as part of an NSF-funded project to develop robust genome annotation methods, tools, and standard training sets for the plethora of plant genomes currently or soon to be sequenced. It has a total of 229 sequence information about *Dendrobium* (http://www.plantgdb.org/search/misc/PublicPlantSeq.php?search=Search&term=Dendrobium&type=organism). The Global Genome Biodiversity Network (GGBN), was formed in October 2011 by 32 representatives of 13 biorepositories, biorepository networks and research organizations in nine countries in an effort to work towards the development of a global network of biorepositories and research organizations that are guided by shared values governing the exchange of samples, technology and information. The DNA Bank Network is one partner of the GGBN and will bring its experiences and knowledge to it. 25 entries were found for the family Orchidaceae (http://www.dnabank-network.org/Query.php?kingdom=Plantae).

Orchid genome databases

Continued improvement in sequencing quality, combined with increasingly sophisticated bioinformatic analysis of sequence data, has increased the relevance of whole genome sequencing to many fields of biological science including the pharmaceutical industry, agriculture, national defence and medicine (Andries *et al.*, 2005). Similarly, the increased availability of sequence data has served to support increasingly complex and informative comparative genomic studies. In both the cases, the enhanced relevance and power of genomic comparisons have, in turn, furthered demand for still more sequence data, with the goal of comparing entire genomes. Although high-throughput sequencing methodologies have been developed to accommodate the demand for sequence output, they consume large amounts of a valuable input: genomic DNA. Less input DNA is required for a genome-wide, microarray-based survey restricted to known mutations, but even this would require some form of amplification step (Syvanen, 1985). Other applications also place a high demand on potentially scarce DNA. Emerging relationships between specific genotypes and risk factors or disease states have focused attention on DNA samples that are of great medical/scientific importance, but of limited supply invasion and discomfort, these methodologies produce far less genomic DNA than less precise, more invasive techniques (Dietmaier, 1999; Hahn *et al.*, 2000; Lasken and Egholm, 2003). Considerable interest also exists in sequencing low abundance DNA from museum or fossil specimens. High rates of consumption, combined with high demand from the scientific community, may result in hard decisions restricting access to these limited or irreplaceable samples.

Genome sequencing is more time-consuming for plants than animals because plastid DNA is about 10 times larger than the mitochondrial DNA used in studying animal genomes. But continued improvements in DNA sequencing technology are now allowing researchers to analyze those larger amounts of data more quickly. In NCBI there are only four complete Orchid genomes (including one hybrid cultiver) available for analysis in its Genome database (http://www.ncbi.nlm.nih.gov/genome? term= Orchidaceae). Orchidstra is a developing database of orchid genome maintained by Agricultural Biotechnology Research Centre, Academia Sinica, National Taiwan University and Yuan Ze University (http:// orchidstra.abrc.sinica.edu.tw/none/). In Orchidstra there are five complete Orchid genomes. The species for which the complete genome sequence is publicly available are:

1. *Erycina pusilla*
2. *Neottia nidus-aris*

3. *Oncidium 'Gower ramsey'*

4. *Phalaenopsis aphrodite*

5. *Phalaenopsis bellina*

6. *Phalaenopsis equestris*

7. *Rhizanthella gardneri*

OrchidBase is another database where high throughput DNA sequences of orchid species by 454 or Solexa are maintained (http://140.116.25.218/EST/releaseSummary.aspx).

Databases for phylogenetic analysis of plants

'TreeBASE' (accessible at http://www.treebase.org) is a relational database containing phylogenetic information from research papers submitted to the Web site (Sanderson *et al.*, 1994; Piel *et al.*, 2002). This site allows users to search the database freely according to various keywords, and see visual representations of the trees. Moreover, it allows the user to gain access to information concerning a tree as well as use comparison tools to learn more about various taxa contained within the tree and their relationships with other taxa within the database. There are about 220 entries in 'TreeBASE' for the genus *Dendrobium* alone (http://treebase.org/treebase-web/search/taxonSearch.html). Attempts were also taken to develop newer database like 'TreeSearch' which were eventually not in work. 'TreeSearch' (once accessible at http://cs.nyu.edu/cs/faculty/shasha/pap ers/treesearch.html) was developed to be applied to phylogeny applications including XML querying (Shasha *et al.*, 2002).

In-silico orchid systematic and classification

Orchidaceae are rapidly becoming one of the best-studied families of the angiosperms in terms of intra-familial phylogenetic relationships. These studies demonstrate that several previous concepts about phylogenertic patterns were incorrect, which make all previous classifications in need of review (Chase *et al.*, 2005).

Genera Orchidacearum is a multi-disciplinary and multi-authored approach in producing a new classification of the orchids that emphasises monophyly. Evidence from molecular analyses is being integrated with morphological, ontogenetic, anatomical, cytological and biochemical evidences to produce a new phylogenetic classification of the orchids. From DNA sequence based phylogeny five subfamilies were recognised: Apostasioideae, Vanilloideae, Cypripedioideae, Orhidoideae are sister to all the rest, followed successively by Vanilloideae, Cypripedioideae and the remainder of the monandrous orchids, Orchidoideae and

Epidendroideae. Tribal, subtribal and generic delimitation has also been substantially revised (Pridgeon *et al.*, 1999; Pridgeon *et al.*, 1999-2005). An outline of the phylogenetic classification based on DNA data is available at 'Orchids' (http://www.orchids.co.in/dna-data-orchidaceae.shtm).

Conclusion

On the basis of current information available on re-evaluation of DNA sequences there is no doubt that the previous classification systems are not only incorrect in many respects but also irrefutable as they were so heavily dependent on the author's intuition. A classification solely based on DNA data may appear to be unwise and undesirable to many botanists. The newly circumscribed taxa may be likely to lack clear, defining variance in DNA sequence of a particular locus. But that same criticism can be made against systems, particularly for groups circumscribed by the lack of apomorphies. Therefore, it may be better to produce a revised classification based at first on DNA data alone. Such classification may be further supported by the collection of additional data (like morphometry) and thus be improved.

Orchids should be considered among one of the premier groups of flowering plants for evolutionary studies. The massive quantities of DNA data that are accumulated in the present time have revolutionized the ideas about orchid taxa. Darwin's next book after 'Origin of Species' was focused on orchids, and the reasons for this are clear. Orchids should be studied more because they optimize evolution in its most dynamic aspect, the rapid production of an incredibly diverse array of species. The challenge is to understand how this has come out, and thus an intensive study of this largest angiosperm family is highly solicited.

Acknowledgement

This work was financially supported by Council of Scientific and Industrial Research, Government of India.

References

Aceto, S., Caputo, P., Cozzolino, S., Gaudio, L. and Moretti, A. 1999. Phylogeny and evolution of *Orchis* and allied genera based on ITS DNA variation: morphological gaps and molecular continuity. Molec. Phyl. Evol. **13**: 67–76.

Andries, K., Verhasselt, P., Guillemont, J., Gohlmann, H.W., Neefs, J.M., Winkler, H., van Gestel, J., Timmerman, P., Zhu, M., Lee, E., *et al.* 2005. A diarylquinoline drug active on the ATP synthase of *Mycobacterium tuberculosis*. Science. **307**: 223–227.

Baldwin, B.G., Sanderson, M.J., Porter, J.M., Wojciechowski, M.F., Campbell, C.S. and Donoghue, M.J. 1995. The ITS region of nuclear ribosomal DNA-A valuable source of evidence on Angiosperm phylogeny. Ann. Mo. Bot. Gard. **82**: 247–277.

Bateman, R..M.., Pridgeon, A..M. and Chase, M.W. 1997. Phylogenetics of subtribe Orchidinae (Orchidoideae, Orchidaceae) based on nuclear ITS sequences. Infrageneric relationships and taxonomic revision to achieve monophyly of *Orchis sensu stricto*. Lindleyana **12**: 113–141.

Bateman, R.M. 2009. Field-based molecular identification within the next decade? J. Hardy Orchid Soc. **6**: 31–34.

Bebeli, P.J., Zhou, Z., Somers, D.J. and Gustafson, J.P. 1997. PCR primed with mini satellite core sequences yields DNA fingerprinting probes in wheat. Theor. Appl. Genet. **95**: 276-283.

Boonsrangsom, T., Pongtongkam, P., Masuthon, S. and Peyachoknagul, S. 2008. Development of microsatellite markers for *Dendrobium* orchids. Thai J. Genetics. **1**: 47-56.

Branton, D., Deamer, D.W., Marziali, A., Bayley, H., Benner, S.A., Butler, T., Ventra, M.D., Garaj, S., Hibbs, A., Huang, X., *et al*. 2008. The potential and challenges of nanopore sequencing. Nat. Biotechnol. **26**: 1146–1153.

Bredemeijer, G.M.M., Arens, P., Wouters, D., Visser, D. and Vosman, B. 1998. The use of semi automated fluorescent microsatellite analysis for tomato cultivar identification. Theor. Appl. Genet. **97**: 584-590.

Brinegar, C., Bruno, D., Kirkbride, R., Glavas, S. and Udranszky, I. 2007. Applications of redwood genotyping by using microsatellite markers. USDA Forest Service Gen. Tech. Rep. PSW-GTR-194, pp. 47-55.

Brow, M.A., Oldenburg, M.C., Lyamichev, V., Heisler, L.M., Lyamicheva, N., Hall. J.G., Eagan, N.J., Olive, D.M., Smith, L.M., Fors, L. and Dahlberg, J.E. 1996. Differentiation of bacterial 16S rRNA genes and intergenic regions and *Mycobacterium tuberculosis* katG genes by structure-specific endonuclease cleavage. J. Clinical Microbiol. **34**: 3129-3137.

Brown, S.M., Hopkins, M.S., Mitchell, S.E., Senior, M.L., Wang, T.Y., Duncan, R.R., Gonzalez-Candelas, F., Kresovich, S. 1996. Multiple methods for the identification of polymorphic simple sequence repeats (SSRs) in sorghum [*Sorghum bicolor* (L.) Moench]. Theor. Appl. Genet. **93**: 190-198.

Brunsfeld, S.J., Soltis, P.S., Soltis, D.E., Gadek, P.A., Quinn, C.J., Strenge, D.D. and Ranker, T.A. 1994. Phylogenetic relationships among the genera of Taxodiaceae and Cupressaceae: evidence from rbcL sequences. Syst. Bot. **19**: 253–262.

Cameron, K.M., Chase, M.W., Whitten, W.M., Kores, P.J., Jarrell, D.C., Albert, V.A., Yukawa, T., Hills, H.G., and Goldman, D.H. 1999. A phylogenetic analysis of the Orchidaceae: evidence from rbcL nucleotide sequences. Am. J. Bot. **86**: 208–224.

Casa, A.M., Brouwer, C., Nagel, A., Wang, L., Zhang, Q., Kresovich, S. and Wessler, S.R. 2000. The MITE family Heartbreaker (Hbr): Molecular markers in maize. Proc. Natl. Acad. Sci. **97**: 10083-10089.

CBOL Plant Working Group 2009. A DNA barcode for land plants. Proc. Natl. Acad. Sci. **106**: 12794 – 12797.

Chang, R.Y., O'Donoughue, L.S. and Bureau, T.E. 2001. Inter-MITE polymorphisms (IMP): a high throughput transposon-based genome mapping and fingerprinting approach. Theor. Appl. Genet. **102**: 773781.

Chase, M.W., Salamin, N., Wilkinson, M., Dunwell, J.M., Kesanakurthi, R.P., Haidar, N. and Savolainen, V. 2005. Land plants and DNA barcodes: Short-term and long-term goals. Phil. Trans. Royal Soc. London, B, Biol. Sci. **360**: 1889 – 1895.

Chattopadhyay, P., Banerjee, N. and Chaudhary, B. 2010. Precise seed micromorphometric markers as a tool for comparative phylogeny of *Dendrobium* (Orchidaceae). Floricul. Ornamental Biotechnol. **4**: 36-44.

Chattopadhyay, P., Banerjee, N. and Chaudhary, B. 2012. Genetic characterization of selected medicinal *Dendrobium* (Orchidaceae) species using molecular markers. Res. J. Biol. **2**: 117-125.

Chaudhary, B., Chattopadhyay, P., Verma, R. and Banerjee, N. 2012. Understanding the phylomorphological implications of pollinia from *Dendrobium* (Orchidaceae). Am. J. Plant Sci. **3**: 816-828.

Chiang, T.Y., Schaal, A.B. and Ching, I.P. 1998. Universal primers for amplification and sequencing a noncoding spacer between the *atpB* and *rbcL* genes of chloroplast DNA. Bot. Bull. Acad. Sin. **39**: 245-250.

Church, G.M., 2006. Genomes for all. Sci. Am. **294**: 46–54.

Clements, M.A. 2003. Molecular phylogenetic systematics of the Dendrobiinae (Orchidaceae), with emphasis on *Dendrobium* section *Pedilonum*. Telopea. **10**: 247–298.

Comes, H.P. and Abbott, R.J. 2001. Molecular phylogeography, reticulation, and lineage sorting in Mediterranean *Senecio* sect. *Senecio* (Asteraceae). Evolution. **55**: 1943–1962.

Coote, T. and Bruford, M.W. 1996. Human microsatellites applicable for analysis of genetic variation in apes and Old World monkeys. J. Heredity **87**: 406-410.

Cozzolino, S., Aceto, S., Caputo, P., Widmer, A. and Dafni, A. 2001. Speciation processes in eastern Mediterranean *Orchis* s.l. species: molecular evidence and the role of pollination biology. Israel J. Plant Sci. **49**: 91– 103.

Cullings, K.W. and Vogler, D.R. 1998. A 5.8S nuclear ribosomal RNA gene sequence database: applications to ecology and evolution. Mol. Ecol. **7**: 919–923.

Dietmaier, W., Hartmann, A., Wallinger, S., Heinmoller, E., Kerner, T., Endl, E., Jauch, K.W., Hofstadter, F. and Ruschoff, J. 1999. Multiple mutation analyses in single tumor cells with improved whole genome amplification. Am. J. Pathol. **154**: 83–95.

Domulin-Lapegue, S., Kremer, A. and Petit, R.J. 1999. Are chloroplast and mitochondrial DNA variation species independent in oaks? Evolution. **53**: 1406-1413.

Douzery, E.J.P., Pridgeon, A.M., Kores, P., Linder, H.P., Kurzweil, H. and Chase, M.W. 1999. Molecular phylogenetics of Diseae (Orchidaceae): a contribution from nuclear ribosomal ITS sequences. Amer. J. Bot. **86**: 887–899.

Drmanac, R., Sparks, A.B., Callow, M.J., Halpern, A.L., Burns, N.L., Kermani, B.G., Carnevali, P., *et al.* 2010. Human genome sequencing using unchained base reads on self-assembling DNA nanoarrays. Science **327**:78–81.

Fan, X.M., Zhang, Y.M., Yao, W.H., Chen, H.M., Tan, J., Xu, C.X., Han, X.L., Luo, L.M. and Kang, M.S. 2009 Classifying maize inbred lines into heterotic groups using a factorial mating design. Agron. J. **101**: 106-112.

Flavell, A.J., Knox, M.R., Pearce, S.R. and Ellis, T.H.N. 1998. Retro transposon-based insertion polymorphisms (RBIP) for high throughput marker analysis. Plant J. **16**: 643-650.

Fu, H., Park, W., Yan, X., Zheng, Z., Shen, B. and Dooner, H.K. 2001. The highly recombinogenic bz locus lies in an unusually gene-rich region of the maize genome. Proc. Natl. Acad. Sci. **98**: 8903–8908.

Gilbert, K. and Robert, M. 1992. Massively parallel, optical, and neural computing in the United States. Moxley.

Godwin, I.D., Aitken, E.A.B. and Smith, L.W. 1997. Application of inter-simple sequence repeat (ISSR) markers to plant genetics. Electrophoresis **18**: 1524–1528.

Govaerts, R. 2010. World Checklist of Ephedraceae and Gnetaceae. The Board of Trustees of the Royal Botanic Gardens, Kew. Published on the Internet; http://www.kew.org/wcsp/.

Gravendeel, B. 2000. Reorganising the Orchid genus *Coelogyne*, a phylogenetic classification based on morphology and molecules. Thesis, Universiteit Leiden.

Gross, B.L., Schwarzbach, A.E. and Rieseberg, L.H. 2003. Origin(s) of the diploid hybrid species *Helianthus deserticola* (Asteraceae). Am. J. Bot. **90**: 1708–1719.

Gu, W., Shan, Y.A., Zhou, J., Jiang, D.P., Zhang, L., Du, D.Y. et al. 2007. Functional significance of gene polymorphisms in the promoter of myeloid differentiation-2. Ann. Surg. **246**: 151–158.

Gu, W.K., Weeden, N.F., Yu, J. and Wallace, D.H. 1995. Large-scale, cost-effective screening of PCR products in marker-assisted selection applications. Theor. Appl. Genet. **91**: 465-470.

Hahn, Y., Lee, Y.J., Yun, J.H., Yang, S.K., Park, C.W., Mita, K., Huh, T., Rhee, M. and Chung, J.H. 2000. Duplication of genes encoding non-clathrin coat protein gamma-COP in vertebrate, insect and plant evolution. FEBS Lett. **482:** 31-36.

Hall, B. 2007. Phylogenetic Trees Made Easy, 3rd edn. Sunderland, MA: Sinauer Associates.

Heinze, B. 2007. A database of PCR primers for the chloroplast genomes of higher plants. Plant Methods **3**: 4.

Heslop-Harrison, J.S. and Schwarzacher, T. 1996. Genomic Southern and *in situ* hybridization for plant genome analysis. In: Jauhar P.P., ed. Methods of genome analysis in plants. Boca Raton: CRC Press, 163-179.

Jordan, S.A. and Humphries, P. 1994. Single nucleotide polymorphism in exon 2 of the BCP gene on 7q31-q35. Human Mol. Gen. **3**: 1915.

Kalendar, R., Grob, T., Regina, M., Suoniemi, A. and Schulman, A. 1999. IRAP and REMAP: two new retrotransposon-based DNA fingerprinting techniques. Theor. Appl. Genet. **98**: 704-711.

Kaushik, P. 1983. Ecological and Anatomical Marvels of the Himalayan Orchids. Today and Tomorrow's Printers and Publishers, New Delhi.

Kawase, D., Yumoto, T., Hayashi, K. and Sato, K. 2007. Molecular phylogenetic analysis of the infraspecific taxa of *Erigeron thunbergii* A. Gray distributed in ultramafic rock sites. Plant Species Biol. **22**: 107-115.

Kochert, G. 1994. RFLP technology. In : Phillips, R. L. and Vasil, I.K., (eds.) DNA – based markers in plants. Kluwer Academic Publishers, Dordrecht, 8-38.

Lasken, R.S. and Egholm, M. 2003. Whole genome amplification: abundant supplies of DNA from precious samples or clinical specimens. Trends Biotechnol. **1**: 531-535.

Lau, N., Lim, L., Weinstein, E. and Bartel, D. 2001. An abundant class of tiny RNAs with probable regulatory roles in *Caenorhabditis elegans*. Science. **294**: 858-862.

Leung, F.C. and Ho, I.S.H. 1998. Isolation of novel repetitive DNA sequences as DNA fingerprinting probes for *Panax ginseng*. Proceedings in the 1st European Ginseng Congress, Marburg, December 1998, pp. 125–132.

Levine, J.M. 2000. Species diversity and biological invasions: relating local process to community pattern. Science. **288**: 852–854.

Li, G. and Quiros 2001. Sequence-related amplified polymorphism (SRAP), a new marker system based on a simple PCR reaction: its application to mapping and gene tagging in *Brassica*. Theor. Appl. Genet. **103**: 455461.

Linder, C.R., Moore, L.A., Jackson, R.B. 2000. A universal molecular method for identifying underground plant parts to species. Mol. Ecol. **9**: 1549-1559.

Lira, C.F., Cardoso, S.R.S., Ferreira, P.C.G. *et al.* 2003. Long-term population isolation in the endangered tropical tree species *Caesalpinia echinata* Lam. revealed by chloroplast microsatellites. Mol. Ecol. **12**: 3219–3225.

Litt, M. and Luty, J.A. 1989. A hypervariable microsatellite revealed by *in vitro* amplification of a dinucleotide repeat within the cardiac muscle actin gene. Am. J. Human Gen. **44**: 397-401.

Lyons-Weiler, J. and Milinkovitch, M.C. 1997. A phylogenetic approach to the problem of differential lineage sorting. Mol. Biol. Evol. **14**: 968–975.

Mardis, E.R. 2008. Next-generation DNA sequencing methods. Annu. Rev. Genomics. Hum. Genet. **9**: 387-402.

Margulies, M., Egholm, M., Altman, W.E., Attiya, S., Bader, J.S., Bemben, L.A., Berka, J., Braverman, M.S., Chen, Y.J., Chen, Z., *et al.* 2005. Genome sequencing in microfabricated high-density picolitre reactors. Nature. **437**: 376–380.

Mitchell, A.D. and Wagstaff, S.J. 1997. Phylogenetic relationships of *Pseudopanax* species (Araliaceae) inferred from parsimony analysis of rDNA sequence data and morphology. Pl. Syst. Evol. **208**: 121-138.

Morgante, M. and Vogel, J. 1994. Compound microsatellite primers for the detection of genetic polymorphisms. U S patent application no. 08/326456.

Ng, Y. A., Zheng, X.A. and Jordan, I.M. 2001. Stable algorithms for link analysis. In Proc. 24th Annual Intl. ACM SIGIR Conference. ACM.

Ngan, F., Shaw, P., But, P.P.H. and Wang, J. 1999. Molecular authentication of *Panax* species. Phytochemistry **50**: 787-791.

Paterson, A. 1996. Genome Mapping in Plants. R.G. Landes, Austin.

Piel, W.H., Donoghue, M.J. and Sanderson, M.J. 2002. TreeBASE: a database of phylogenetic knowledge. Pp. 41-47. In: **Shimura, J., K. L. Wilson, and D. Gordon** (eds.) *To the interoperable "Catalog of Life" with partners Species 2000 Asia Oceanea.* Research Report from the National Institute for Environmental Studies No. 171, Tsukuba, Japan.

Porreca, R., Drulhe, S., de Jong, H. and Ferrari-Trecate, G. 2008. Structural identification of piecewise-linear models of genetic regulatory networks. J. Comput. Biol. **15**: 1365–1380.

Pridgeon, A.M.., Bateman, R..M., Cox, A.V., Hapeman, J.R. and Chase, M.W. 1997. Phylogenetics of the subtribe Orchidinae (Orchidoideae,Orchidaceae) based on nuclear ITS sequences. 1. Intergeneric relationshipsand polyphyly of *Orchis sensu lato*. Lindleyana. **12**: 89–109.

Pridgeon, A.M., Cribb, P.L., Chase, M.W. and Rasmussen, F.N. (eds.) (1999– 2005). Genera *Orchidacearum*, vols 1–4. Oxford University Press, Oxford.

Roder, M.S., Korzun, V., Wendehake, K., Plaschke, J., Tixier, M.H., Leroy, P., Ganal, M.W. 1998. A microsatellite map of wheat. Genetics. **149**: 2007–2023.

Sanchez, I. and Kron, K.A. 2008. Phylogenetics of Polygonaceae with an emphasis on the Evolution of Eriogonoideae. Syst. Bot. **33**: 87-96.

Sanderson, M.J., Donoghue, M.J., Piel, W. and Eriksson, T. 1994. TreeBASE: a prototype database of phylogenetic analyses and an interactive tool for browsing the phylogeny of life. Am. J. Bot. **81**: 183.

Sanger, F. and Coulson, A.R. 1975. A rapid method for determining sequences in DNA by primed synthesis with DNA polymerase. J. Mol. Biol. **94**: 441–448.

Schuster, S.C. 2008. Next generation DNA sequencing transforms today's biology. Nat. Methods. **5**: 16–18.

Shasha, D., Wang, J. and Giugno, R. 2002. Algorithmics and applications of tree and graph searching. In Proc. 21th ACM Symp. on Principles of Database Systems (PODS'02). 39–52.

Singh, K.P., Phukan, S. and Bujarbarua, P. 2001. Orchidaceae. (eds.) Singh, N.P. and Singh, D.K. *in* Floristic Diversity and Conservation Strategies in India (Botanical Survey of India, Kolkata), Vol. IV, pp. 1735-1846.

Soliva, M., Kocyan, A. and Widmer, A. 2001. Molecular phylogenetics of the sexually deceptive orchid genus *Ophrys* (Orchidaceae) based on nuclear and chloroplast DNA sequences. Mol. Phylogenet. Evol. **20**: 78-88.

Stevens, P.F. 2001. Angiosperm Phylogeny Website. Version 9, June 2008 (and more or less continuously updated since). http://www.mobot.org/MOBOT/research/APweb/.

Syvanen, M. 1985. Cross-species Gene Transfer; Implications for a New Theory of Evolution. J. Theor. Biol. **112**: 333-343.

Ten Bosch, J.R. and Grody, W.W. 2008. Keeping Up with the Next Generation. J. Mol. Diagnos. **10**: 484–492.

Tsumura, T. and Suyama, Y. 1998. Differentiation of mitochondrial DNA polymorphism in populations of five Japanese *Abies*. Evolution. **52**: 1031-1042.

Tucker, T., Marra, M. and Friedman, J.M. 2009. Massively Parallel Sequencing: The next big Thing in genetic medicine. The Ame. J. Human Gen. **85**: 142–154.

van den Berg, C. 2000. Molecular phylogenetics of Epidendreae with emphasis on subtribe Laeliinae (Orchidaceae). Ph.D. dissertation, University of Reading, Reading, UK.

van der Wurff, A.W.G., Chan, Y.L., van Straalen, N.M. and Schouten, J. 2000. TE-AFLP: combining rapidity and robustness in DNA fingerprinting. Nucleic Acids Res. **28**: 105-109.

Vos, P., Hogers, R., Bleeker, R., Reijans, M., van de Lee, T., Homes, M., Frijters, A., Pot, J., Peleman, J., Kupier, M. and Zabeau, M. 1995. AFLP: a new technique for DNA fingerprinting. Nucleic Acids Res. **23**: 4407–4414.

Wang, H.Z., Feng, S.G., Lu, J.J., Shi, N.N. and Liu, J.J. 2009. Phylogenetic study and molecular identification of 31 *Dendrobium* species using intersimple sequence repeat (ISSR) markers. Sci. Hort., **122**: 440-447.

Waugh, R., McLean, K., Flavell, A.J., Pearce, S.R., Kumar, A., Thomas, B.T., Powell, W. 1997. Genetic distribution of BARE-1 retro transposable elements in the barley genome revealed by sequence-specific amplification polymorphisms (S-SAP). Mol. Gen. Genet. **253**: 687 694.

Welsh, J. and McClelland, M. 1990. Fingerprinting genomes using PCR with arbitrary primers. Nucleic Acids Res. **18**:7213-7218.

Wicke, S. and Quandt, D. 2009. Universal primers for the amplification of the plastid *trnK/matK* region in land plants. Anales del Jardín Botánico de Madrid. **66**: 285-288.

Williams, J.G.K., Kublik, A.R., Livak, K.J., Rafalski, J.A. and Tingey, S.V. 1990. DNA polymorphism amplified by arbitrary primers are useful as genetic markers. Nucleic Acids Res. **18**: 6531-6535.

Wojciechowski, M.F., Sanderson, M.J. and Hu, J.M. 1999. Evidence on the monophyly of *Astragalus* (Fabaceae) and its major subgroups based on nuclear ribosomal DNA ITS and chloroplast DNA *trnL* intron data. Syst. Bot. **24**: 409–437.

Wolf, P.G., Murray, R.A., Sipes, S.D. 1997. Species-independent, geographical structuring of chloroplast DNA haplotypes in a mountain herb *Ipomopsis* (Polemonieaceae). Molec. Ecol. **6**: 283-291.

Wood, H. 2006. The *Dendrobiums*: A.R.G. Ganter Verlag, Ruggell/Liechtenstein.

Xu, Z., Lu, C., Hiou, F. and Ding. 2011. Development of novel chloroplast microsatellite markers for *Dendrobium officinale* and cross-amplification in other *Dendrobium* species (Orchidaceae). Sci. Hortic. **128(4)**: 5.

Yu, Y., Yuan, D.J., Liang, S.G., Li, X.M., Wang, X.Q. *et al.* 2011.Genome structure of cotton revealed by a genome-wide SSR genetic map constructed from a BC1 population between *Gossypium hirsutum* and *G. barbadense*. BMC Genomics **12**: 15.

Yuan, Z.L., Chen, Y.C. and Yang, Y. 2009. Diverse non-mycorrhizal fungal endophytes inhabiting an epiphytic, medicinal orchid (*Dendrobium nobile*): estimation and characterization. World J. Microbiol. Biotechnol. **25**: 295-303.

Zhang, J., Wheeler, D.A., Yakub, I., Wei, S., Sood, R., Rowe, W., Liu, P.P., Gibbs, R.A. and Buetow, K.H. 2005. SNP detector: a software tool for sensitive and accurate SNP detection. PLoS Comput. Biol. **1**: 0395-0404.

Zietkiewicz, E., Rafalski, A. and Labuda, D. 1994. Genomic fingerprinting by simple sequence repeat (SSR)-anchored polymerase chain reaction amplification. Genomics **20**: 176-183.

Chapter 4

Effective Nutritional Requirements and an *in vitro* Approach for Asymbiotic Seed Germination and Seedling Growth of Some Terrestrial Orchids

Madhumita Majumder and Nirmalya Banerjee

Introduction

Orchidaceae is the largest and most diverse family of the flowering plants, consisting of ~35,000 species under 800 genera. Due to poor hybridization barrier, more than 1,50,000 hybrids already have been produced and registered, many of which are multigeneric. Exquisite and perpetual flower of orchids made them doyen among ornamentals. Evident by recent increases in world floriculture trade, orchids became the second most popular cut flowers as well as potted floriculture crop with high market value. Due to continuous destruction of natural habitats, unauthorized trade and ruthless collection by orchid lovers, many orchid species in nature are disappearing at an alarming rate. Continuing loss of native orchid habitat has led to an increased emphasis on orchid conservation. Furthermore, their high commercial demand has undoubtedly led to an increased emphasis on mass propagation and conservation of important orchids. Major obstacles for mass propagation of economically important orchids for commercial purposes as well as conservation are: (1) unavailability of efficient and reliable protocols for seed germination, (2) obligate mycorrhizal association for natural seed germination, (3) a clear understanding of early seedling growth and development and, (4) high mortality of seedlings during transplantation. Use of *in vitro* protocols has been foreseen as a successful approach for *ex-situ* conservation and reintroduction of endangered orchids. Plants regenerated from seeds have a broader genetic background than those developed by clonal propagation methods. Therefore, the former meet the goals of a reintroduction programme better, in the sense of warranting sufficient genetic resources in the reintroduced population to undergo adaptive evolutionary change. Ever since the development of a method for asymbiotic germination of orchid seeds by Knudson (1922), the techniques have been used routinely for large scale propagation of a number of orchid species and their hybrids, but a very few studies critically investigated the peculiarities of seed germination and protocorm development. Keeping this in mind, the present study was under taken with a view to i) develop efficient protocol for *in vitro* asymbiotic germination of seeds, ii) study the effects of nutritional requirements, activated charcoal and cultural conditions on seed germination and growth of seedlings. These findings also offer an opportunity to commercial nurseries for large-scale propagation as well as for *ex situ* conservation of several genera.

Orchids are one of the largest and most diverse groups among angiosperms. The family Orchidaceae includes 800 genera and 25,000 species (Atwood, 1986; Freudenstein and Rasmussen, 1999; Stewart and Griffiths, 1995; Ng and Hew, 2000). The family is considered as an ancient group, with their present distribution pattern suggesting an origin that probably predates the break-up of Gondwana land about 125 million years ago. Yet, orchids have all the earmarks of a group in active evolution – species, genera, tribes and subtribes are all difficult to delimit (Dressler and Dodson, 1960).

Orchids are well known for their long lasting beautiful flowers. The immense floral variations within the species along with their long shelf lives make orchids one of the most important floricultural crops of present day (Dressler, 1993). These culminate one of the evolutionary lines of monocots and are still in the process of active speciation. A complicated floral make up, a specialized pollination mechanism and dependence on a suitable mycorrhizal association for seed germination are some of their adapted features. The orchids are essentially out-breeders, and as a consequence free exchange of gene pools accounts for an endless series of continuous morphological variability in many taxa. The extreme degree of morphological variability in orchids is attributed to genetic drift. They have a world-wide distribution, occurring in various ecological habitats as saprophytes, terrestrials or epiphytes (Dressler, 1981; Batty *et al.*, 2002). More than 70% of orchid species are epiphytic and mostly inhabit in the tropic (Atwood, 1986). The most fascinating fact that in India alone about 1229 species have been found that are distributed in 184 genera. Among them more than 300 species are endemic. The main orchid- rich belts in the country are Peninsular India, North-Eastern India, Eastern and the Western Himalayas.

The flowers of orchids have always fascinated botanists and horticulturists which are of various shapes, sizes and colours. For this unusual aesthetic beauty, these are quite popular among the professional and amateur orchid lovers in many parts of the world. According to Benzing (1986a, 1986b) the floral structure of orchids is rather stereotyped in so far as the number and organization of floral parts are concerned. But the real diversity is found in the size, shape and other structural details. These variations in floral morphology are associated with the specialized and bizarre mechanisms of insect pollination (Dressler, 1981; Batty *et al.*, 2002). Orchids generally do not allure the insect visitors with such rewards as nectar or food pollen. Instead they have developed different structural modifications of their floral parts, especially the labellum and the column, for the enticement of insect pollinators. Therefore, the relationship between orchids and their insect pollinators is not based on mutualism and the orchids are regarded as parasites on behaviour patterns of the pollinators (Barth, 1985).

The economic importance of orchids lies mainly in their ornamental value and horticultural uses. They provide cut blooms which keep fresh for long, make pretty corsages and add to the variety of floral arrangements and potted plants (Laws, 1995). A large number of our native species have been used for long in Europe and the USA as progenitors for the production of some of the famous hybrids and even today are in great demand by orchid dealers abroad. In addition to their ornamental value, orchids are also well known for their medicinal usage especially in the traditional folk medicine. The medicinal use of orchids was first reported in China (Bulpitt, 2005). The Chinese pharmacopoeia, 'the Sang Nung Pen Tsao Ching", illustrated *Dendrobium* as a source of tonic, astringent, analgesic, and anti-inflammatory compounds as far back as 200 BC and also since the Vedic period in India (Singh and Tiwari, 2007). Orchids are used in traditional medicine as they are rich in active compounds including several alkaloids (Okamoto *et al.*, 1966; Lawler and Slaytor, 1969; Elander *et al.*, 1973; Nurhayati *et al.*, 2009). In the Ayurvedic branch of traditional medicine, a group of eight drugs, known as "Ashtavarga", provides important ingredients for different types of tonics. Dried pseudo-bulbs of *Malaxis acuminata* serve as important sources of 'Astavarga' utilized in the preparation of the Ayurvedic tonic 'Chyavanprash'.The latter is one of the most widely used Ayurvedic preparations for promoting human health and preventing disease (Uniyal, 1975; Govindarajan *et al.*, 2007). A large number of orchids such as *Acampe papillosa, Dendrobium alpestre, D.chrysotoxum, D.nobile, Eulophia nuda, Vanda tessellata* are reported to have medicinal importance (Handa, 1986; Soon, 1989; Hegde, 1996). Cured capsules of *Vanilla planifolia, V. pompona* and *V. tahitensis* are popularly used as the source of essence of vanilla.

However, inspite of their manifold utility, for a long time the cultivation of orchids for commercial purposes was considered very difficult as most orchids show extremely slow rate of vegetative multiplication. Noel Bernard (1899,1902, 1903, 1904, 1906, 1909) and later Lewis Knudson (1921,1922,1924,1925,1926,1927,1929,1930) demonstrated for the first time that orchid seeds could be germinated in large numbers in artificial cultural condition following both symbiotic and asymbiotic pathways. This advent had facilitated the commercial production of economically important orchids. The micropropagation methods for orchids were worked out by Rotor (1949) using flower- stalk nodes (Rotor, 1949; Ernst, 1994) and shoot meristem (Thomale, 1957; Arditti and Ernst, 1993). But this technology was not commercially utilized in many orchid-rich countries until recent time. In Asia, commercial propagation of orchids using *in vitro* culture methods began in the 1970s (Gavinlartvatana and Prutpongse, 1993). Therefore, the orchid dealers still mainly depend on collection of plants from forests to meet a large part of their foreign and local demands. Owing to their indiscriminate collection from nature, rare orchids are now confronting the inevitable danger of depletion. As a result large number of orchids has become rare, threatened or endangered in its natural habitat due to direct and indirect effect of human activities (Batty *et al.*, 2002). Some other causes for the disappearance of orchid population are the natural calamities such as seasonal flood, landslide and erosion, shifting cultivation (jhuming), forest fires and deforestation. The bewitching beauty of orchid flowers has attracted the attention of indiscriminate collectors who exploited the plants commercially, endangering even their survival. The orchid flora is perhaps the most sensitive indicator of habitat change. The decline of species both in number and diversity can be used as an indication of ecosystem degradation. The cumulative result of such causes has now necessitated strict conservation procedures to protect the orchids from further extinction and has highlighted the need for updating basic data and eventually organizing their hybridization and multiplication with native species as they are important for us as potential material for commercial cultivation and exploitation. According to the recent figures of IUCN (International Union for Conservation of Nature and Natural resources) Red List of threatened plants, 1779 species of orchids are threatened with extinction (Walter and Gillett, 1998). However it should be noted that this number is at a global scale. Numerous other species are threatened at national or regional level, which have increased the demand for Red Lists at subglobal scales (Gärdenfors, 2001). In India, 215 species have been declared as endangered and 14 nearly extinct which comprises more than 20% of its total orchid flora (Hegde, 1996). These include species like *Vanda coerulea, Dendrobium chrysotoxum, D. wardianum, D. densiflorum, D.nobile, Thunia alba, Cymbidium elegans, Paphiopedilum* sp., *Renanthera imschootiana.* The IUCN has a species survival commission (SSC) which is active in the preservation programme. The Convention on International Trade in Endangered species of Wild Flora and Fauna (CITES) has agreed to restrict international trade in rare species of the flora of India, including the orchids. Accordingly, all species of the Orchidaceae are now on Appendix II of the convention (Hegde, 1996) and this restricts their export.

Hegde (1996) has mentioned various strategies for *ex situ* and *in situ* conservation of orchids in India. *In situ* conservation is the most efficient practice to conserve orchids which exists as an integral part of the ecosystem. Efficient management of natural habitat, formation of sanctuaries and national parks are some of the most important tool for the conservation of threatened taxa. *Ex situ* conservation is primarily aimed at creation and maintenance of germplasm pool outside the natural habitat. Although species conservation is achieved most effectively through the management of wild populations and natural habitats (*in situ*), *ex situ* technique can be used to complement the *in situ* methods and in some cases, may be the only option for some species (Sarasan *et al.*, 2006).

With the advent of germination of the orchid seeds in artificial condition using different culture media (Bernard, 1904, 1906, 1909; Knudson, 1921, 1922, 1925), the technique has been variously modified and utilized for different orchid species. Earlier studies on orchid seed germination have demonstrated that different species require different and often species-specific medium composition to obtain optimum germination and seedling growth (Arditti, 1967b; Oliva and Arditti, 1984; De Pauw *et al.*, 1995). Therefore, it is necessary to formulate the appropriate *in vitro* seed culture protocol for every species, specially having commercial importance. Large number of minute seeds lacking any storage tissues is the characteristics of orchids. Orchid seeds also favour the expression of genetic variability (Batty *et al.*, 2002). Thus, asymbiotic orchid seed propagation technique has been applied for the production of commercially important orchid, and has been regarded as an efficient tool for the conservation of endangered and threatened orchid taxa and may be useful in the re-introduction purposes.

Impressive results concerning asymbiotic germination and reintroduction of rare and endangered orchids were obtained at the Royal Botanic Gardens, Kew (Ramsay and Stewart, 1998). From conservation point of view, in contrast to clonal propagation, the *in vitro* seed culture technique would be more suitable. Since it provides the maintenance of optimum genetic diversity. In the present investigation, the main objective is to produce the large number of seedlings within a short time which would decrease the collection pressures on already endangered populations.

Apart from the difficulties of seed germination, in natural conditions, the life cycle of orchids is very long. It takes about 5-10 years for a plant to bloom and produce fertile seeds. Vegetative propagation of orchids is also an extremely slow process. Therefore, mass scale production of orchids depends heavily on different *in vitro* methods. Such methods could be effectively used for conservation purposes in both direct and indirect ways. The *in vitro* propagation techniques, developed since 1960 (Morel, 1960, 1964) has become most conducive approach for commercial purposes. Conventional propagation through seeds is less desirable especially for horticultural exploitation due to the long juvenile period before flowering (Decruse *et al.*, 2003). Effective, reliable method of clonal propagation has long been in demand and successively various micropropagation methods have been developed to establish effective clonal propagation for the development of orchid industries in various countries.

For the last few decades *in vitro* techniques for rapid propagation of plants have been essential tools for the horticultural industries, particularly for plants like orchids whose multiplication rates are extremely slow through conventional *ex situ* methods (Rao, 2004; Mukhopadhyay and Roy, 1994). In addition, the significance of artificial propagation for conservation of endangered/rare species has been emphasized for various angiospermic plants, including orchids (Seeni and Latha, 1992; Arditti and Ernst, 1993; De Pauw *et al.*, 1995; Ramsay and Stewart, 1998). Although, the techniques using meristem, shoot-tip or nodal culture offer true-to-type clonal propagation of selected genotypes, their application in the conservation of rare species is limited. This is primarily because extensive use of these techniques would result in undesirable genetic pauperization of natural population (Arditti and Ernst, 1993; Millner *et al.*, 2008). Moreover, the development of these protocols involves the sacrifice of large number of individual plants, adding extra pressure on the depleted wild populations. Therefore, a seed culture-based technique (Shimura and Koda, 2004) which provides the maintenance of optimum genetic diversity with least damage to the natural population, would be most suitable for conservation purpose. Therefore, micropropagation could be used as an indirect approach of *ex situ* conservation.

Materials and Methods

Pollination and capsule collection

Fig. 4.1: Fruits of Orchids

With the help of a needle, the anther cap of a flower was removed and the pollinium was first taken out and then transferred to the sticky stigmatic surface of another flower, whose pollinium was already removed. After this, the pollen receiving flower (female parent) was labelled with a tag containing the date of pollination and the name of the pollen-donor species. The cross-pollinated flower was then left as such for the development of capsule. This procedure was applied for both siblings (crossing two plants of the same species) and hybrids (crossing of two plants of different species). After several months the undehisced fruits were harvested while they were still green and immediately taken to the laboratory for the culture of seeds. The age of the fruits of each species and hybrid would be mentioned individually in later sections.

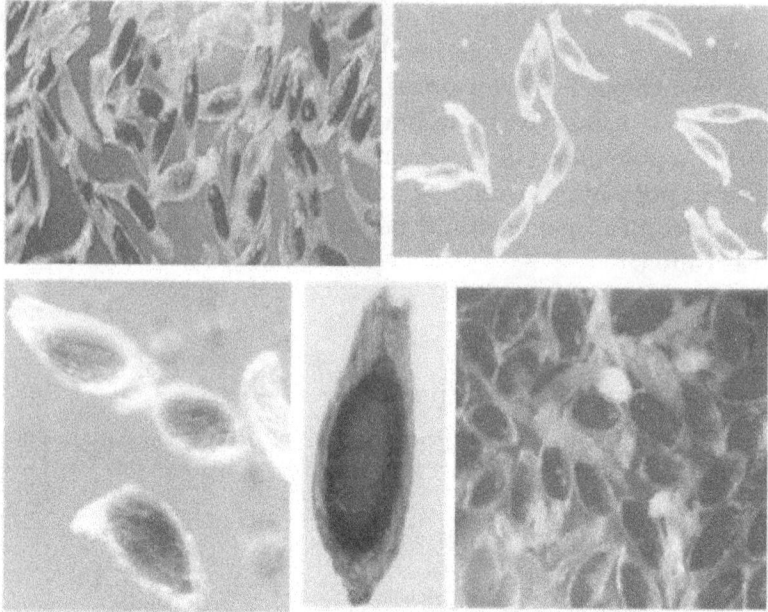

Fig. 4.2. A-E. Qualitative testing of viability through TTC porofile for seeds of : A. *Dendrobium nobile* (10x); B. *D. chrysanthum* (10x) C. *D.* farmeri (10x); D. *D. parishii* (10x); E. *Hygrochita parishii* (10x)

Determination of Seed Viability

The viability of the seeds was tested with 1% (w/v) TTC (2, 3, and 5-triphenyl tetrazolium chloride) solution, as described by Shivanna and Johri (1985) for the angiosperm pollen grains. The solution was prepared in 0.15M Tris-HCl buffer (pH 7.8) and stored in a dark bottle under refrigeration. The test was performed on cavity slides. First, a thin film of petroleum jelly was applied around the cavity of the slide and a drop of 1% TTC solution was added inside the cavity. The seeds were then added in the solution and a cover glass was placed above. Care was taken to prevent entrapment of air bubbles inside the cavity chamber. For each fruit sample three replicate slides were prepared. The slides were then placed inside a pair of Petri plates lined with moistened filter paper and kept under dark condition at 30±2°C for 48 hour.

After the completion of the treatment period the slides were observed under a light microscope. The seeds showed a gradation in response, i.e. the colour of seeds varied from dark red to orange and white. The seeds with white embryo could be easily recognized as nonviable ones, but in other cases (with various shades of red and orange) it was difficult to fix a cut-off point. Therefore, the test was used only for the qualitative assessment of seed conditions, i.e., whether majority of the seeds were viable (with red to orange colouration) or not.

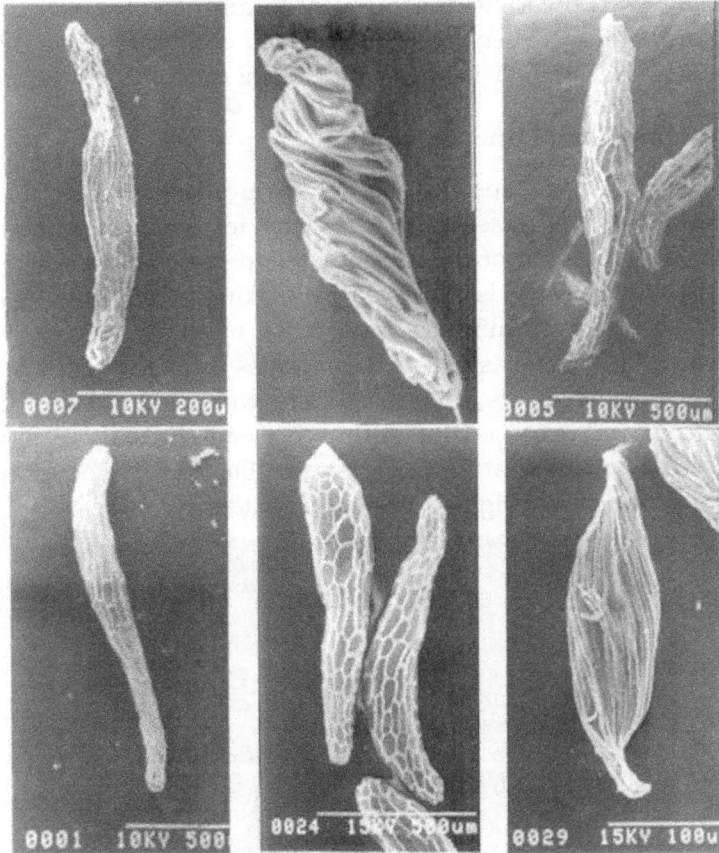

Fig. 4.3. Structure of seeds of different species of orchids under SEM

Study of Seed Structure

The mature seeds of different species were freshly collected from the naturally dehiscing capsules. The seed structure of all the above taxa was studied through light microscopy as well as by scanning electron microscopy (SEM). For each species 25 randomly selected seeds were examined under light microscope. Length and width were also measured with the help of stage micrometer at the longest and widest axis of the seed. Seeds exhibit different forms. Therefore, seed volumes were calculated using the formula $2[(W/2)2\ (1/2\ L)\ (1.047)]$, where, $1.047 = \pi/3$, W=Width and L= Seed length (Arditti *et al.*, 1980). Orchid embryos are elliptical in cross section and therefore their volume was calculated by using the formula: $4/3\ \pi\ ab^2$ where a= ½ its length and b= ½ its width.

For SEM the seeds were directly (without any pre-treatment) attached to the aluminum stubs using double- sided adhesive tape and surrounded the specimens with silver conducting paint to form an electrical connection to the ground. The specimens were then coated with gold using the sputter coating unit and examined in the SEM Hitachi model S530.

Culture Establishment and Incubation

Depending upon the species, the pods were harvested from the plant and the capsules were rinsed with tap water for about 30 minutes. After that surface disinfestation of the capsules was performed by rinsing in 95% (v/v) ethanol for 60 seconds, followed by soaking in 0.1% (w/v) mercuric chloride solution for 15 minutes and finally washing thrice with sterile distilled water and swift passing of the capsules through the flame. Surface sterilized capsules were dissected aseptically with a scalpel and the seeds were taken out and sown on the surface of semisolid culture media. The cultures were maintained at 25 ± 2°C under 10h photoperiod provided by Philips white fluorescent lights of 3000 lux intensity.

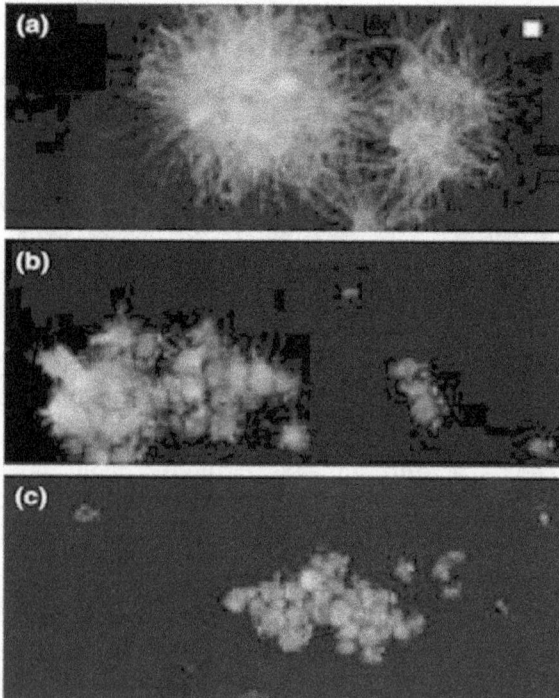

Fig. 4.4: a. Protocorm development with numerous rhizoids; b. and c. Protocorm development and morphology and lack of rhizoids

Nutrient Medium

Knudson C (Arditti, 1982 modified after Knudson, 1946) and MS media, in various formulations, have been used extensively for temperate terrestrial orchid seed germination (Arditti, 1982; Clements, 1982; Fast, 1982; Rasmussen, 1995; Michel, 2002). Arditti (1982) solidified Knudson C medium with 12–15 g agar, whilst the modifications of Stoutamire (1964) and Anderson (1990) employed 8 and 6 g l^{-1} agar, respectively. In this study both Knudson C and MS media were solidified with 8 g l^{-1} agar to regulate diffusion, absorption and water potential effects (George, 1993). Media were solidified in Petri dishes (15 ml media), glass culture tubes (ø 25 mm; 10 ml media) and culture jars (25 ml media). Liquid cultures were continuously agitated by means of an orbital shaker operating at 50 cycles min^{-1} (Oliva and Arditti, 1984; Chu and Mudge, 1994). Decontaminated capsules were cut into a minimum of six transverse sections (2–4 mm thick, depending on the length of the capsule), with the two end sections being discarded to reduce contamination. Capsule sections were placed, cut surface down, onto the germination media. Sowing density was two capsule sections per tube and four sections per jar culture. Knudson C and MS media, solidified with agar and supplemented with vitamins, were used as control. Mature seeds were tested with various media additives, inorganic salt concentrations and media viscosity manipulations as well as a range of illumination and temperature regimes. Treatments were applied independently. 'Green-pod' cultures involved charcoal, inorganic salt concentration and media viscosity manipulation, which were applied as for the mature seed. Each species investigated did not experience all possible media and culture condition manipulations.

The germination media were based on inorganic salts of Knudson's C with some modification. The iron source prescribed for KC media was replaced by the iron –EDTA as described by Murashige and Skoog (1962). The medium was also modified by the inclusion of 0.1% peptone and 2% (w/v) sucrose served as carbon source. The detailed composition of the culture media is given in Table 1. In all instances, the pH of the medium was adjusted to 5.2, prior to autoclaving with 0.05(N) KOH and 0.05(N) HCl. The media were solidified with 0.9% (w/v) agar (Merck Ltd. Mumbai, India). The culture vessels were plugged with non-absorbent cotton. The sterilization of culture media was performed by autoclaving at 1.05 $kgcm^{-2}$ for 20 min.

Media Modifications

Seeds were inoculated into culture tubes each containing 20 ml of KCP basal medium. In addition to the constituents shown in the Table 4.1, different organic substances like peptone (.05%, 0.1%, 0.2% and 0.4%), yeast extract (0.00625%, 0.025%, 0.0125% and 0.05%) and coconut water (5%, 10%, 20% and 40%), different plant growth regulators like auxin (Indole acetic acid, ∝-napthalene acetic acid, 2,4- dichlorophenoxyacetic acid) and cytokinin (6- benzylaminopurine) were also added to medium in various concentrations.

Table 4.1: Composition of KC basal medium

$Ca(NO_3)_2.4H_2O$ (mg l^{-1})	1000.00
$(NH_4)_2SO_4$ (mg l^{-1})	500.00
H_2PO_4 (mg l^{-1})	250.00
$MgSO_4.7H_2O$ (mg l^{-1})	250.00
Na_2EDTA (mg l^{-1})	37.25
$FeSO_4.7H_2O$ (mg l^{-1})	27.25
$MnSO_4.4H_2O$ (mg l^{-1})	7.50
Sucrose (g l^{-1})	20.00
Water	1.00 litre

Germination and seedling growth

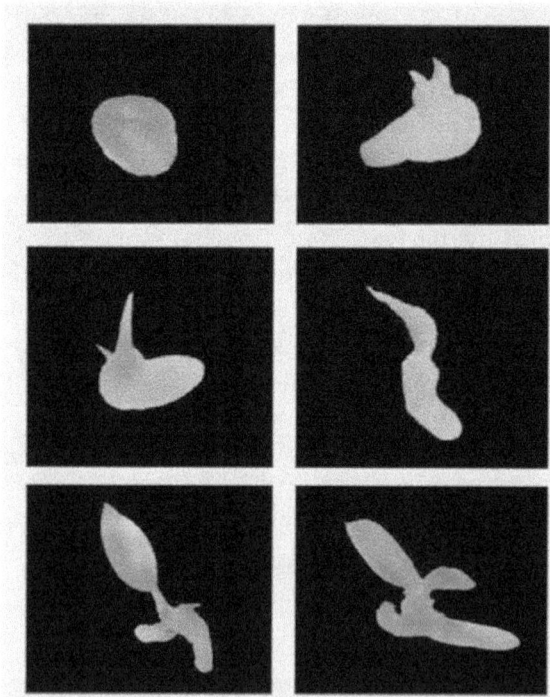

Fig. 4.5. Developmental stages of Protocorms after Germination of Seeds of Orchids

Germination responses were graded according to modifications of the broad categories employed by Warcup (1975), Oliva and Arditti (1984), Smreciu and Currah (1989) and LeRoux *et al.* (1997): (i) no germination; (ii) testa split and embryo swollen; (iii) embryo enlarged, rhizoids present; (iv) protocorm considerably larger than testa; (v) differentiation of the apical meristem; and (vi) greening of the primary leaf tissue. Germination and early seedling development were recorded fortnightly up to a maximum of 18 months, where primary seedling development was concluded by the concomitant initiation of primary leaves and root initials. Percentages represent mean counts for a minimum of 100 randomly chosen seeds from at least five cultures. Germination of seeds was considered to have occurred when the embryo emerged from the ruptured seed coat (De Pauw *et al.* 1995). The growth of the seedling in different treatments was expressed through quantitative assessment of the relative stage of seedling development, such as seedlings with or without visible shoot apex, seedlings with expanded leaves and seedlings with roots. In the latter case the growth was expressed as a percentage of the total live seedlings at the end of the three-month culture period.

General Accounts of Orchid Seeds

Morphology and its Consequence

Orchid seeds are unique in all respects. The structure and size of the seeds are among the most interesting characteristics of the family orchidaceae (Arditti and Ghani, 2000). They differ from other angiospermic plant and resemble the so-called 'dust seeds' of other plants (Fleischer, 1929, 1930; Ziegenspeck, 1936; Rauh *et al.*, 1975; Rasmussen, 1995).

Number and Size

Orchid seeds are generally produced in large number and the number of seeds per fruit ranges from 20-4,000,000. From this vast number of seeds, sometimes only one may germinate and grow to blooming size (Arditti, 1967b). Orchid seeds are very small (Beer, 1863; Ziegenspeck, 1936; Poddubnaya-Arnoldi and Selezneva, 1953, 1957a, b; Arditti, 1967b, 1979, 1992; Rauh *et al.*, 1975; Arditti, 1982; Arditti and Ernst, 1984; Rasmussen, 1995; Arditti and Ghani, 2000; Yam and Arditti, 2009) and the variation of size occurs in the family, in genera and even within species. The size of the seeds ranged between 0.05mm to 6.0 mm in length (Hallé, 1977) and 0.01mm- 0.93mm in width (Benzing and Clements, 1991; Barthlott and Ziegler, 1981; Dressler, 1993). However, the statement that the seeds of epiphytic species are smaller than those of terrestrial ones (Rasmussen, 1995) is not true (Arditti and Ghani, 2000).

Embryo and seed coat

The orchid seeds consist of a small embryo (Rasmussen, 1995) usually without any cotyledon or an endosperm (Arditti, 1967b; Arditti and Earnst, 1984; Arditti and Ghani, 2000). The embryo remains suspended within a membranous, transparent and pigmented seed coat (Burgeff, 1936; Darwin, 1888; Knudson, 1929; Arditti, 1967b). The seed coat normally consists of dry membranous interlaced with heavily thickened cell walls (Arditti, 1967b). The coat may be thin and with reticulate and sclerotic (Rosso, 1966; Arditti, 1967b). In addition to being small, because of their large internal air space orchid seeds are also very light, so they are very buoyant in both air and water.

Viability

The viability of orchid seeds is also variable. Some seeds may lose viability within 2 months (Brummitt, 1962) or less (Lindquist, 1965) while others remain viable for long (Arditti, 1967b, 1979, 1992, 1993). When dried and kept inside desiccators at 0°C (Kano, 1965), it remains viable even up to 18 years. According to Whigham *et al.* (2006) *in situ* study on temperate terrestrial orchids indicated that most seeds of *Goodyera pubescens* germinated within one year, whereas 4 other species continued to germinate sparsely during the 4 year observation period, and after almost 7 years, many seeds were still viable (Whigham *et al.*, 2006).

Seed dispersal

Seed dispersal may occur a) by air b) by water and c) by land animals and birds (Arditti and Ghani, 2000). High number as well as physical characteristics of orchid seeds facilitate extensive coverage of areas around the seed parent and wider dispersion further away.

Reserve nutrients

Structurally, two major groups of orchid seeds are there (Arditti, 1967b). A few species may have relatively differentiated embryos with a rudimentary cotyledon and are rather easy to germinate (Burgeff, 1936). Another group, that includes majority of species, has relatively undifferentiated seeds and contains no cotyledons and no endosperm (Arditti, 1967b). This has been described as being a characteristic of higher groups. For germination, it is claimed that orchid seeds require a symbiotic association with fungi (Rasmussen, 1995). Orchid species vary in the degree to which they depend on fungi and it is quite amenable that their reliance on fungi in germination may not be due to resource limitation but for their inability to metabolize nutrient reserves rapidly (Harrison, 1977; Rasmussen,

1995). Many temperate terrestrial species, such as *Dactylorhiza* and *Orchis* can germinate in water and remain alive for some days to weeks, without receiving any external nutrients (Rasmussen, 1995).

Since orchid seeds generally have no endosperm and no cotyledons, lipid droplets serve as the principle food reserve (Knudson, 1922, 1927; Poddubnaya-Arnoldi and Zinger, 1961; Arditti, 1967b, 1979; Arditti and Ernst, 1984). *Cattleya* seeds reportedly contained 32% fat, 1.2% sugar and no starch (Knudson, 1922; Carlson, 1940; Arditti, 1967b). The same has been verified for several other species (Knudson, 1927; Arditti, 1967b).

Many temperate terrestrial species, such as *Dactylorhiza* and *Orchis*, can germinate in water and remain alive for some days to weeks, without receiving any external nutrients (Rasmussen, 1995). The self-sustaining capacity of these species indicate that, they can mobilize at least some part of their food-reserve, immediately after the embryo cells get hydrated, so that the actual germination takes place prior to, and without the fungal infection (Rasmussen, 1995).

In vitro studies with other species indicated that, addition of some external nutrients is necessary, so as to germinate without fungal aid. This requirement is highly variable among species. The requirement includes, soluble sugars, some minerals, generally organic undefined nitrogen forms, but sometimes specific amino acids or inorganic nitrogen; some specific vitamins (like glycine, tryptophan etc) or a vitamin mixture; some undefined organic additives; some definite cytokinin or other growth factors. Some less exacting species, although capable of germination without the fungal intervention, show an increase in germination frequency in presence of some additives. Also seeds from the same batch may differ in their nutritional requirement (Rasmussen, 1995), that in turn, pinpoint the genetic variability among themselves.

Germination of Orchid Seeds: A Prologue

The germination of orchid seeds has long been recognized as difficult and uncertain of attainment (Knudson, 1922). It appeared that there were some environmental factors that crippled the process of germination. Further, it was pointed out that the inherent characteristics of the seeds rendered the germination refractory (Knudson 1922).

A. Brief account on stages of germination and seedling growth

Orchid embryo is undifferentiated in most cases, except that the cells at the apical region are smaller, densely granulated, constituting the meristematic region, while rest cells at the basal region are large. At the base, a delicate suspensor is present in some species. The embryo remains enclosed within a transparent integument with an opening at the lower end through which the suspensor, when present, may protrude.

During germination, the embryo first absorbs water and enlarges in transverse direction till a small spherule stage is reached. Initially, these remain white, even in presence of light and in most cases, chlorophyll appears soon (Arditti *et al.*, 1981; Oliva and Arditti, 1984), being more pronounced at the meristematic region (Bernard, 1909; Knudson, 1922). But, in some species, chlorophyll may be present from onset, as in *Cypripedium* (Oliva and Arditti, 1984). Then the embryo ruptures the integumemt, which is accompanied by emergence of absorbing hairs out of the epidermis. The embryo further enlarges, more absorbing hairs develop near the basal region, and larger spherule, a somewhat top-shaped structure is attained. Bernard (1909) termed it as the 'protocorm' (for its similarity with pteridophytic protocorm body) which is characterized by a marked depression or notch at the apical part. Next, approximately at the middle of the depression, the first leaf point emerges out, which subsequently develops to form the first leaf. Followed by the first leaf, a second and a third leaf may unfold; alongside, elongation continues, and a distinct stem becomes apparent. The first root may arise either from the protocorm or from the stem below the second or the third leaf (Bernard, 1909; Knudson, 1922). However, the roots thus formed are adventitious, since these are not formed out of a root meristem, which is absent among orchids. Thus an orchid plantlet is now formed (Oliva and Arditti, 1984). As is reported, in most of the *Cypripedium* species, roots and shoots appear together; but in *C. acaule*, shoots are formed after the roots, while in *C. californicum*, that occurs in reverse order (Oliva and Arditti, 1984). The period required to attain these developments for *in vitro* seedlings may vary and can take 4-6 months (Bernard, 1909; Knudson, 1922; Oliva and Arditti, 1984), even up to 30 months (for some *Cypripedium* spp.; Oliva and Arditti, 1984).

In a number of terrestrial species, protocorm does not directly give rise to leaf or aerial shoot; rather forms an intermediate rhizome stage, which after attaining certain length (Roy and Banerjee, 2002), give rise to the upright aerial shoot, quite similar to that described above. The rhizome tip grows vertically (as observed in *Geodorum densiflorum*, *G. citrinum* etc.), forms leaves and the root emerges from the base. In contrast, some seedlings may directly form shoots before the appearance of rhizomes, as in seedlings of *Epipactis*, *Goodyera*, *Piperia*, *Platanthera* (Arditti *et al.*, 1981).

It is evident that seed germination of orchids differs from that of other seeds (Arditti *et al.*, 1981), because of the absence of an endosperm, radicle, leaf rudiments and root meristem (Arditti, 1967b; Arditti *et al.*, 1981). So, for orchids, the germination is defined as the appearance of green or white protocorm; and development means, the appearance of chlorophyll, absorbing hairs, rhizomes (in some species), shoots and roots (Arditti, 1967b, 1979; Arditti *et al.*, 1981; Oliva and Arditti, 1984).

B. Early enquiries on germination

The process of orchid seed germination escaped beyond detection for many years. Orchid seeds were believed as not to be viable (Constantin, 1913; Bouriquet, 1947) or at least incapable of germination and they multiply by means of bud or gemma-like structures which undergo a series of metamorphoses prior to the formation of another mature plant (Fabre, 1855, 1856; Moran, 1890; Arditti, 1967b).

In 1840, Link presented an ambiguous graphical indication for the presence of fungi in root cells of *Goodyera procera* (Link, 1840; Ramsbottom, 1922a; Yam and Arditti, 2009). Later, universal occurrence of the endophyte within the orchid roots was firmly established after careful examination of the roots of 500 orchid species from all parts of the world and was identified as species of *Nectria* that was further verified by others (Prillieux, 1856, 1860; Prillieux and Rivière, 1856; Fabre, 1856; Arditti, 1967b). Frank first used the term "mycorrhiza" to denote root-fungus and pointed out the possibility of a symbiotic association (Frank, 1892; Ramsbottom, 1922a). MacDougal (1899a, b) suggested that the fungus might benefit the plant, but its exact role was still obscure. During 19th century, commercial demand for orchids was increased but the growers had no method for germination of the orchid seeds. When spontaneous seedlings of terrestrial orchids have been observed, they are most often found close to roots of adult plants (Fabre, 1856; Ames, 1922; Rasmussen and Whigham, 1998a, b). John Harris observed that orchid seeds when scattered at the base of the mother plant will germinate (Anonymous, 1893; Arditti, 1967b; Arditti and Ernst, 1993). The precise requirements for germination was first revealed by Noel Bernard, who quite coincidentally, during a stroll in the forest, noticed an array of developing seedlings of *Neottia nidus-avis*, and detected that all the seedlings were infected by a fungus. Hereby, he envisaged the possible role of the fungus in orchid seed germination (Bernard, 1899; Arditti, 1967b).

C. Period of debate over 'Obligate Symbiosis'

Bernard successfully isolated the fungus from the infected orchid seedlings and advocated that the germination of orchid seeds and subsequent growth of the seedlings took place only upon infection with some strains of fungi those generally found living in the orchid roots (Bernard, 1903,

1904, 1906, 1909; Arditti, 1967b). Bernard pioneered a method of germination known as 'symbiotic germination', in which orchid seeds and fungus were co-cultured and this was probably the first ever *in vitro* technique for propagation (Arditti and Ernst, 1993). Burgeff (1909) also believed on obligate symbiosis and classified those endophytes as a separate group called Orcheomyces (Burgeff, 1909, 1911, 1932, 1936, 1959; Wynd, 1933b). However, Burgeff (1909) failed to recognize the importance of his own experiment in which *Laelio-Cattleya* seeds germinated on 0.33% sucrose solution in the dark; the plants lived for 10 months, beyond which further development was impossible without the fungus either in the light or dark (Wynd, 1933b). It was believed that the fungus excreted some enzymes that would digest the starch within the embryo, which in turn, would cause an increase in the cell-sap, thereby inducing the germination and the formation of protocorm (Bernard, 1909). It was also pointed out that the fungus can invert sucrose and this may occur in the embryo (Bernard, 1909). The partisans of obligate symbiosis (Bernard, 1909; Burgeff, 1909, 1911, 1936; Constantin, 1917; Constantin and Magrou, 1922; Ramsbottom, 1922b, 1927) believed symbiosis to be a pre-requisite for "normal" development of orchids. Also, they advocated that the requirement of fungal strain is stringent (Bernard, 1909; Burgeff, 1909).

D. Requirements for asymbiotic seed germination and protocorm growth of orchids

During 19th century, commercial demand for orchid was increased but the growers had no method for germination of the orchid seeds. The history of *in vitro* culture of orchids began with the attempts to germinate orchid seeds that normally show extremely poor germination under natural condition. This fact together with the slow rate of vegetative propagation has always hindered commercial production of orchids. Previously, John Harris observed that orchid seeds could occasionally germinate when scattered at the base of a mature plant (Anonymous, 1893; Arditti, 1967a; Arditti and Ernst, 1993), but the exact requirements for germination were unknown. The precise requirement for orchid seed germination was first noticed by Noel Bernard, who found the association of a fungal infection with the developing orchid seedlings (*Neottia nidus-avis*). Hereby, he envisaged about the possible role of the fungus in the germination of seeds (Bernard, 1899; Harley, 1959; Arditti, 1967a). Later Bernard successfully isolated the fungus from the infected orchid seedlings and was able to establish the fact that, symbiotic association with a specific mycorrhizal partner is a pre-requisite for orchid seed germination (Bernard, 1903, 1904, 1906, 1909; Harley, 1959; Arditti, 1967a,b). Although the exact role of mycorrhizal fungi is still questionable, it is generally considered that the associated endophyte provides soluble carbohydrates, essential minerals, water, enzyme precursors and even hormones for germinating embryos and protocorms (Smith, 1966; Arditti, 1967a; Alexnder *et al.*, 1984; Alexnder and Hadley, 1985; Rasmussen, 1992). The works of Bernard was probably the first *in vitro* technique for the propagation of any plant (Arditti and Ernst, 1993). As described by Bernard (1909), Burgeff (1909, 1911, 1932), Rasmussen (1992), the method of symbiotic germination involved the inoculation of the orchid seeds with the fungi in a suitable culture medium. But this method requires elaborate procedure and the frequency of germination is often very low (Bernard, 1909; Harley, 1959).

Lewis Knudson noted that the presence of fungi was not absolutely essential for the *in vitro* germination of orchid seeds and demonstrated that germination could occur in a sugar and mineral medium (Knudson, 1921, 1922; Harley, 1959; Arditti, 1967a). During germination, the embryo first absorbs water and enlarges in transverse direction till a small spherule stage is reached. Initially these remain white, even in presence of light and in most cases; chlorophyll appears soon (Arditti *et al.*, 1981; Oliva and Arditti, 1984). Being more pronounced at the meristematic region (Bernard, 1909; Knudson, 1922), the embryo ruptures the integument, which is accompanied by emergence of absorbing hairs out of the epidermis. The embryo further enlarges and a larger spherule, a somewhat top-shaped structure is attained. Bernard (1909) termed it as the 'protocorm' which is characterized by a marked depression; the first leaf point emerges out, which subsequently develops to form the first leaf. Followed by the first leaf, second and a third leaves unfold and a distinct stem becomes apparent and later roots develop from the protocorm. Thus, an orchid plantlet is formed (Oliva and Arditti, 1984). Knudson was able to raise seedlings of a number of orchids, namely *Cattleya, Laelia, Cymbidium, Odontoglossum, Phalaenopsis, Ophyrus* and *Dendrobium* (Knudson, 1921, 1922, 1925, 1946), through the asymbiotic method.

In a number of terrestrial species, protocorm does not directly give rise to aerial shoot rather forms an intermediary rhizome stage, which after attaining certain length (Roy and Banerjee, 2002), give rise to the upright aerial shoots. In contrast, some seedlings may directly form aerial shoots before the appearance of rhizomes, as in seedlings of *Epipactis, Goodyera, Piperia, Platanthera* (Arditti *et al.*, 1981)

In contrast, upon culturing in nutrient media containing all essential salts but lacking sugar, embryos became green, beyond which no development was noted (Knudson, 1924). The role of orchid symbiont was now questioned as whether symbiosis was a necessity for germination or merely incidental and not of any significance (Knudson, 1925). He also advocated that the presence of sugar was essential only in the primary stages of germination. Following the pioneering work of Knudson a great deal of work has been done on the asymbiotic germination of orchid seeds (Clement, 1924a, 1926; Ballion and Ballion, 1924; Bultel, 1926; La Garde, 1929). These works demonstrated that different species of orchids showed different cultural requirements. However, it has been stated that the symbiotically germinated seedlings are of good quality (Wynd, 1933b) and generally more robust, less subject to fungal attack and have a better chance of survival (Blowers, 1966). As a result, several workers devised different nutrient media. Many of these were based on Knudson's medium. While others made significant changes in the composition. (Burgeff, 1936, 1954; Withner, 1947, 1959b; Vacin and Went, 1949; Yates and Curtis, 1949; Hager, 1954; Mc Ewan, 1961; Raghavan and Torrey, 1964; Mitra *et al.*, 1976; Harvais, 1982). However, a large array of *in vitro* studies on germination of orchid seeds and subsequent development of protocorm indicate that, the requirement for germination as well as growth differs among the species. The factors required for *in vitro* germination of orchid seeds may be divided into 3 categories, viz.

(a) Media components

(b) Culture environment and its components (viz. light, temperature, humidity, and the type of the culture vessel)

Pre-treatment with NaOCl or $(Ca)_2OCl$ and/or cold or stratification or soaking to break the dormancy of the seeds, when soaked.

Effects of carbohydrates

The success on asymbiotic germination of orchid seeds (Knudson, 1922, 1924, 1925, 1927; Clement, 1924a, b, 1926, 1929, 1932; Ballion and Ballion, 1924, 1928; Bultel, 1924-25, 1926) followed by normal flowering (Knudson, 1930) on the asymbiotically raised plant confirmed that appropriate sugar is required for germination. However, it remained for Knudson (1922, 1924, 1925, 1926, 1927) to point out that the true significance of the mycorrhizal condition was in furnishing a source of carbohydrate to the orchid embryo and in maintaining a favourable degree of acidity (Wynd, 1933b).

It has been mentioned earlier that various plant extracts (Knudson, 1922, 1925) and sugars, such as glucose, fructose, mannose, sucrose (Knudson, 1922, 1924, 1925; Clement, 1924a, b, 1926, 1929, 1932; Ballion and Ballion, 1924, 1928; Bultel, 1924-25, 1926) were all successful to cause germination of orchid seeds. Regarding efficiency for germination and growth of orchid seedlings, Knudson (1922) found: Sucrose> fructose> glucose. Smith (1932) found good growth with sucrose, glucose and maltose, singly or in different combinations and found no apparent difference in the growth of the seedlings. Wynd (1933b) reported the order of excellence over a series of sugars added to three different inorganic media of Knudson (1922), Shive (1915) and La Grade (1929).

Good germination was also obtained with sucrose combined with glucose (Mariat, 1951; Liddell, 1953b); sucrose along with organic additive (Tsukamoto *et al.*, 1963), or mixture of glucose and fructose (Stoutamire, 1963) and starch (Bouriquet, 1947). Raghavan and Torrey (1964) reported good germination and growth with sucrose, on ammonium containing media. Kano (1965) and Ernst reported that orchid seedlings grew best on 0.117M sucrose solution, but appeared not to be harmed by concentration in the range of 0.015- 0.233M. Sometimes, some orchid-extract (Bouriquet, 1947) or fungal extract (Downie, 1940, 1941, 1943, 1949a, b) stimulated germination in presence of sugars.

Regarding utilization, sugar present in medium that does not support growth of the seedlings (e.g. Lactose) is not hydrolysed (Knudson, 1924; Ernst and Arditti, 1990) and may be toxic in some cases (Ernst, 1967). In most instances, a wide variety of sugars act as C sources for germination of orchid seeds and subsequent seedling growth, whereas organic acids are of little value (Ernst, 1966a; Arditti, 1967b). It is evident that orchid embryos and seedlings are not only unable to utilize their own food reserves, but also are unable to efficiently hydrolyse and utilize larger polysaccharides. All these necessitate the fungal symbiosis to be the pre-requisite for germination of orchids that functions so as to provide exogenous source of C (Knudson, 1921, 1922) and also aid to breakdown such larger molecules and transport simpler sugars (Harley, 1969; Smith, 1966).

In *Phalaenopsis* hybrids, Ernst and Arditti (1990) found that germinating seeds and developing seedlings can utilize glucose, maltose, maltotriose, maltotetraose, maltopentose and maltohexose as carbon sources, but an inverse relationship of oligomer length with survival and growth parameters was observed.

Seeds of *Cattleya aurantiaca* asymbiotically grown on a sucrose-containing media developed leaf-bearing plants, but could not differentiate to go beyond protocorm stage in the absence of sucrose (Harrison and Arditti, 1978).

Nakamura (1982) observed that for germination and seedling growth of *Galeola septentrionalis*, glucose, glycerol, fructose, inositol, mannitol, mannose, ribose, sorbitol and xylose were utilized as C sources while arabinose, rhamnose and sorbose were not, and galactose inhibited growth.

Smith (1966,1967) by using [^{14}C] glucose as the C source, demonstrated the translocation of fungal carbohydrate 'trehalose' by the fungal hyphae into the seedlings of *Dactylorhiza purpurella* which soon broke down into glucose, a form that could be readily utilized by orchid embryos (Knudson, 1925). For asymbiotic germination of *D. purpurella* and *Bletilla hyacinthina*, trehalose and glucose both acted as satisfactory C source for growth of non-photosynthetic protocorms but mannitol did not (Smith, 1973).

In *Dactylorhiza* spp. seedling development was stimulated with glucose and sucrose at concentration of 10gdm^{-3} each but each species showed significant difference in shoot and root development depending on sugar and PGR combinations (Wotanová *et al.*, 2007).

Orchid mycorrhizal fungi, can utilize many carbon sources including complex polysaccharides (Harvais and Hadley, 1967) and cellulose appears to be the normal carbohydrate source in nature for protocorm growth in a variety of orchid-fungus systems including both temperate and tropical orchids (Hadley, 1969).

Van Waes (1984) noted that for *Orchis morio*, higher germination occurred with 29 and 58 mM sucrose than in control without sucrose, but further higher concentration of sucrose (87 and 116mM) inhibited germination to a level inferior to those of the sugar- free control. Similar inhibition result was also reported by Mead and Bulard (1979) in germination of *Orchis laxiflora* and *Dactylorhiza majalis* (Rasmussen, 2000).

Another study with *in vitro* grown protocorms, plantlets and field plants of *Goodyera repens* revealed that C moved only from fungus to orchid and this movement ceased when the host reached certain stage of development. With advanced leafy protocorms of a *Cattleya* hybrid, external C resulted in increased dry weight and root growth, but decreased shoot/root ratio and there was little evidence of nutrient translocation by the mycorrhizal fungus from the nutrient to the non-nutrient side in the absence of carbohydrates.

In situ studies on *Tipularia discolor* revealed that seed germination of a terrestrial orchid could be enhanced by the presence of decomposing wood (Rasmussen and Whigham, 1998b). McKendrick *et al.* (2000) also indicated that in *Corallorhiza trifida* the rate of seedling growth could be determined by the ability of the fungal symbionts to transfer C from their ectomycorrhizal co-associates.

Based on nutritional modes at mature stage, orchids are categorized largely as (i) obligate autotrophic and (ii) obligate myco-heterotrophic (Leake,

1994; Leake *et al.*, 2004; Smith and Read, 1997). The myco-heterotrophic orchids are known to acquire their nitrogen and carbon from the mycorrhizal fungi (Leake, 1994; Leake *et al.*, 2004; Smith and Read, 1997; Gebauer and Meyer, 2003). Many green orchids may have the capacity for mixotrophy where they can obtain C through photosynthesis and heterotrophically from their fungal association that enable them to grow at or below the compensation point (Julou *et al.*, 2005).

Cameron *et al.* (2006) indicated that when extra radical mycelia system of *Goodyera repens* was employed with double-labelled [^{13}C-^{15}N] glycine; both ^{13}C and ^{15}N were assimilated by the fungus and transferred to the roots. Thus again indicating a fungus dependent pathway for organic nitrogen acquisition by the orchid. Subsequent study revealed net plant-to-fungus C flux and the rapidity of bidirectional C flux is indicative of dynamic transfer at an interfacial apoplast as opposed to reliance on digestion of fungal peptons (Cameron *et al.*, 2008).

Rasmussen (2000) advocated that the need for exogenous sugars arises after the seeds have germinated. Some species can germinate equally well in water and in weak solutions of sucrose or an equal molarity of mannitol or in distilled water (Harvais, 1972). Prior to development of photosynthetic apparatus, carbohydrates may be obtained either from the nutrients stored in the embryo or from those acquired from or made accessible by the infecting fungus (Rasmussen, 2000).

Thus, germination of orchid seeds is enhanced by mycorrhizal fungi (Masuhara and Katsuya, 1994; Zettler and Hofer, 1998), which is, for *in vitro* cultivations, substituted by suitable sugars (Wotavová-Novotná *et al.*, 2007). While polysaccharides (starch or cellulose) are suitable for symbiotic germination (Smith and Read, 1997), mono or disaccharides (sucrose, glucose or fructose) are preferred for asymbiotic way (Harvais, 1973).

Effects of mineral ions and different media

For germination of orchid seeds, Knudson (1922) used Pfeffer's solution and introduced his "solution B" which is the pioneer media of its kind. Following his works (Knudson, 1921, 1922, 1925, 1946), a considerable number of other media have been devised (La Garde 1929; Wynd, 1933a, b; Burgeff, 1936; Knudson, 1946; Yates and Curtis, 1949; Vacin and Went, 1949; Vacin, 1950a, b; Sideris, 1950; Withner, 1959a, b; Kano, 1965; Arditti, 1967b). Some of these are mere modifications of Knudson's B or C (1922, 1946) media whereas others include considerable changes as an effort to improve seed germination and seedling growth, especially in species which are difficult to germination (Arditti, 1967b).

Although wild population of orchid (terrestrials) often grow in soil that has a low content of accessible mineral ions, a high water potential is generally desirable in artificial germination substrates (Rasmussen, 2000). Though mineral ions may be required for the young seedlings, complete absence of minerals may often be advantageous for the actual germination process. Rasmussen (2000) observed that for *Spiranthes magnicamporum*, water-agar medium induced 99% germination compared with only 45% on mineral medium and 25% on an oat medium. This negative result with minerals may be due to the osmotic effects, since none of the mineral ions which are generally added to media are known to inhibit germination to any great extent. Moreover it appears that orchid seeds and seedlings can adapt easily to a wide variety of inorganic salt combinations and concentrations from as low as 102 p.p.m. total salt content to many times that amount (Curtis, 1947c; Withner, 1959a; Arditti, 1967b).

Iron deficiencies may occur in orchid seed cultures due to the ease with which this metal may precipitate (Sideris, 1950; Vacin, 1950b; Yamada, 1963). In attempts to increase iron availability various iron salts have been used. Various organic and inorganic acids, such as tartaric acid (Vacin, 1950b; Sideris, 1950), citric acid (Burgeff, 1936; Sideris, 1950; Liddell, 1953a, b) etc have been used in an effort to assure iron availability (Arditti, 1967b). Tomato and pineapple juice, potassium tartarate and sodium hexametaphosphate are also suggested as good iron stabilizers (Sideris, 1950; Arditti, 1967b). However, chelated iron preparations have been used successfully (Yamada, 1963).

Addition of manganese to orchid seedling cultures improved growth and development (Noggle and Wynd, 1943; Knudson, 1946; Yates and Curtis, 1949; Sideris, 1950; Vacin, 1950b; Arditti, 1967b), stimulated root growth but had little or no effect on shoot growth; enhanced seedling color to a richer green but caused chlorosis at higher concentrations (Yates and Curtis, 1949).

Orchid tubers and seeds contain very little calcium (Tienken, 1947) and may therefore have a low calcium requirement (Arditti, 1967b) and it may precipitate as calcium phosphate.

Potassium may play an important role in the germination of the orchid *Galeola septentrionalis* (Nakamura, 1962, 1963). Among various potassium salts, KH_2PO_4 did not improve germination whereas K_2CO_3, KI and KCOOH inhibited it (Arditti, 1967b).

Several microelement solutions are routinely used in orchid seed culture media. Boron, copper, molybdenum and zinc are normal components of such solutions. In a comparative study, growth was increased in presence of boron, iron, molybdenum, copper, manganese and zinc to a number of iron and manganese lacking basal media (Arditti, 1967b). Iodine is recommended for immature seed cultures (Tsuchiya, 1954). Copper is recommended to be used with great caution, since orchids are very sensitive to this metal.

Seeds of *Aplectrum hyemale* germinated very poorly on full and half strength Curtis medium (1936). In a modifications that contained urea in Curtis medium, showed improved growth. Thus, this species may require or at least benefit from nitrogen in this form (Oliva and Arditti, 1984).

Seeds of *Spiranthes sinensis* germinated on KC and Karasawa media (Nishimura, 1982). *S. cernua* germinated on a modified KC medium (Stoutamire, 1964). In another study, *S. gracilis* and *S. romanzoffiana* germinated and developed well on full strength Curtis (1936) medium (Oliva and Arditti, 1984). *S. cernua* germinated only in media that contained peptone (Stoutamire 1964), thus indicating its special requirement for yet another or more organic compounds (Oliva and Arditti, 1984).

Alam *et al.* (2002) stated that, seeds of *Dendrobium transparens* germinated in different media in the following order: MS>Hyponex>KC>OKF1 medium. On the other hand, seeds of *Geodorum densiflorum* germinated and formed light green globular structure on both MS and PM media; these structures proliferated and developed into rhizome on PM medium while those on MS medium directly produced tiny seedlings (Bhadra and Hossain, 2003). In case of *Arachnis labrosa*, germination occurred in different media in the order of Mitra *et al.*>MS>KC (Temjensangba and Deb, 2005).

Pedroza-Manrique and Mican Gutiérrez (2006) described the effect of different media on seed germination of *Odontoglossum gloriosum*. It was found that better and rapid germination was obtained on medium containing Hydro-coljap salts and 2.68 μM NAA, in comparison to MS or KC salts. Among six asymbiotic media (ML, MS, LM, VW, MM, KC), seeds of *Habenaria macroceratitis* showed highest germination on both LM and KC media while protocorm development was enhanced on MM media (Stewart and Kane, 2006).

Among 3 *Vanda* hybrids, germination and protocorm development varied with the media (KC, HMS or PT) used under different photoperiods (Johnson and Kane, 2007). According to Johnson *et al.* (2007) seeds of *Eulophia alta* showed better germination and protocorm development on PT medium than on KC, MM, HMS or VW medium. For *Epidendrum ibaguense*, maximum seed germination was achieved on Mitra (1976) and phytamax (PM) fortified with peptone, over MS and KC medium (Hossain, 2008).

For asymbiotic germination of six populations of *Calopogon tuberosus* var. *tuberosus* (ectotypic variant), Kauth *et al.* (2008) stated that different culture media (BM-1, KC, MM, HMS, P723 and VW) had little effect on the germination of Michigan and Florida populations, but germination of South Carolina seeds was high on media with higher calcium and magnesium.

In summary, it may be stated that the mineral requirements of germinating orchid seeds and developing young seedlings are not substantially different from those of most other flowering plants, but some preferences are not uncommon among species.

Effect of nitrogen

Both ammonium and nitrate ions may be utilized by orchids (Curtis and Spoerl, 1948) and appears to be absorbed at approximately same rate (Sideris, 1950; Arditti, 1967b). Both ammonium and nitrate forms can be utilized by *Cattleya mollie, C. trianae, Cymbidium* hybrid and *Vanda tricolor* (Curtis, 1947b; Raghavan and Torrey, 1964; Nakamura, 1982), while growth of *Paphiopedilum insigne* and *Vanilla planifolia* is suppressed by both of these forms (Lugo-Lugo, 1955a, b; Nakamura, 1982).

Ammonical nitrogen is required for *Paphiopedilum* seeds (Burgeff, 1936) and beneficial for *Cymbidium, Cattleya* (Spoerl, 1948). It is more beneficial than the organic nitrogen for *Laeliocattleya* (Burgeff, 1936; Magrou *et al.*, 1949; Arditti, 1967b). *Cattleya labiata* seeds readily germinated on ammonium nitrogen, but drastically inhibited when nitrate was used as the only source of nitrogen (Raghavan and Torrey 1963, 1964). Superiority of ammonical nitrogen over nitric nitrogen is also observed on germination of *Vanilla planifolia* (Lugo-Lugo, 1955a, b). NH_4NO_3 induced better growth of *Dactylorhiza* seedlings than the nitrateform (Malmgren, 1996). But ammoniacal nitrogen is inferior for *Vanda* (Curtis and Spoerl 1948) and may even cause death of the protocorms of *Orchis latifolia* (Mead and Bullard, 1979), when used as the sole N source.

According to Harvais (1973) *Cypripedium reginae* could not tolerate casein hydrolysate (CH) or yeast extract supplements. In contrast, germination and growth of *Dactylorhiza purpurella* in the mineral medium containing KNO_3 were markedly superior when casamino acid was added and was further augmented with addition of yeast extract in the media (Harvais, 1972; Mead and Bullard, 1979). Similar enhancement in protocorm development was noted in *Orchis laxiflora* and *Ophrys sphegodes* with NH_4NO_3 in presence of organic nitrogen and vitamins (Mead and Bullard, 1975). However, CH was found to be replaced by a reconstitution of its amino acids or by glutamine alone; and thiamine proved to be necessary for optimal development of *Orchis laxiflora*. Organic nitrogen when used as the only N source has been variously reported to support good growth of *Orchis laiflora*, such as with casamino acids and edamine (Zeigler *et al.*, 1967). The mineral N source, in contrast, when used alone, delayed development, or even caused death of the protocorms (Mead and Bullard, 1979).

On modified KC medium, seeds of *Aplectrum hyemale* failed to germinate (Stoutamire 1964; Oliva and Arditti 1984); Poor germination occurred on full and half strength Curtis medium, but that was further improved with urea supplements, thus indicating that the species either require or benefit from N in this form (Oliva and Arditti, 1984) like that of many others (Burgeff 1936). Certain compounds related to ornithine cycle (e.g. arginine, ornithine, urea) efficiently replaced NH_4NO_3 and enhanced growth (Raghavan and Torrey, 1963, 1964).

Nakamura (1982) reported that growth of *Cymbidium* hybrids and *Epidendrum o'brienianum* is stimulated by amino-acid mixture in the form of protein hydrolysate, but growth of *Cattleya* sp. and *Epidendrum cochleatum* is inhibited when amino acids are added to the nutrient media (Spoerl, 1948; Raghavan and Torrey, 1964; Nakamura, 1982). Almost all amino acids have been investigated for their ability to promote growth and provide N with several species (Knudson, 1932; Burgeff, 1936, 1954; Schaffstein, 1941; Curtis, 1947a; Spoerl, 1948; Magrou *et al.*, 1949; Withner, 1955; Nakamura, 1962; Raghavan, 1964; Raghavan and Torrey, 1964; Kano, 1965; Arditti, 1967b) and the results are quite variable.

For both germination and growth of *Cymbidium mastersii*, dl-tryptophan, l-aspartic acid and l-asparagine were beneficial while dl-histidine HCl, l-arginine HCl and l-cysteine were inhibitory (Prasad and Mitra, 1975). No single amino acid alone was as good as urea for *Galeola septentrionalis*, but a mixture of amino acids proved to be comparable or superior over urea (Nakamura, 1982).

In *Catasetum fimbriatum* (Majerowicz *et al.*, 2000) high frequency of biomass accumulation was found in plants supplied with glutamine while no significant difference was observed in plants incubated in the presence of inorganic nitrogen form (NO_3^- : NH_4^+ ratio). Enzymatic activities on shoot and root indicated that organic nitrogen and NH_4^+ to be the most important N sources for this plant.

Several experiments revealed that both nucleic acids and nitrogen bases show variable responses regarding germination and growth. Nucleic acids appear to be more stimulating in *Phalaenopsis*, *Vanda*, *Cymbidium* and *Laeliocattleya* than $(NH_4)_2SO_4$ and NH_4NO_3 (Burgeff, 1936, Arditti, 1967b) while of little or no effect (Withner, 1951; Arditti, 1967b) or inhibitory (Schaffstein, 1941, Magrou *et al.*, 1949; Arditti, 1967b) in others. Among the nitrogen bases, adenine induced growth promotion in *Cattleya* either alone or along with ribose and pyruvate (Withner, 1942, 1943, 1951), while had no effect, or even inhibitory for germination and growth of the others (Kano, 1965, Arditti, 1967b; Ernst, 1966b). Guanine and guanine hydrochloride inhibited the growth of *Phalaenopsis, Vanda, Cymbidium* and *Laeliocattleya* (Burgeff, 1936, Arditti, 1967b).

Dijk (1990) described the effect of different mineral ions in orchid seed culture. It was found that in juvenile seedlings of *Orchis morio* and *Dactylorhiza majalis*, compatible host/fungus combinations facilitated growth at low nitrogen levels whereas, *Ceratobasidium* caused growth depression of *D. majalis* at higher nitrogen levels. Dijk and Eck (1995) pointed out that axenic culture of 3-month old protocorms of two different populations of *Dactylorhiza incarnata* (coastal and inland) conveyed a differential response to N forms in that, coastal populations showed a positive response to nitrate ions, whereas the inland population did not. Instead, both showed ammonium toxicity that was in turn reduced in alkaline conditions, the yield reached its optimum at pH 8.0. Beyrle *et al.* (1995) demonstrated that when the fungal hyphae invade the orchid tissues, the balance between the symbionts is affected amongst other things by the source of nitrogen so that a low supply coinciding with a high supply of N in any combination resulted in parasitism and tended to increase fungal virulence.

According to Lee *et al.* (1997) and Rasmussen (2002) *in vitro* symbiotic seedlings reach a higher nitrogen concentration in their tissues than asymbiotic controls which confirmed that the mycobionts assist in nutrient uptake for the plants. So, numerous orchid species are unable to use inorganic nitrogen. This fact can thus be explained by mycorrhizal symbiosis, where their requirements for amino compounds are satisfied by a symbiotic fungus.

Extraradical mycelia systems of *Goodyera repens* was employed with double-labelled [^{13}C-^{15}N] glycine, both ^{13}C and ^{15}N were assimilated by the fungus and transferred to the roots, thus again indicating a fungus-dependent pathway for organic N acquisition by the orchid (Cameron *et al.*, 2006).

Effects of vitamins

Apart from sugars and nitrogen, some other factors may be required for germination and growth of orchid seedlings, at least sometimes, and the mycorrhizal fungi may make this available to the host (Knudson, 1922, 1924, 1925; Curtis, 1939; Noggle and Wynd, 1943). Such a factor supplied by the mycorrhizal symbiont, may be a vitamin (Schaffstein, 1938, 1941;

Noggle and Wynd, 1943; Burgeff, 1959) and this may be the inherent cause of rapid and more vigorous growth (Bultel, 1924-25, Wynd, 1933b) of the orchid seedlings in symbiotic culture (Noggle and Wynd, 1943).

It has been demonstrated that vitamin C to be necessary for germination of *Cymbidium* seeds which were also found to be difficult to germinate by asymbiotic means (Curtis, 1943). Presence of vitamin C in the medium resulted in increased germination and accelerated growth of *Cattleya labiata autumnalis* and *Oncidium pulvinatum* (Pollacci and Berganischi, 1940), while it exerted no positive effect on *Goodyera repens* (Downie, 1940), *Dendrobium nobile* (Schaffstein, 1941), *Cattleya* hybrids (Withner, 1942, Noggle and Wynd, 1943) and *Epidendrum tempense* (Withner, 1942). Bouriquet (1947) pointed out that in *Vanilla planifolia*, meager germination occurred with vitamin but the protocorms soon died. No effect of vitamins B1, B2, B6, C niacin, biotin, pantothenic acid and inositol was found on *Cattlya* hybrid and *Epidendrum tempense* (Withner, 1942). Similarly, *Vanilla planifolia* showed growth promotion with the application of either mixture of vitamins B1, B6, niacin, five amino acids and IBA; or vitamins and IBA; or vitamin B12 alone (Withner, 1955).

Noggle and Wynd (1943) demonstrated that the *Cattleya* hybrid germinated and grew well in un-purified maltose media, but no germination occurred in purified maltose media, even in presence of thiamin HCl, ascorbic acid or calcium pantothenate. However, the latter media exerted meager germination and slow development with riboflavin, good germination but poor development with pyridoxine and both good germination and excellent development with nicotinic acid (P-P factor).

Vitamin B1, pyrimidine along with thiazole, niacin, biotin, vitamin B6 and thiamin was favourable for germination and growth of *Cattleya* seedlings (Mariat 1948, 1949, 1952). Inositol, folic acid and para-amino benzoic acid stimulated germination, while vitamins B2 and B6 were effective for differentiation of the plants already at the leaf-point stage. In many instances, B1 was either ineffective or at times inhibitory (Schaffstein, 1938, 1941; Downie, 1940, 1943; Withner, 1942; Noggle and Wynd, 1943), while it had a favourable effect on some other species (Arditti, 1967b).

Seed germination and growth of *Cymbidium mastersii* was enhanced by thimine HCl, pyridoxine HCl, biotin and folic acid, but was inhibited with riboflavin and completely suppressed by niacin (Prasad and Mitra, 1975). For *Acampe praemorsa*, thiamine HCl showed slight enhancement of germination, than that of niacin (Krishnamohan and Jorapur, 1986).

Germination and growth of *Orchis laxiflora* occurred in nutrient media containing inorganic and organic nitrogen along with vitamins (thiamine, pyridoxine, nicotinic acid and biotin) among which, thiamine proved to be necessary for the formation of chlorophyll and further development, rather than for germination itself (Mead and Bullard, 1975). Considerable enhancement was also noted on *Dendrobium* with thiamine and "Panvitan" (Kano, 1965; Arditti, 1967b).

Numerous isolates of the form-genus *Rhizoconia* which are obtained as orchid mycobionts (Warcup, 1975) are apparently less exacting for their nutrients (Hadley and Ong, 1978). Some of these are reportedly heterotrophic for thiamine and biotin (Vermeulen, 1947), or at least partially so for thiamine and n-aminobenzoic acid (Stephen and Fung, 1971).

Vitamin heterotrophy in orchid fungi may confirm a degree of evolutionary adaptation to the root-infecting habit (Hadley and Ong, 1978), and a parallel situation occurs in ectomycorrhizal fungi and other higher basidiomycetes (Harley, 1969).

Effect of Different Organic Additives

Effect of peptones

Peptone is a complex additive that contains many amino acids, several amides and a number of vitamins (Powell and Arditti, 1975; Oliva and Arditti, 1984). The reports of the application of peptone in orchid culture media are less common and its effects are not yet well established (Arditti and Ernst, 1993; Mukhopadhyay and Roy, 1994). The major amino acids that have been found in peptone are glycine (23%), glutamic acid (11%), arginine (8%), aspartic acid (59%), lysine (4.3%), leucine (3.5%), valine (3.2%), tyrosine (2.3%) and phenyl alanine (2.3%). Peptone also contains various minerals (e.g., P, K, Na, Mg, Ca, Cl, Mn, Pb, As, Cu and Zn) and vitamins (e.g. pyridoxine, biotin, thiamine, niacin and riboflavin).

Previous observations showed that the effects of peptone on orchid seed germination varied with the species. Peptone enhanced the rate of germination and growth of protocorms in *Cattleya* (Curtis, 1943; Alberts, 1953; Stoutamire, 1964; Harvais, 1973, 1974; Linden, 1980; Oliva and Arditti, 1984), *Cypripedium* (Liddell, 1953a), *Dendrobium* (Alberts, 1953), *Vanilla* (Bouriquet, 1947) and various European terrestrial orchids. Stoutamire (1964) pointed out that seeds of *Spiranthes cernua* could germinate only in medium supplemented with peptone. According to Prasad and Mitra (1975) the use of 1% peptone proved effective for both seed germination and better growth of protocorms. Seeds of *Acampe praemorsa* showed better germination in presence of 0.05 and 0.1% peptone. Again for *Vanda*, peptone is ineffective for early stages of germination (Mathews and Rao, 1980; Mitra, 1986). Similar improving effect of peptone in protocorm growth and development was noted in *Dendrobium* (Alberts, 1953, Morel, 1974) and in *Vanda* (Lami, 1927, Morel, 1974, Mathews and Rao, 1980; Mitra, 1986). Such promoting effects have also been reported on *Cyperipedium* (Withner, 1947), *Dendrobium*, *Cypripedium reginae*, *Cymbidium virescens*, *Cymbidium* hybrid, *Paphiopedilium insigne* (Kano, 1965) and *Phalaenopsis* hybrids (Ernst, 1966b).

Effect of coconut water

Coconut water (CW), liquid endosperm of *Cocos nucifera* fruit, is an undefined complex organic substance. It has been successfully employed for the culture of different plant species (Arditti and Ernst, 1993; Sutter, 1996). It has been shown that, coconut water contains a large number of amino acids, like alanine, arginine, aspartic acid, glutamic acid, glycine, leucine etc. and other substances, like (e.g., citric acid and malic acid), vitamins (biotin, folic acid, niacin, pantothenic acid, riboflavin, pyridoxine and thiamine) and several plant hormones like auxins, cytokinins and gibberellins (Raghavan, 1966, 1976; Arditti and Ernst, 1993). Van Overbeek *et al.* (1941, 1942) and Arditti and Ernst (1993) used coconut water for the culture of *Datura* embryos and showed its beneficial effects. Earlier Mariat (1951) used CW in an orchid culture medium, where it was shown to inhibit germination of *Cattleya lawrenceoma* seeds. In orchid seed culture the effect of CW was shown to be species specific. According to Kotomori and Murashige (1965), germination of *Dendrobium* seeds was unaffected by its application, but the growth of protocorms was inhibited by its presence. However, improved germination as well as growth was reported in *Cymbidium mastersii* (Prasad and Mitra, 1975), *Acampe praemorsa* (Krishnamohan and Jorapur, 1986) and *Spathoglottis plicata* (Chennaveeraiah and Patil, 1975). In *Arundina bambusifolia*, rate of germination and growth of seedlings enhanced by the application of coconut water (Kanjilal *et al.*, 1998a). Coconut water exerted a stimulatory effect in seed culture of *Paphiopedilum, Vanilla* and *Cattleya* (Hegarty, 1955; Arditti, 1967a). For *Cattleya*, it was excellent when applied in a mixture with tomato and pineapple juice plus banana pulp and vitamin (Lawrence and Arditti, 1964a, b). The effects of coconut water is very species specific and in addition as it is a natural product, the constituents as well as activity of liquid endosperm may depend on factors like age, variety, seasons and means of *in vitro* sterilization (Arditti and Ernst, 1993).

Effects of yeast-extract

Yeast-extract (YE) contains water-soluble components of the yeast cell, like carbohydrates, amino acids, peptides vitamins and salts. In many orchid species yeast-extract has been successfully used for seed germination as well as in protocorm development (Curtis, 1947a; Flamce, 1978; Mariat, 1948; Mathews and Rao, 1980). In the present study germination of different species of *Dendrobium* was found to be stimulated by 0.05% YE, but remains unaffected with further increase in the concentration. Previously, it has been shown that the effect of YE on germination of *Dendrobium* could be both stimulatory as well as inhibitory (Kano, 1965; Ito, 1955). This result partly contradicts the present finding on *Dendrobium* species. Schaffstein (1938) and Tonnier (1954) also reported that application of YE in the media enhanced the germination and the present observation also contradicts this report.

Effect of banana extracts

Banana extracts stimulated germination and growth of various species (Withner, 1955; Lawrence and Arditti, 1964a, b; Ernst, 1966b; Arditti, 1967b). Seeds of *Anoectochilus formosanus* germinated on HMS medium supplemented with 0.2% activated charcoal and 8% banana homogenate and the germinated seedlings were cultured in HMS liquid medium with BAP and subsequently on HMS medium supplemented with banana homogenate along with NAA, BAP and activated charcoal (Shiau *et al.*, 2002). Seeds of *Spathoglottis plicata* germinated well in VW medium that was further improved with ripe banana extract, followed with ripe tomato extract and yogurt water (Roy and Biswas, 2005). Also onion extracts (Ito, 1955), fresh apple juice (Tsukamoto *et al.*, 1963, Kano, 1965), dried apple (Kano, 1965), fresh Sphagnum beds (Lindquist, 1965) etc induced germination and protocorm growth of different orchid species with variable degrees of excellence.

Effect of plant growth regulators

Several studies indicated the requirement of exogenous supplies one or more growth factors in addition to mineral nutrients and carbohydrates for seed germination and/or growth of the protocorms of orchids, since auxin is known to be produced by many fungi (Gruen, 1959; Ulrich 1960) and cytokinins have been identified in the culture media of the mycorrhizal fungus *Rhizopogon roseolus*. It has been suggested that the growth factors produced by symbiotic fungi may be involved in the development of orchid embryos (Hadley and Harvais, 1968). However it should be remembered that only traces of auxin have been found in *Cypripedium* seeds more at all in *Calanthe* and *Dendrobium* seeds (Poddubnaya-Arnoldi and Zinger, 1961; Arditti, 1967b).

Stimulation in both germination and seedling growth was obtained in various orchid species when either of Knudson's, Burgeff's, and Sladden's or tomato juice media was supplemented with "Hortomone A" (Arditti, 1967b). Both germination and growth was promoted by NAA in *Cattleya* (Withner, 1951), *Cattleya* hybrid (Mariat, 1952) and *Epidendrum nocturnum* (Yates and Curtis, 1949) while no significant effect was observed on seedling growth of *Euanthe*, *Phalaenopsis* and *Vanda* (Burgeff, 1936). Root growth was stimulated by NAA in *Epidendrum nocturnum* (Yates and Curtis, 1949). IBA, on the other hand stimulated germination and/or seedling growth of various species (Withner, 1951) and had no effect on some other (Burgeff, 1936). At low concentration, IBA had no effect on *Dendrobium* and *Brassolaeliocattleya* and stimulated growth of *Vanilla planifolia,* while at higher concentration, it was inhibitory for these three species (Withner, 1955; Kano, 1965). IAA, however slightly promoted seedling growth of *Cattleya* (Withner, 1951); increased fresh weight of *Phalaenopsis* hybrid seedlings (Ernst, 1966b) and slightly inhibited the growth of *Dendrobium* (Kano, 1965).

Gibberellic acid and gibberellins have been shown to break seed dormancy and the effects of gibberellins on cell division in the stem apex and subsequent cell elongation, are well known (Paleg, 1965; Hadley and Harvais, 1968). Gibberellic acid promoted germination in *Cattleya, Cymbidium, Cypripedium* and *Odontoglossum* (Humphreys, 1958). Enhanced seedling growth was recorded in various species and hybrids with gibberellins and/or gibberellates (Moir, 1957; Smith, 1958). Gibberellin induced growth promotion primarily involves early growth of the protocorms and the effect diminishes as the seedlings grow older, such that the control itself becomes equal or exceeds the treated ones, as in *Cymbidium, Laeliocattleya* (Smith 1958). However, the negative effect of gibberellins on germinating orchid embryos is also common (Kano, 1965; Ernst, 1966b; Arditti, 1967b). Gibberellic acid had no stimulatory effect on germination of *Orchis purpurella* whether alone or in combination with any of IAA, adenine or kinetin (Hadley and Harvais, 1968). It induced abnormal extension growth of shoots and protocorms, and etiolation of leaf initials (Hadley and Harvais, 1968; Hadley, 1970).

On the other hand, auxin and gibberellins may interact in controlling cell enlargement while the ratio of auxin to cytokinin is recognized as important in controlling cell division and the initiation of buds and roots (Kefford and Goldacre, 1961). Mixtures of IAA with either GA or adenine were without any beneficial effects on germination or protocorm growth of *Orchis purpurella* (Hadley and Harvais, 1968). However, mixtures of IAA and kinetin enhanced protocorm growth considerably and the initial elongation characteristic of IAA soon gave way to a more proportionate development (Hadley and Harvais, 1968).

Seeds of *Spathoglottis plicata* germinated on a modified White's (1934) medium containing CH, CW, 2,4-D, IAA, NAA and kinetin, either singly or in various combinations (Chennaveeraiah and Patil, 1975). Organic supplements enhanced germination when alone but were inhibitory in presence of PGRs.

Seed germination and protocorm growth of *Cymbidium mastersii* were enhanced with adenine and kinetin with the former being more effective, whereas, the beneficial effect of IAA was noticed only in seed germination (Prasad and Mitra, 1975).

Dendrobium fimbriatum seeds, when grown on a specially formulated medium, normal development of seedlings were accompanied by direct development of some seeds into a callus mass that showed differentiation of PLBs in the absence of exogenous PGRs (Mitra *et al.*, 1976). The addition of adenine in the media containing BA increased the number of PLBs (Oyamada and Takano, 1985).

Effect of different isolates of orchid-associated bacteria (OAB) on mycorrhiza-assisted germination of the terrestrial orchid *Pterostylis vittata* revealed that symbiotic germination was enhanced by IAA, inhibited by gibberellic acid and suppressed by kinetin (Wilkinson *et al.*, 1994). Such enhancement may have resulted either from the production of IAA by the OAB and/or by the induction of endogenous hormones in the orchid by the metabolites of the bacteria and/or mycorrhizal fungus (Wilkinson *et al.*, 1994).

Among various cytoknins tested, *in vitro* germination of *Cypripedium candidum* was enhanced by BAP and 2-iP while the effect of kinetin was not significantly different from that of control (De Pauw *et al.*, 1995). The protocorm morpho-types found in the kinetin and control treatments differed from those observed in the presence of BAP and 2-iP and might have resulted via a different pathway. Also, BA and zeatin in low concentrations enhanced germination and protocorm growth of *Cypripedium macranthos* and *Dactylorhiza aristata*, while kinetin did not enhance germination (Tomita, 2002).

Shimura and Koda (2004) reported that seeds of *Cypripedium macranthos* var. *rebunense* germinated in Hyponex-peptone medium supplemented with both NAA and cytokinin forming slow-growing protocorms, but plantlets were easily regenerated on PGR-free medium.

Seeds of *Geodorum densiflorum* germinated on both MS and PM media; seedling (in MS media) grew with NAA along with BAP while the rhizome tips formed multiple shoot buds with PGRs (NAA, IAA, IBA, piclorum, BAP and Zeatin) (Bhadra and Hossain, 2003). On the other hand, immature seeds of *Cymbidium faberi* formed rhizomes on PGR-free medium and multiplied 5 times with 1mg l^{-1} NAA and more than 90% of the rhizomes initiated shoots on the media containing 0.5 or 1mg/L NAA+ 2or 5 mg/L BA (Chen *et al.* 2005).

Among four exogenous cytokinins (BA, Zeatin, Kinetin, 2-iP), both zeatin and kinetin at 1µM enhanced seed germination of *Habenaria macroceratitis* (Stewart and Kane, 2006). Shoot growth rate and shoot length of *Dactylorhiza* was enhanced by cytokinins (2-iP or BA) and their combination with auxin (IBA) while the root growth rate and root length of seedlings were increased in presence of IBA and NAA (Wotavova-Novotna *et al.*, 2007). Seedlings grown from *Rhyncostylis retusa* seeds attained maximum length on MS medium + 6 µM BA+ 0.2 µM NAA+ 1g/L AC (Thomas and Michael, 2007). Among kinetin and TDZ, maximum multiple shoot number was observed with 2 µM TDZ in presence of AC and the shoots were rooted on HMS medium supplemented with 2 µM IBA. Thus PGRs appear to have a low importance for germination but are important for subsequent development of the protocorms (De Pauw *et al.* 1995). Their influence often depends on the nature and concentration of the PGRs and varies among the species (Hadley 1970; Van Waes and Debergh 1986a; Wotavova-Novotna *et al.* 2007).

Wilkinson *et al.* (1994) described the effects of different isolates of orchid-associate bacteria (OAB) on mycorrhiza-assisted germination of the terrestrial orchid *Pterostylis vittata*. It was revealed that symbiotic germination was enhanced by IAA, but inhibited by gibberellic acid and suppressed by kinetin. Such enhancement may have resulted either from the production of IAA by the OAB and/or by the induction of endogenous hormones in the orchid by the metabolites of the bacteria and mycorrhizal fungus.

De Pauw *et al.* (1995) stated that PGRs appear to have less influence on seed germination but are important for subsequent development of the protocorms. He studied the role of various cytokinins on the germination of *Cypripedium candidum*. Among them BAP and 2-iP enhanced seed germination, though the effect of kinetin was not significantly different from the control. However, protocorm development was much faster in the presence of all the cytokinins. In different treatments nine morphologically different types of protocorms were recognized and the type of protocorms found in the kinetin and control treatments differed from those observed in the presence of BAP and 2-iP. At low concentrations, BA and Zeatin showed increased rate of germination and protocorm growth of *Cypripedium macranthos* and *Dactylorhiza aristata*, while kinetin did not promote germination (Tomita, 2002). On the contrary, Shimura and Koda (2004) described the effect of PGR on seed germination of *Cypripedium macranthos* var. *rebunense*. It was found that seeds germinated in Hyponex-peptone medium supplemented with both NAA and cytokinin forming slow growing protocorms (mentioned as PLBs), but plantlets were easily regenerated on PGR-free medium.

Studies on the *in vitro* seedling production were also reported in *Laelia purpurata* (Stancato *et al.*, 1998), *Encyclia boothiana* var. *erthronioides* (Stenberg and Kane, 1998) and *Cymbidium aloifolium* (Dasgupta and Bhadra, 1998;

Buzarbarua, 1999). Buzarbarua (1999) showed that in PGR-free medium the protocorms failed to produce shoot and root, but with the application of higher concentrations of NAA, IBA and kinetin this could be easily achieved.

Shiau *et al.* (2002) described the effect of growth regulators in HMS medium on germination. It was found that, seeds of *Anoectochilus formosanus* germinated on basal HMS medium supplemented with 0.2% activated charcoal and 8% banana homogenate and the germinated seedlings were cultured in HMS liquid medium with BAP and subsequently on HMS medium supplemented with banana homogenate along with NAA, BAP and activated charcoal. Stimulation of germination in seeds of *Spathoglottis plicata* occurred in VW medium, that was further improved with banana extract, followed with ripe tomato extract and yogurt water (Roy and Biswas, 2005).

Lee and Lee (2003) showed that in *Dicentra spectabilis* embryogenic callus was obtained from seeds, which were cultured on Murashige and Skoog medium supplemented with various concentrations of 2,4-D, under dark condition.

Shiau *et al.* (2005) developed an *in vitro* propagation protocol for *Haemaria discolor* var. *dawsoniana* by artificial cross-pollination and asymbiotic germination of seeds. Germination of seeds obtained on half-strength Murashige and Skoog's medium supplemented with 3% sucrose, 0.2% activated charcoal, banana homogenate TDZ and NAA. Survival frequency of plantlets was 96%.

Effects of surface adsorbants

Activated charcoal can be produced from wood, wood waste, paper-mill waste liquors and peat and is variously used in orchid culture media (Pan and Staden, 1998). Its addition to culture medium may promote or inhibit *in vitro* growth, depending on species and tissues used. The consequences of addition of activated charcoal may be attributed to establishing a darkened environment; adsorption of undesirable/inhibitor substances; adsorption of growth regulators and other organic compounds, or the release of growth promoting substances present in or adsorbed by activated charcoal (Pan and Staden, 1998).

For germination and further development of Western European orchids, addition of activated charcoal to the sowing medium lowered the germination rate and slowed down the seedling development, but its addition to the transplantation medium was stimulatory for several species with a high release of polyphenols in the sowing medium and failed to stimulate the species with no visible release of polyphenols in the medium (Van Waes, 1987). Its addition to these growing medium of *Epidendrum radicans* improved *ex-vitro* establishment of the seedlings (Pateli *et al.* 2003), but failed to stimulate either germination or the development of *Odontoglossum gloriosum* seedlings (Pedroza-Manrique and Mican-Gutièrrez, 2006).

In *Cattleta labiata,* after an initial period of growth in NH_4NO_3, the seedlings overcame the inhibitory effect of nitrate and this ability to utilize nitrate reductase (Raghavan and Torrey, 1963, 1964; Arditti, 1967b; Hew *et al.,* 1988). So an age-dependent change in tolerance level becomes evident.

Due to over collection and loss of natural habitat, different species of *Cypripedium* have become extremely rare and many of them are hardly found in natural habitat. In an effort to save these species from extinction, several workers have made an effort to develop an efficient system of seedling production through *in vitro* seed germination technique Ramsay and Stewart (1998), under the Sainbury Orchid Conservation Project, carried out a successful attempt to raise seedlings of *C. calceolus* and re-establish them in their natural habitat. From the works of Light and Mac Conaill (1998), it was found that chilling treatment was a necessary factor for the germination of mature *C. calceolus* var. *pubescens* seeds, and this event could be overcome if the seeds were harvested prematurely (7-8 weeks after pollination). In other studies, *in vitro* seed germination was performed on *C. reginae* (Frosch, 1986; Ballard, 1987) and *C. acaule* (St-Arnaud *et al.,* 1992).

Malmgren (1996) stated that Ammonium nitrate (NH_4NO_3) induced better growth of *Dactylorhiza* seedlings than the nitrate form.

Kanjilal *et al.* (1998a) described the effects of medium composition on the germination of *Arundina bambusifolia* seeds, which showed that germination could occur in both Knudson's C (Knudson, 1946) and Vacin and Went (1949) media, with or without CW, tomato juice or banana extract. Among these supplements, only CW promoted germination of seeds, while both tomato juice and banana extract proved inhibitory. In further studies, Kanjilal *et al.* (1998b) stated that the *in vitro* raised seedlings were shown to contain aneuploid variation with 2n=36 and 2n=38, beside the normal one 2n=40.

Roy and Banerjee (2001) studied *in vitro* seed germination and protocorm development of *Geodorum densiflorum* in three basal media (KC, VW and ½ strength MS). Seed germination was very high (up to 96%) in KC medium; ½ strength MS being slightly more productive than VW.

For germination and further development of *Epidendrum radicans,* the addition of activated charcoal to the growing medium improved *ex vitro* establishment of the seedlings (Pateli *et al.,* 2003), but activated charcoal failed to stimulate either germination or the development of *Odontoglossum gloriosum* seedlings (Pedroza-Manrique and Mican-Gutiérrez, 2006).

In *Coelogyne mossiae* Sebatinraj *et al.* (2006) studied the effects of different nutrient solutions, organic supplements, and plant growth regulators on *in vitro* seed germination and subsequent plantlet development.

For asymbiotic seed germination of a threatened North America native terrestrial orchid *Bletia purpurea,* six asymbiotic orchid seed germination media (KC, VW, P723,VW, ½ MS and MM) showed an effective germination and protocorm development in either a 0/24h or 16/8h L/D photoperiod (Dutra *et al.,* 2008).

Dutra *et al.* (2009) examined the growth promoting effect of five asymbiotic orchid seed germination media (1/2 MS, KC, MM, P723 and VW) on *Cyrtopodium punctatum,* an endangered Florida native orchid under different photoperiods.

Effect of Light and Temperature

Optimal *in vitro* temperature and light parameters are subject to specific variation and reflect taxonomic affiliation and/or differences in habitat. Optimum incubation temperatures for Orchidaceae of tropical origin were reported in the range 22–25°C (Borris, 1969; Stoutamire, 1974, 1990; Arditti, 1982; Oliva and Arditti, 1984; Rasmussen, 1995; Michel, 2002). Alternatively, Rasmussen *et al.* (1990) reported that temperate orchids required temperatures somewhat lower (< 20°C) than those reported elsewhere. Data are also available that have revealed the positive effects of cold and warm stratification seed treatments prior to long-term incubation (Withner, 1959a; Nakamura, 1976; Hadley, 1982; Fast, 1982; Pritchard, 1985; Ballard, 1987; Van der Kinderen, 1987; Van Waes, 1987; Coke, 1990; Stoutamire, 1990; Yanetti, 1990; Rasmussen, 1992; De Pauw and Remphrey, 1993; Miyoshi and Mii, 1998). Many terrestrial species benefit from incubation in continuous darkness for a minimum defined period following sowing (Harvais and Hadley, 1967; Fast, 1982; Van Waes and Debergh, 1986b; Ballard, 1987; Van Waes, 1987; Rasmussen *et al.,* 1990; Rasmussen and Rasmussen, 1991; Rasmussen, 1992, 1995). However, variants include continuous low-light treatments and a range of lowlight: dark photoperiods (Knudson, 1943; Stoutamire, 1963, 1964, 1974; Arditti, 1967a, 1979; Harvais, 1972, 1973; Clements, 1982; Anderson, 1991; Rasmussen and Rasmussen, 1991; Zettler and McInnis,1992).

Effect of Seed Maturity

Seed maturity also influences *in vitro* germination success. Immature seed ('green-podding') is widely used for germinating terrestrial orchids (Stoutamire, 1964, 1974; Borris, 1969; Sauleda, 1976; Linde´n 1980, 1992; Vogelpoel, 1980, 1987; Arditti, 1982; Clements, 1982; Fast, 1982; Oliva and

Arditti, 1984; Pritchard, 1985; Anderson, 1990, 1991; Rasmussen *et al.*, 1990; Rasmussen and Rasmussen, 1991; Zettler and McInnis, 1992; Rasmussen, 1995; LaCroix and LaCroix, 1997; Wodrich, 1997). However, the appropriate developmental stage for excision must be assessed for each species, with Michel (2002) reporting optimal harvest times in the range 21–360 days post-pollination for tropical terrestrial species. By contrast, Rasmussen (1995) reported a maximum of 63 days for Holarctic terrestrials. 'Green-podding' also removes the need for seed decontamination, which is desirable since some orchid seed displays sensitivity to decontaminants (Bergman, 1995, 1996; Rasmussen, 1995). However, exposure is not necessarily disadvantageous as chemical scarification increases the percentage germination in several Holarctic terrestrials (Stoutamire, 1964; Harvais and Hadley,1967; Purves and Hadley, 1976; Harvais, 1982; Linde´n, 1980; Arditti, 1982; Van Waes and Debergh, 1986a, b; Rasmussen, 1992, 1995; Miyoshi and Mii, 1995).

In vitro asymbiotic seed germination of *Dendrobium nobile* varied significantly with fruit harvesting time and seed culturing medium. Immature seeds that were harvested 96 and 116 days after pollination (DAP) showed lower germination response. But mature seeds harvested 188 DAP showed higher frequency at stage 5 (emergence of 1st leaf) on different growth media indicating the absence of testa imposed dormancy in this endangered epiphytic orchid (Vasudevan and Staden, 2010).

Conclusion

These findings provide several clues regarding the germination requirements of several tropical orchids. They are clearly divided—as attested to by the *in vitro* germinability indices, with regard to the germination requirements of individuals. Germination of different seeds incubated with plant growth regulators or undefined media additives need not be interpreted as a lack of response to these supplements. The *in vitro* studies of different species showed that the nutritional requirements for various phases of seedling development differ considerably. The results of *in vitro* studies on germination of seeds of different orchid species indicated that, a simple nutrient medium Knudson's C (1946), is efficient for germination. The effect of different organic additives like CW, yeast extract and nitrogenous source like peptone was negligible. Similarly, different PGR modulators and inhibitors of endogenous auxins hardly affected germination. Seedling necrosis was common occurrence during seed culture which in most cases effectively occurred in presence of different auxins. Seedling necrosis could be overcome with the inclusion of different organic additives in the germination media. However, the nutritional requirements of the seedlings of different species varied considerably at different phases of their development. As a consequence, different organic supplements and PGRs differentially affected the post-germination growth and developmental events in various species. Bearing in mind that the utilization of land for agriculture, grazing or urbanization has devastated many native orchid populations and the protocol described above is a useful tool for a massive production of the species studied for re-introduction to threatened areas. Moreover, because of the robust character of orchids as well as their beautiful, brightly variegated and long-living flowers this investigation could provide an impulse to the use of these species as garden or potted plants. The improvement of the germination protocol in order to reduce the time needed until plantlet development, Zettler (1997) concluded that if an orchid is critically dependent on a compatible mycorrhiza for germination, the loss of that fungus *in situ* will ultimately result in the inability of that species to establish new stands. An additional benefit of culturing orchid seeds symbiotically is that the resulting seedlings can serve as both plant material and inoculum for conservation efforts (Batty *et al.*, 2002). The isolation of a suitable mycobiont is a promising step forward in ongoing efforts to develop reintroduction and conservation protocols for several endangered and threatened orchid species. Continued research should focus on improving the efficiency of orchid symbiotic seed germination, acclimatization, and *in situ* establishment methodologies to further progress in rare orchid conservation techniques.

References

Alam, M.K., Rashid, M.H. Hossain, M.S. Salam, M.A. and Rouf, M.A. 2002. *In vitro* seed propagation of *Dendrobium* (*Dendrobium transparens*) orchid as influenced by different media. Biotech. **1(2-4):** 111-115.

Alberts, A.A. 1953. Use of fish emulsion for the germination of orchid seed. Orch. J. **2(10):** 464-466.

Alexander, C., Alexander, I.J. and Hadley, G. 1984. Phosphate uptake in relation to mycorrhizal infection. New Phytol. **97:** 401-411.

Alexander, C. and Hadley, G. 1985. Carbon movement between host and mycorrhizal endophyte Oliva during the development of the orchid *Goodyera repens* Br. New Phytol. **101:** 657-665.

Ames, O. 1922. Notes on New England orchids. II. The mycorrhiza of *Goodyera pubescens*. Rhodora. **24 (270):** 37-46.

Anderson, A.B. 1990. Asymbiotic germination of seeds of some North American orchids. In: C.E. Sawyers (eds.). North American native terrestrial orchid propagation and production. Chadds Ford Press, Pennsylvania, USA.

Anderson, A.B. 1991. Symbiotic and asymbiotic germination and growth of *Spiranthes magnicamporum* (Orchidaceae). Lindleyana **6:** 183–186.

Anonymous. 1893. The history of orchid hybridization. Orch. Rev. **1(1):** 3-6.

Arditti, J. 1967. Factors affecting the germination of orchid seeds. Bot. Rev. **33:** 1-97.

Arditti, J. 1979. Aspects of the physiology of orchids. Adv. Bot. Res. **7:** 241-665.

Arditti J, Michaud, J.D. and Healey, P.L. 1980. Morphometry of orchid seeds. II. Native California and related species of *Calypso, Cephalanthera, Corallorhiza and Epipactis.* Am. J. Bot. **67:**347–360.

Arditti, J. Michaud, J.D. and Oliva, A.P. 1981. Seed germination of North American orchids. I. Native California and related species of *Calypso, Epipactis, Goodyera, Piperia* and *Platanthera.* Bot. Gaz. **142(4):** 442-453.

Arditti, J. 1982. Orchid seed germination and seedling culture-a manual. Introduction, general outline, tropical orchids (epiphytic and terrestrial) and North American terrestrial orchids. In: J.Arditti (ed.). Orchid Biology: Reviews and perspectives. Vol.1. Cornell Univ. Press, Ithaca, N.Y, pp. 243-293.

Arditti, J. and Ernst, R. 1984. Physiology of germinating orchid seeds. In: J. Arditti (eds.). Orchid Biology reviews and perspectives, Vol. 3., Ithaca, NY, USA: Cornell University Press, pp. 179-222.

Arditti, J. 1992. Fundamentals of orchid biology. John Wiley & Sons Inc., NY: 691.

Arditti, J. 1993. Storage and longevity of orchid seeds. Malay. Orch. Rev. (Singapore) **27:** 59-63, 82-87.

Arditti, J. and Ernst, R. 1993. Micropropagation of orchids. John Wiley & sons, NY.

Arditti, J. and Ghani, A.K.A. 2000. Numerical and physical properties of orchid seeds and their biological implications. New Phytol. **145:** 367-421.

Atwood, J.T. 1986. The size of the orchidaceae and the systematic distribution of epiphytic orchids. Selbyana **7:** 171-186.

Ballard, W.W. 1987. Sterile propagation of *Cypripedium reginae* from seeds. Amer. Orch. Soc. Bull. **56:** 935-946.

Ballion, G. and Ballion, M. 1924. The non-symbiotic germination of orchid seed in Belgium. Orch. Rev **32:** 305-309.

Ballion, G. and Ballion, M. 1928. Asymbiotic germination of orchid seed. Ibid. **36:** 103-112.

Barth, F.G. 1985. Insects and flowers: The biology of a partnership. George Allen & Unwin, London.

Barthlott, W. and Ziegler, B. 1981. Mikromorphologie der Samenschalen als systematisch Merkmal bei orchideen. Berichte der Deutsche Botanische Gesellschaft **94:** 267-273.

Batty, A.L. Dixon, K.W. Brundrett, M.C. and Sivasithamparam, K. 2002. Orchid conservation and mycorrhizal associations. Chapter 7. In: K.Sivasithamparam, K.W. Dixon and R. L. Barrett (eds.), Microorganisms in Plant conservation and Biodiversity: Kluwer Academic Publishers, pp. 195-226.

Beer, J.G. 1863. Beitrage zur morphologie und biologie der familie der orchidéen. Vienna, Austria: Druck und Verlag von CarlmGerold's Sohn.

Benzing, D.H. 1986a. The genesis of ochid diversity: emphasis on floral biology leads to misconception. Lindleyana 1: 73-89.

Benzing, D.H. 1986b. The vegetative basis of vascular epiphytism. Selbyana 9: 23-43.

Benzing, D.H. and Clements, M.A. 1991. Dispersal of the orchid *Dendrobium insigne* by the ant *Iridomyrmex cordatus* in Papua New Guinea. Biotropica 23: 604-607.

Bergman, F.J. 1995. The disinfection of orchid seed. Orchid Rev. 103: 217–219.

Bergman, F.J. 1996. Orchid seed disinfection—the effect of pH and temperature. Orchid Rev 104: 245–247.

Bernard, N. 1899. Sur la germination du *Neottia nidus-avis*. Compt. Rend. Acad. Sci. Paris 128: 1253-1255.

Bernard, N. 1902. Études sur la tubérisation. Revue Généale de Botanique 13: 1-92.

Bernard, N. 1903. La germination des orchidées. Comp. Rend. Acad. Sci. Paris 137: 483-485.

Bernard, N. 1904. Recherches expérimentalis sur les Orchidées. Rev. Gén. Bot. 16: 405-451, 458-476.

Bernard, N. 1906. Symbiosis d'orchidées et de divers champignons endophytes. Comp. Rend. Acad. Sci. Paris 142: 52-54.

Bernard, N. 1909. L'evolution dans la symbiosie, les orchidées et leur champignons commensaux. Annales des Sciences Naturelles, Botanique 9 (9): 1-196.

Beyrle, H.F. Smith, S.E. Peterson, R.L. and Franco, C.M.M. 1995. Colonisation of *Orchis morio* protocorms by a mycorrhizal fungus: Effects of nitrogen nutrition and glyphosate in modifying the responses. Can. J. Bot. 73: 1128-1140.

Bhadra, S.K. and Hossain, M.M. 2003. *In vitro* germination and micropropagation of *Geodorum densiflorum* (Lam.) Schltr., an endangered orchid species. Plant Cell Tiss. Org. Cult. 13(2): 165-171.

Blowers, J.W. 1966. Orchids. How important is mycorrhiza. Gard. Chron. 159(13): 293.

Borris, H. 1969. Samenvermehrung und anzucht Europaischer erdorchideen. In: Proceedings of the 2nd European Orchid Congress. Paris, France.

Bouriquet, G. 1947. Sur la germination des grains de vanilier (*Vanilla planifolia* And). L'Agron. Tropicale 2: 150-164.

Brummitt, L.W. 1962. *Cyperipediums*. Orch. Rev. 70 (828): 181-183; (830): 249-251; (834): 384-387.

Bulpitt, C.J. 2005. The uses and misuses of orchids in medicine. QJM: An Int J Medicine 98: 625–631.

Bultel G. 1924-25. Germination aseptiques d'orchidbes. Cultures symbiotique et asymbiotique. Rev. Hort. Paris 96: 268-271, 1924. 97:318-321, 359-363. 1925.

Bultel, G. 1926. Les orchidBes germBes sans champignons sont des plantes normales. Ibid.98: 155.

Burgeff, H. 1909. Die Wurzelpilze der Orchideen, ihre Kulter und ihre Leben in der Pflanze. G. Fischer Verlag, Jena.

Burgeff, H. 1911. Die Anzucht tropischer Orchideen aus Samen.G. Fischer Verlag, Jena.

Burgeff, H. 1932. Saprophytismus und symbiose. Studien a tropischen orchideen. Jena, Germany: Gustav Fischer.

Burgeff, H. 1936. Samenkeimung der orchidéen und Entwicklung ihrer Keimp anzen. Jena, Germany: Verlag von Gustav Fischer.

Burgeff, H. 1954. Samenkeimung und Kultur Europaischer Erdorchieen. G. Fischer Verlag, Stuttgart.

Burgeff, H. 1959. Mycorrhiza of orchids. In: C.L. Withner, (ed.). The Orchids. The Ronald Press C., New York, 361-395.

Buzarbarua, A. 1999. Effect of auxins and kinetin on germination of *Cymbidium aloifolium* Sw. seeds. Indian J. Plant. Physiol. **4(1):** 46-48.

Cameron, D.D., Johnson, I. and David, J.R. 2006. Mutualistic mycorrhiza in orchids: evidence from plant-fungus carbon and nitrogen transfers in the green leaved terrestrial orchid *Goodyera repens*. New Phytol. **171:** 405-416.

Cameron, D.D. Johnson, I. Read, D.J. and Leake, J.R. 2008. Giving and receiving: measuring the carbon cost of mycorrhizas in the green orchid, *Goodyera repens*. *New Phytol.* **180:** 176 –184.

Carlson, M.C. 1940. Formation of the seed of *Cypripedium parviflorum*. Bot. Gaz. **102(2):** 295-300.

Chen, Y., Liu, X. and Liu, Y. 2005. *In vitro* plant regeneration from the immature seeds of *Cymbidium faberi*. Plant cell Tiss. Org. Cult. **81:** 247-251.

Chennaveeraiah, M.S. and Patil, S.J. 1975. Morphogenesis in seed cultures of *Spathoglottis*. Curr. Sci. **44(2):** 68.

Chu, C-C. Mudge, K.W.1994. Effects of prechilling and liquid suspension culture on seed germination of the yellow Lady's slipper orchid (*Cypripedium calceolus* var. *pubescens*). Lindleyana 9:153–159.

Clement, E. 1924a. Germination of *Odontoglossum* and other seed without fungal aid. Orch. Rev. **32:** 233-238.

Clement, E. 1924b. The non-symbiotic germination of orchid seeds. Ibid. 359-365.

Clement, E. 1926. The non-symbiotic and symbiotic germination of orchid seeds. Ibid. **34:** 165-169.

Clement, E. 1929. Non-symbiotic and symbiotic germination of orchid seed. Ibid.37:68-75.

Clement E. 1932. Raising orchid seedlings. Ibid. **40:** 195-206.

Clements, M.A. 1982. Orchid seed germination and seedling culture- a manual: Australian native orchids (epiphytic and terrestrial). In: Arditti, J. (ed.). Orchid biology: reviews and perspectives.II. Cornell University Press, Ithaca, New York, 295-303.

Coke, J.L. 1990. Aseptic germination and growth of some terrestrial orchids. In: Sawyers, C.E. (ed.) North American native terrestrial orchid propagation and production. Chadds Ford Press, Pennsylvania, USA.

Constantin, J. 1913. The development of orchid cultivation and its bearing on evolutionary theories. Smithson, Inst. Ann. Rep. **1:** 345-358.

Constantin, J. 1917. La vie des Ochidèes. 188. Paris.

Constantin, J. and Magrou, J. 1922. Applications industrielles d'une grande dècouverte Francaise. Actualitès Biol. Ann. Sci. Nat. Bot. IV. **10;** 1-34.

Curtis, J. 1936. The germination of some native orchid seeds. Amer. Orch. Soc. Bull. **5:** 42-47.

Curtis, J.T. 1939. The relation of specificity of orchid mycorrhizal fungi to the problem of symbiosis. Amer. J. Bot. **26:** 390-399.

Curtis, J.T. 1943. Germination and seedling development in five species of *Cypripedium* L. Amer. J. Bot. **30:** 199-206.

Curtis, J.T. 1947a. Studies on the nitrogen nutrition of orchid embryos. I. Complex nitrogen sources. Amer Orch. Soc. Bull. **16(12):** 654-660.

Curtis, J.T. 1947b. Undifferentiated growth of orchid embryos on media containing barbiturates. Science **105**(2718):128.

Curtis, J.T. 1947c. Ecological observations on orchids of Haiti. Amer. Orch. Soc. Bull. **16:** 262-269.

Curtis, J.T. and Spoerl, E. 1948. Studies on the nitrogen nutrition of orchid embryos. Amer. Orch. Soc. Bull. **17(2):** 111-114.

Darwin, C. 1888. The various contrivances by which orchids are fertilized by insects. 2nd ed. John Murray, London: 300.

Dasgupta, P. and Bhadra, S.K. 1998. *In vitro* production of seedlings in *Cymbidium aloifolium* (L.) Sw. Plant Cell Tiss. Org. Cult. **8:** 177-182.

Decruse, S.W., Gangaprasad, A., Seeni, S. and Menon, V.S. 2003. Micropropagation and ecorestoration of *Vanda spathulata*, an exquisite orchid. Plant Cell Tiss. Org. Cult. **72:** 199-202.

De Pauw, M.A. and Remphrey, W.R. 1993. *In vitro* germination of three *Cypripedium* species in relation to time of seed collection, media and cold treatment. Can. J. Bot. **71:** 879–885.

De Pauw, M.A., Rempherey, W.R. and Palmer, C.E. 1995. The cytokinin preference for *in vitro* germination and protocorm growth of *Cypripedium candidum*. Ann. Bot. **75:** 267-275.

Dijk E. 1990. Effects of mycorrhizal fungi on *in vitro* nitrogen response of juvenile orchids. Agri. Ecosys. & Env. **29**(1-4): 91-97.

Dijk, E. and Eck, N. 1995. Axenic *in vitro* nitrogen and phosphorous responses of some Dutch marsh orchids. New Phytol. **131:** 353-359.

Downie, D.G. 1940. On the germination and growth of *Goodyera repens*. Trans. Proc. Bot. Soc. Edinb. **33(1):** 36-51.

Downie, D.G. 1941. Notes on the germination of some British orchids. Trans. Proc. Bot. Soc. Edinb. **33(2):** 94-103.

Downie, D.G. 1943. Notes on the germination of *Corallorhiza inaata*. Trans. Proc. Bot. Soc. Edinb. **33**(4):380-382.

Downie, D.G. 1949a.The germination of *Goodyera repens* (L.). R. Br. in fungal extract. Trans. & Proc. Bot. Soc. Edinb. **35(2):** 120-125.

Downie, D.G. 1949b. The germination of *Listera ovata* (L.). R. Br. in fungal extract. Trans. & Proc. Bot. Soc. Edinb. **35(2):** 126-130.

Dressler, R.L. and Dodson, C.H. 1960. Classification and phylogeny in the Orchidaceae. Ann. Miss. Bot. Gard. **47(1):** 25-68.

Dressler, R.L. 1981. The orchids: natural history and classification. Cambridge, Mass., Harvard Univ. Press.

Dressler, R.L. 1993. Phylogeny and classification of the orchid family. Dioscorides press, Portland, Oregon, USA.

Dutra, D., Johnson, T.R., Kauth, P.J., Stewart, S.L., Kane, M.E. and Richardson, L. 2008. Asymbiotic seed germination, *in vitro* seedling development and greenhouse acclimatization of the threatened terrestrial orchid *Bletia purpurea.* Plant Cell Tiss. Org. Cult. **94:**11-21.

Dutra, D., Kane, M.E. and Richardson, L. 2009. Asymbiotic seed germination and *in vitro* seedling development of *Cyrtopodium punctatum*: a propagation protocol for an endangered Florida native orchid. Plant Cell Tiss. Org. Cult. **96(3):** 235-243.

Elander, M., Leander, K., Rosenbloom, J. and Ruusa, E. 1973. Studies on orchidaceae alkaloids. XXXII. Crepidine, crepidamine and dendrocrepine from *Dendrobium crepidatum* Lindl. Acta Chem. Scand. **27**: 1907–1913.

Ernst, R. 1966a. The growth response to variations of molecular structure of carbohydrates with freshly germinated *Phalaenopsis* and *Dendobium* seed. Amer. Orch. Soc. Bull. **36**: 1068-1073.

Ernst, R. 1966b. Process for producing acyclic surfactant sulfobetaines. United States Patent 3,280,179.

Ernst, R. 1967. Effect of carbohydrate selection on the growth rate of freshly germinated *Phalaenopsis* and *Dendrobium* seed. Am. Orchid Soc. Bull. **3b:**1068-1073.

Ernst, R. 1994. Effect of Thidiazurone on *in vitro* propagation of *Phalaenopsis* and *Doritaenopsis*. Plant Cell Tiss. Org. Cult. **39:** 273-275.

Ernst, R. and Arditti, J. 1990. Carbohydrate physiology of orchid seedlings. III. Hydrolysis of maltooligosaccharides by *Phalaenopsis* (Orchidaceae) seedlings. Amer. J. Bot. **77(2):** 188-195.

Fabre, J.H. 1855. Recharches sur le tubercules de l'*Himantoglossum hircinum.* Ann. Sci. Nat. Bot. **4(3):** 253-291.

Fabre, J.H. 1856. De la germination des Ophrydees et de la natur de leur tubercules. Ann. Sci. Nat. Bot. **4(5):** 163-186.

Fast, G. 1982. Orchid seed germination and seedling culture- a manual: European terrestrial orchids (symbiotic and asymbiotic methods). In: J. Arditti (eds.). Orchid biology: reviews and persrectives. II. Cornell University Press, Ithaca, New York, pp. 309-326.

Flamce, M. 1978. Influence of selected media and supplements on the germination and growth of *Paphiopedilum* seedlings. Amer. Orch. Soc. Bull. **47**: 419-423.

Fleischer, E. 1929. Zur Biologie feilspanf rmiger Samen. Botanisches Archiv. **26**: 86-132.

Fleischer, E. 1930. Ophic Zur Biologie feilspanf rmiger Samen. Botanisches Zentralblatt **158**: 91-92.

Frank, A.B. 1892. Lehrbuch der Botanik. Bd. I. W. Engelmann, Leipiz: 264.

Freudenstein, J.V. and Rasmussen, F.N. 1999. What does morphology tell us about orchid relationships?-a cladistic analysis. Amer. J. Bot. **86(2)**: 225-248.

Frosch, W. 1986. Asymbiotic propagation of *Cypripedium reginae*. Amer. Orch. Soc. Bull. **55**: 14-15.

Gärdenfors, U. 2001. Classifying threatened species at national versus global levels. Trends in ecology and Evolution **16(9)**: 511-516.

Gavinlarvatana, P. and Prutpongse, P. 1993. Commercial micropropagation in Asia. In: P.C. Debergh and R.H. Zimmermann (ed.). Micropropagation: Technology and Application. Kluwer Acad., Dordrecht, pp. 181-189.

Gebauer, G. and Meyer, M. 2003. ^{15}N and ^{13}C natural abundance of autotrophic and mycoheterotrophic orchids provides insight into nitrogen and carbon gain from fungal association. New Phytol. **160**: 209-223.

George, E.F. 1993. Plant propagation by tissue culture. Part 1. The Tecnology. Exegetics Ltd., London.

Govindarajan, R., Singh, D.P. and Rawat, A.K.S. 2007. High-performance liquid chromatographic method for the quantification of phenolics in 'Chyavanprash' a potent Ayurvedic drug. J. Pharm. Biomed. Anal. **43**: 527-532.

Gruen, M.E. 1959. Auxins and fungi. Annual Rev. Plant Physiol. **10**:404.

Hadley, G. 1969. Cellulose as a cabon source for orchid mycorrhiza. New Phytol. **68**: 933-939.

Hadley, G. 1970. The interaction of kinetin, auxin and other factors in the development of north temperate orchids. New Phytol. **69(3)**:549-555.

Hadley, G. 1982. Orchid mycorrhiza. In: Arditti, J. (ed.) Orchid biology—reviews and perspectives II. Cornell University Press, New York, USA.

Hadley, G. and Harvais, G. 1968. The effect of certain growth substances on asymbiotic germination and development of *Orchis purpurella*. New Phytol. **67**:441-445.

Hadley, G. and Ong, S.H. 1978. Nutritional requirements of orchid endophytes. New Phytol. **81(3)**: 561-569.

Hager, H. 1954. Growing Cattleyas from seed to flower in $2^1/_2$ years. Amer. Orch. Soc. Bull. **23(2)**: 78-81.

Hallé, N. 1977. Flore de la Nouvelle-Calédonie et dependances.8. Orchidacées. Paris, France: Musuém National D'Historie Naturelle.

Handa, S.S. 1986. Orchids for drugs and chemicals. In: Vij, S.P., ed. Biology, Conservation and culture of Orchids. Affiliated East-West Press Private Ltd., New Delhi, pp. 89-100.

Harley, J.L. 1959. The biology of mycorrhiza. Leonard Hill, London.

Harley, J.L. 1969. The biology of mycorrhiza. 2nd ed. Leonard Hill Limited, London.

Harrison, C.R. 1977. Ultastructural and histochemical changes during the germination of *Cattleya aurantiaca* (Orchidaceae). Bot. Gaz. **138(1)**: 41-45

Harrison, J.L. and Arditti, J. 1978. Physiological changes during the germination of *Cattleya aurantiaca* (Orchidaceae). Bot. Gaz. **139(2):** 180-189.

Harvais, G. 1972. The development and growth requirements of *Dactylorhiza purpurella* in asymbiotic cultures. Can. J. Bot. 50:1223–1229.

Harvais, G. 1973. Growth requirements and development of *Cypripedium reginae* in axenic culture. Can. J. Bot. **51:** 327-332.

Harvais, G. 1974. Notes on the biology of some native orchids of Thunder Bay, their endophytes and symbionts.Can. J. Bot. **52:** 451–460.

Harvais, G. 1982. An improved culture medium for growing the orchid *Cypripedium reginae* axenically. Can. J. Bot. **60:** 2574-2555.

Harvais, G. and Hadley, G. 1967. The relation between host and endophyte in orchid mychorriza. New Phytol. **66:** 205–215.

Hegarty, C.P. 1955. Observations on the germination of orchid seed. Amer. Orch. Soc. Bull. **24(7):** 457-464.

Hegde, S.N. 1996. Orchid wealth of India. Arunachal Forest News **14:** 6-19.

Hew, C.S., Ting, S.K. and Chia, T.F. 1988. Substrate utilization by *Dendrobium* tissues. Bot Gaz. **149(2):** 153-157.

Hossain, M.M. 2008. Asymbiotic seed germination and *in vitro* seedling development of *Epidendrum ibaguense* Kunth. (Orchidaceae). Afr. J. Biotech. **7(20):** 3614-3619.

Humpheys, J.L. 1958. Gibberellins for orchids. Orch. Rev. **66(782):** 176.

Ito, I. 1955. Germination of seeds from immature pods and subsequent growth of seedlings in *Dendrobium nobile* Lindl. Sci. Rep. Saikyo Univ. Agr. **7:** 35-42.

Johnson, T.R. and Kane, M.E. 2007. Asymbiotic germination of ornamental *Vanda*: *In vitro* germination and development of three hybrids. Plant Cell Tiss Org. Cult. **91:** 251-261.

Johnson, T.R., Stewart, S.L., Dutra, D., Kane, M.E. and Richardson, L. 2007. Asymbiotic and symbiotic seed germination of *Eulophia alta* (Orchidaceae)-preliminary evidence for the symbiotic culture advantage. Plant Cell Tiss. Org. Cult. **90:** 313-323.

Julou, T., Burghardt, B. Gebauer, G. Berveiller, D. Damesin, C. and Selosse, M. 2005. Mixotrophy in orchids: insight from a comparative study of green individuals and non-photosynthetic individuals of *Cephalanthera damasonium*. New Phytol. **166:** 639-653.

Kanjilal, B., Datta, K.B. and De Sarkar, D. 1998a. Conservation studies on a rare and endangered orchid of Terai hills (West Bengal). Cytol. Genet. **9:** 319-324.

Kanjilal, B. Datta, K.B. and De Sarkar, D. 1998b. Aneuploid lines in *Arundina graminifolia* (Don) Hocher. (Orchidaceae) obtained through *in vitro* micropropagation. Cytol. Genet. **9:** 417-421.

Kano, K. 1965. Studies on the media for orchid seed germination. Memoirs of the faculty of Agriculture Kagawa University **20:** 1-79.

Kauth, P. Kane, M.E. Vendrame, W.A. and Reinhardt-Adams, C. 2008. Asymbiotic germination response to photoperiod and nutritional media in six populations of *Calopogon tuberosus* var. *tuberosus* (Orchidaceae): evidence for ecotypic differentiation. Ann. Bot. **102(5):** 783-793.

Kefford, N.P. and Goldacre, P.L. 1961. The changing concept of auxin. Amer J. Bot. **48**: 643.

Knudson, L. 1921. La germinacion no simbibtica de las semillas de orquideas. Bol. Real. Soc. Espafiola Hist. Nat. **21**: 250-260.

Knudson, L. 1922. Non symbiotic germination of orchid seeds. Bot. Gaz. **73(1):** 1-25.

Knudson, L. 1924. Further observations on non-symbiotic germination of orchid seeds. Bot. Gaz. **77(2):** 212-219.

Knudson, L. 1925. Physiological study of the symbiotic germination of orchid seeds. Bot. Gaz. **79(4):** 345-380.

Knudson, L. 1926. Physiological investigations on orchid seed germination. Int. Cong. Plant Sci. Ithaca, Proc. 1183-1189.

Knudson, L. 1927. Symbiosis and asymbiosis relative to orchids. New Phytol. **26(5):** 328-336.

Knudson, L. 1929. Physiological investigations on orchid seed germination. Procedings of the International congress of Plant Sci. **2**: 1183-1189.

Knudson, L. 1930. Flower production by orchids grown non-symbiotically. Bot. Gaz. **89:** 192-199.

Knudson, L. 1932. Direct absorption and utilization of amino acids by plants. NY (Cornell) Agr. Exp. Sta., Ann. Rep. 45: iii.

Knudson, L .1943. Nutrient solutions for orchid seed germination. Amer. Orch. Soc. Bull **12**:77–79.

Knudson, L. 1946. A new nutrient solution for the germination of orchid seeds. Amer. Orch. Soc. Bull. **14:** 214-217.

Kotomori, S. and Murashige, T. 1965. Some aspects of propagation of orchids. Am. Orch. Soc. Bull. **34**: 484-489.

Krishnamohan, P.T. and Jorapur, S.M. 1986. *In vitro* seed culture of *Acampe praemorsa* (Roxb.) Blatt.& McC. In: S.P.Vij (ed.), Biology, Conservation and Culture of Orchids. Affiliated East-West press Pvt. Ltd., New Delhi, pp. 437-439.

LaCroix, I.F. and LaCroix, E. 1997 African orchids in the wild and in cultivation. Timber Press, Portland, USA.

La Garde, R.V. 1929. Non-symbiotic germination of orchids. Ann. Mo. Bot. Gard. **16**: 499-514.

Lami, R. 1927. Influence d'une peptone sur la germination de quelques Vandées. Compt. Rend. Acad. Sci. Paris. 184: 1579-1581.

Lawler, L.J. and Slaytor, M. 1969. The distribution of alkaloids in New South Wales and Queensland Orchidaceae. Phytochemistry 8: 1959–1962.

Lawrence, D. and Arditti, J. 1964a. A new medium for the germination of orchid seeds. Amer. Orch. Soc. Bull. **33(9):** 766-768.

Lawrence, D. and Arditti, J. 1964b. On the effect of Gro-Lux lamps on the growth of orchid seedlings following transflasking. Amer. Orch. Soc. Bull. **33(11):** 948.

Laws. 1995. Cut orchids in the world market. Flora Cult. Int. **5**: 12-15.

Leake, J.R. 1994. The biology of myco-heterotrophic (Saprophytic) plants. New Phytol. **127**: 171-216.

Leake, J.R. McKendrick, S.L. Bidartondo, M. and Read, D.J. 2004. Symbiotic germination and development of the myco-heterotroph *Monotropa hypopitys* in nature and its requirement for locally distributed *Tricoloma* spp. New Phytol. **163**: 405-423.

Lee, S. Park, S. Kim, T. Paek, K. Lee, S.S. Park, S.S. Kim, T.J. and Paek, K.Y. 1997. Effect of orchid habitat soil on growth of tissue cultured *Cymbidium kanran* and *C. goeringii* and root infection by orchid mycorrhizal fungus. J. Kor. Soc. Hort. Sci. **38**: 176-182.

Lee, Y.L. and Lee, N. 2003. Plant regeneration from protocorm-derived callus of *Cypripedium formosanum*. *In vitro* Cell Dev. Bio-Plant. **39**: 475-479.

LeRoux, G. Barabe, ´ D. and Vieth, J. 1997. Morphogenesis of the protocorm of *Cypripedium acaule* (Orchidaceae). Plant Syst. Evol. **205**:53–72.

Liddell, R.W. 1953a. Notes on the germinating *Cypripedium* seed. Amer. Orch. Soc Bull. **22(8)**: 580-582.

Liddell, R.W. 1953b. Notes on the germinating *Cypripedium* seed-II. Amer. Orch. Soc Bull. **22(8): 580-582.**

Light, M.H.S. and Mac Conaill, M. 1998. Factors affecting germinable seed yield in *Cypripedium calceolus* var. *pubescens* (Willd.) Correll and *Epipectis helleborine* (L.) Crantz (Orchidaceae). Bot. J. Linn. Soc. **126**: 3-26.

Linde´n, B. 1980. Aseptic germination of seeds of northern terrestrial orchids. Ann. Bot. Fennici. **17**: 174-182.

Linde´n, B. 1992. Two new methods for pre-treatment of seeds of northern orchids to improve germination in axenic culture. Ann. Bot. Fenn. **29**: 305–313.

Lindquist, B. 1965. The raising of *Disa uniflora* seedlings in Gothenburg. Amer. Orch. Soc. Bull. **34 (4):** 317-319.

Link, H.F. 1840. Icones selectee anatomico-botanicae, Fasc. II. Ausgewählte Anatomisch botanische Abildungen, vol 2. Lûderitz, Berlin.

Lugo-Lugo, H. 1955a. Effects of nitrogen on the germination of *Vanilla planifolia* seeds. Amer. Orch. Soc. Bull. **24**: 309-312.

Lugo-Lugo, H. 1955b. The effect of nitrogen on the germination of *Vanilla planifolia*. Amer. J. Bot. **42(7):** 679-684.

Mac Dougal, D.T. 1899a. Symbiotic saprophytism. Ann. Bot. **13(49)**:1-47.

Mac Dougal, D.T. 1899b. Symbiosis and saprophytism. Bot. Gaz. **28(3):** 220-222.

Magrou, J. Mariat, F. and Rose, H. 1949. Sur la nutrition azotée des orchidées. Compt. Rend. Acad. Sci. Paris. **229**: 665-688.

Majerowicz, N. Kerbauy, G.B. Nievola, C.C. and Suzuki, R.M. 2000. Growth and nitrogen metabolism of *Catasetum fimbriatum* (Orchidaceae) growth with different nitrogen sources. Env. Exp. Bot. **44 (3)**: 195-206.

Malmgren, S. 1996. Orchid Propagation: theory and practice. In: C. Allen, (Ed.). Proceedings of the North American Native Terrestrial Orchids Propagations and Production Conference. National Arboretum, Washington, pp. 63-71.

Mariat, F. 1948. Influence des facteurs de croissance sur le development et la differenciation des embryons d'orchidées. Rev. Gén. Bot. **55**: 229-243.

Mariat, F. 1949. Action de l'acide nicotinique sur la germination et le development des embryons e *Cattleya*. Bull. Soc. Bot. France **98(7-9)**: 260-263.

Mariat, F. 1951. Influence du lait coco et du coprah sur le development d jeunes plantules de Cattleya. Comp. Rend. Acad. Sci. (Paris) **229**: 1355-1357.

Mariat, F. 1952. Recherches sur la physiologie des embryons d'orchidee. Rev. Gen. Bot. **59**: 324-377.

Masuhara, G. and Katsuya, K. 1994. *In situ* and *in vitro* specificity between *Rhizoctonia* spp. and *Spiranthes sinensis* (Persoon) Ames. var. *amoena* (*M. biebersteini*) Hara (Orchidaceae). New Phytol. **127**: 711-718.

Mathews, V.H. and Rao, P.S. 1980. *In vitro* multiplication of *Vanda* hybrids through tissue culture technique. Plant Sci. **17**: 383-389.

Mc Ewan, B. 1961. Stock solution medium for flasking. Orch. Soc. Southern Caliornia Rev. **3(5)**: 7-9.

McKendrick, S.L. Leake, J.R. Taylor, D.L. and Read, D.J. 2000. Symbiotic germination and development of myco-heterotrophic plants in nature: Ontogeny of *Corallohiza trifida* and characterization of its mycorrhizal fungi. New Phytol. **145**: 523-537.

Mead, J.W. and Bulard, C. 1975. Effects of vitamins and nitrogen sources on asymbiotic germination and development of *Orchis laxiflora* and *Ophrys sphegodes*. New Phytol. **74**: 33-40.

Mead, J.W. and Bulard, C. 1979. Vitamins and nitrogen requirements of *Orchis laxiflora* Lamk. New Phytol. **83(1)**: 129-136.

Michel, E.E. 2002. Asymbiotic propagation of tropical terrestrial orchid species. In: J.Clark, WM. Elliott, G.Tingley, J.Biro (eds.). Proceedings of the 16th World Orchid Conference. Vancouver Orchid Society, Vancouver, Canada.

Millner, H.J., Obeng, A., Mccrea, A.R. and Baldwin, T.C.2008. Axenic seed germination and *in vitro* seedling development of *Restrpia brachypus* (Orchidaceae). J. Torrey Bot. Soc. **135**: 497-505.

Mitra, G.C., Prasad, R.N. and Roy Chowdhury, A. 1976. Inorganic salts and differentiation of protocorms in seed-callus of an orchid and co-related changes in its free amino acid content. Ind. J. Exp. Biol. **14**: 350-351.

Mitra, G.C. 1986. *In vitro* culture of orchid seeds for obtaining seedlings. In: Vij, S.P., (ed.) Biology, Conservation and culture of orchids. Affiliated East-West Press Pvt. Ltd., New Delhi, 401-412.

Miyoshi, K. and Mii, M. 1995. Enhancement of seed germination and protocorm formation in *Calanthe discolor* (Orchideaceae) by NaOCl and polyphenol treatments. Plant Tiss. Cult. Lett. **12**: 267–272.

Miyoshi, K. and Mii, M. 1998. Stimulatory effects of sodium and calcium hypochlorite, pre-chilling and cytokinins on the germination of *Cypripedium macranthos* seed *in vitro*. Physiol. Plant. **102**: 481–486.

Moir, W.W.G. 1957. Orchids and Gibberellin compounds. Pac. Orch. Soc. Hawaii Bull. **14(3-4):** 92-96.

Moran, C De. 1890. Les orchidèes et l'engrais. J. Orch. **1(17):** 271-273.

Morel, G.R. 1960. Producing virus-free *Cymbidiums*. Amer. Orch. Soc. Bull. **293:** 495-497.

Morel, G. 1964. Tissue culture-a new means of clonal propagation of orchids. Amer. Orch. Soc. Bull. **33:** 473-478.

Morel, G.M. 1974. Clonal multiplication of ochids. In: Withner, C.L. (ed.), The Orchids: Scientific studies, Wiley, NY.

Mukhopadhyay, K. and Roy, S.C. 1994. *In vitro* induction of 'runner' – a quick method of micropropagation in orchid. Sci Hortic. **56:** 331-337.

Murashige, T. and Skoog, F. 1962. A revised medium for rapid growth and bio assays with Tobacco tissue cultures. Physiol. Plant. **15:** 473-497.

Nakamura, S.J. 1962. Zur Skmenkeimung einer Chlorophyllfreien Erdorchidee *Galeola septentrionalis* Reicb.f.Z. Fur Botanik **50:** 487-497.

Nakamura, S.J. 1963. The accelerating influences of potassium ions on the non-symbiotic seed germination of the achlorophyllous terrestric orchid *Galeola septentrionalis*. Int. Symp. Physiol. Ecol. Biochem. Germination, AIII, 9:1-4.

Nakamura, S.I. 1976. Atmospheric conditions required for the growth of *Galeola septentrionalis* seedlings. Bot. Mag. (Tokyo) **89:** 211–218.

Nakamura, S.I. 1982. Nutritional conditions required for the non-symbiotic culture of an achlorophyllous orchid *Galeola septentrionalis*. New Phytol. **90:**701–715.

Ng, C.K.Y. and Hew, C.S. 2000. Orchid pseudobulbs- 'false' bulbs with a genuine importance in orchid growth and survival. Sci. Hortic. **83:** 165-172.

Nishimura, G. 1982. Orchid seed germination and seedling cultue- a mutual:Japanese orchids. In: Arditti. J. (ed.), Orchid biology-reviews and perspective. Vol.2. Cornell Univ. Press, Ithaca, N.Y. 331-346.

Noggle, G.R. and Wynd, F.L. 1943. Effects of vitamins on germination and growth of orchids. Bot. Gaz. **104(3):** 455-459.

Nurhayati, N. Gonde, D. and Ober, D. 2009. Evolution of pyrrolizidine alkaloids in *Phalaenopsis* orchids and other monocotyledons: identification of deoxyhypusine synthase, homospermidine synthase and related pseudogenes. Phytochemistry **70:** 508–516.

Oliva, A.P. and Arditti, J. 1984. Seed germination of North American orchids. 11. Native California and related species of *Aplectrum, Cypripedium* and *Spiranthes*. Bot. Gaz. **145** (4): 495-501.

Okamoto, T. Natsume, M. Onaka, T. Uchmaru, F. and Shimizu, M. 1966. The structure of dendramine (6-oxydendrobine) and 6-oxydendroxine. The fourth and fifth alkaloid from *Dendrobium nobile*. Chem. Pharm. Bull. **14:** 676–680

Oyamada, T. and Takano, T. 1985. Investigations on the macro-inorganic nutrients in special reference to ratios of NO_3/SO_4 and K/Ca in tissue culture media of orchids. **21:** 1-9.

Paleg, L.G. 1965. Physiological effects of gibberellins. Annual Rev. Plant Physiol. **16**: 291.

Pan, M.J. and Van Staden, J. 1998. The use of charcoal in *in vitro* culture- A review. Plant Growth Regul. **26**: 155-163.

Pateli, P. Papafotiou, M. and Chronopoulos, J. 2003. Influence of *in vitro* culture medium on *Epidendrum radicans* seed germination, seedling growth and *ex vitro* establishment. Acta Hort. (ISHS) **616**: 189-192.

Pedroza-Manrique and Mican-Gutiérrez. 2006. Asymbiotic germination of *Odontoglossum gloriosum* Rchb.f. (Orchidaceae) under *in vitro* conditions. *In vitro* Cell Dev. Bio.-Plant. **42(6)**: 543-547.

Poddubnaya-Arnoldi, V.A. and Selezneva, V.A. 1953. Viyrazivanye orchidei iz semyan (Cultivation of orchids from seeds). Trudy Glavyi Botanicheskii Sad, **3**: 106-124.

Poddubnaya-Arnoldi, V.A. and Selezneva, V.A. 1957a. Metodika semenogo razmonozenya orhidei (Methods for propagation of orchids from seeds). Biuleten Glavnyi Botanicheskii Sad. **27**: 33-40.

Poddubnaya-Arnoldi, V.A. and Selezneva, V.A. 1957b. Orchidie irih kultura (Orchids and their cultivation). Moscow, Russia: Akademia Nauki.

Poddubnaya-Arnoldi, V.A. and Zinger. 1961. Application of histochemical technique to the study of embryonic processes in some orchids. Recent advances in Botany (Univ. Toronto Press) **8**: 711-714.

Pollacci, G. and Berganischi. 1940. Azione... Orchidee (Effect of vitamins on seeds of orchids). Boll. Soc. Ital. Biol. Sper. **15**: 326-327.

Powell, K.B. and Arditti, J. 1975. Growth requirements of *Rhizoctonia repens* M32. Mycopathologia. **55**: 163-167.

Prasad, R.N. and Mitra, G.C. 1975. Nutrient requirements for germination of seeds and development of protocorms and seedlings of *Cymbidium* in Aseptic cultures. Indian J. Exp. Biol. **13**: 123-126.

Prillieux, E. 1856. De la structure anatomique et du mode de vegetation du *Neottia nidus-avis*. Ann. Sci. Nat. **4 (Bot. 5)**: 267-282.

Prillieux, E. 1860. Observation sur la germination du *Miltonia spectabilis* et de divers autres orchidées. Ann. Sci. Nat. **4(Bot. 13)**: 288-296.

Prillieux, E. and Rivière, A. 1856. Observations sur la germination et le dèvelopment d'une orchidèes (*Anagraecum maculatum*). Ann. Sci. Nat. **4(Bot, 5)**: 119-136.

Pritchard, H.W. 1985. Growth and storage of orchid seeds. In: K.W. Tan (ed.). Proceedings of the 11th World Orchid Conference. World Orchid Conference Inc, Miami, USA.

Purves, S. and Hadley, G. 1976. The physiology of symbiosis in *Goodyera repens*. New Phytol **77**: 689–696.

Raghavan, V. 1964. Effects of certain organic nitrogen compounds on growth *in vitro* of seedlings of *Cattleya*. Bot. Gaz. **125(4)**: 260-267.

Raghavan, V. 1966. Nutrition, growth and morphogenesis of plant embryos. Biol. Rev. **41**: 1-58.

Raghavan, V. 1976. Experimental embryo genesis in vascular plants. Acad. Press, London.

Raghavan, V. and Torrey, J.G. 1963. Inorganic nitrogen nutrition of the embryos of the orchid, *Cattleya*. Amer. J. Bot. **50(6):** 617.

Raghavan, V. and Torrey, J.G. 1964. Inorganic nitrogen nutrition of the seedlings of the orchid *Cattleya*. Amer. J. Bot. **51(3):** 264-274.

Ramsay, M.M. and Stewart, J. 1998. Re-establishment of the lady's slipper orchid (*Cypripedium calceolus* L.) in Britain. Bot. J. Linn. Soc. **126(1):** 173-181.

Ramsbottom, J. 1922a. Orchid mycorrhiza. Charlesworth and Co. England.

Ramsbottom, J. 1922b. The germination of orchid seed. Orch. Rev. **30(342):** 197-202.

Ramsbottom, J. 1927. Orchid mycorrhiza. Charlesworth and Co. England.

Rao, N.K. 2004. Plant genetic resources: Advancing conservation and use through biotechnology. Afr. J. Biotech. **3(2):** 136-145.

Rasmussen, H.N. 1992. Seed dormancy patterns in *Epipactis palustris* (Orchidaceae): requirements for germination and establishment of mycorrhiza. Physiol Plant **86:**161–167.

Rasmussen, H.N. 1995. Terrestrial orchids from seed to mycotrophic plant. Cambridge, UK: Cambridge University Press.

Rasmussen, H.N. 2000. Ins and outs of orchid phylogeny. In: Wilson, K.L. and Morrison D.A. (eds.), Monocots; systematics and evolution. Melbourne, Australia: CSIRO: 430-435.

Rasmussen, H.N. 2002. Recent development in the study of orchid mycorrhiza. J. Ecol. **90(6):** 1002-1008.

Rasmussen, H.N., Andersen, T.F. and Johansen, B. 1990. Temperature sensitivity of *in vitro* germination and seedling development of *Dactylorhiza majalis* (Orchidaceae) with and without a mycorrhizal fungus. Plant Cell Environ **13:**171–177.

Rasmussen H.N. and Rasmussen F.N. 1991. Climatic and seasonal regulation of seed plant establishment in *Dactylorhiza majalis* inferred from symbiotic experiments *in vitro*. Lindleyana 6:221–227.

Rasmussen, H.N. and Whigham, D.F. 1998a. The underground phase: a special challenge in studies of terrestrial orchid populations. Bot. J. Linn. Soc. **126:** 49-64.

Rasmussen, H.N. and Whigham, D.F. 1998b. Importance of woody debris in seed germination of *Tipularia discolor* (Orchidaceae). Amer. J. Bot. **85:** 829-834.

Rauh, W. Barthlott, W and Ehler, N. 1975. Morphologie und funktion der testa staubf rmiger flugsamen. Botanische Jahbucher fur Systematik, Pflanzengeschichte und Pflanzengeographie 96: 353-374.

Rosso, S. 1966. The vegetative anatomy of the *Cyperipedioideae* (Orchidaceae). J. Linn. Soc. (Bot.) 59: 309-341.

Rotor, G Jr. 1949. A method for vegetative propagation of *Phalaenopsis* species and hybrids. Amer. Orch. Soc. Bull. **18:** 738-739.

Roy, J. and Banerjee, N. 2001. Cultural requirements for *in vitro* seed germination,protocorm growth and seedling development of *Geodorum densiflorum* (Lam.) Schltr. Ind. J. Exp. Biol. **39**: 1041-1047.

Roy, J. and Banerjee, N. 2002. Rhizome and shoot development during *in vitro* propagation of *Geodorum densiflorum* (Lam.) Schltr. Sci.Hortic. **94**: 181-192.

Roy, S. and Biswas, A.K. 2005. Isolation of a white flowered mutant through seed culture in *Spathoglottis plicata* Blume. Cytologia **70**: 1-6.

Sarasan, V., Cripps, R., Ramsay, M.M., Atherton, C., McMichen, M., Prendergast, G. and Rowntree, J.K. 2006. Conservation *in vitro* of threatened plants-Progress in the past decade. *In vitro* Cell. Dev. Biol. –Plant **42**: 206-214.

Sauleda, R.P. 1976. Harvesting times of orchid seed capsules for the green pod culture process. Am. Orchid Soc. Bull. **45**: 305–309.

Schaffstein, G. 1938. Untersuchungen uber die Avitaminose der Orchideenkeimlinge. Jahrb. Wiss. Bot. **86**: 720-752.

Schaffstein, G. 1941. Die hvitaminose der Orchideenkeimlinge. Jahrb. Wiss. Bot. **90**:141-198.

Sebastinraj, J. Britto, J.S., Robinson, P.J., Kumar, V.D. and Kumar SS. 2006. *In vitro* seed germination and plantlet regeneration of *Coelogyne mossiae* Rolfe. J. Biol. Res. **5**: 79-84.

Seeni, S. and Latha, P.G. 1992. Folier regeneration of the endangered Red Vanda, *Renanthera imschootiana* Rolfe (Orchidaceae). Plant Cell Tiss. Org. Cult. **29**: 167-172.

Shiau, Y.J., Sagare, A.P., Chen, U.C., Yang, S.R. and Tsay, H.S. 2002. Conservation of *Anoectochilus formosanus* Hayata by artificial cross-pollination and *in vitro* culture of seeds. Bot. Bull. Acad. Sin. **43**: 123-130.

Shiau, Y.J., Nalawade, S.M., Hsia, C.N., Mulabagal, V. and Tsay, H.S. 2005. *In vitro* propagation of the Chinese medicinal plant. *Dendrobium candidum* Wall.ex Lindl. from axenic nodal segments. *In Vitro* Cell Dev Biol-Plant **41**: 666–670.

Shimura, H. and Koda, Y. 2004. Enhanced symbiotic seed germination of *Cypripedium macranthos* var. *rebunense* following inoculation after cold treatment. Plant Cell Tiss. Org. Cult. **78**: 273-276.

Shivana, K.R. and Johri, B.M. 1985. The Angiosperm Pollen: Structure and Function. Wiley Eastern Ltd., New Delhi.

Shive, J.W. 1915. A three salt nutrient solution for plants. Am. J. Bot. **2**: 157-160.

Sideris, C.P. 1950. A nutrient solution for germination of orchid seeds. Bull. Pac. Orch. Soc. Hawaii **8(4)**:337-339.

Singh, A.K.R. and Tiwari, C. 2007. Harnessing the economic potential of Orchids in Uttaranchal. Envis Bull. Hima. Ecol. **14**: 1–3.

Smith, F. 1932. Raising orchid seedlings asymbiotically. Gard. Chron. **91**(2349): 9-11.

Smith, D. E. 1958. Effect of gibberellins on certain ochids. Amer. Orch. Soc. Bull. **27(11)**:742- 747.

Smith, S.E. 1966. Physiology and ecology of orchid mycorrhizal fungi with reference to seedling nutrition. New Phytol. **65:** 488-499.

Smith, S.E. 1967. Carbohyrate translocation in orchid mycorrhizal fungi. New Phytol. 66: 371.

Smith, S.E. 1973. Asymbiotic germination of orchid seeds on carbohydrates of fungal origin. New Phytol. **72(3):** 497-499.

Smith, S.E. and Read, D.J. 1997. Mycorrhizal Symbiosis, 2nd edn.San Diego, CA, USA: Acad. Press.

Smreciu, E.A. and Currah, R.S. 1989. Symbiotic germination of seeds of terrestrial orchids of North America and Europe. Lindleyana **1:** 6–15.

Soon, T.E. 1989. Orchids of Asia. Times Books Intl., Singapore.

Spoerl, E. 1948. Amino acids as sources of nitrogen for orchid embryos. Amer. J. Bot. **35(2):**88-95.

Stancato, G.C., Chagas, E.P. and Mazzafera, P. 1998. Development and germination of seeds of *Laelia purpurata* (Orchidaceae). Lindleyana. **13:** 97-100.

St-Arnaud, M. Lauzer, D. and Barabe, D. 1992. *In vitro* germination and early growth of seedlings of *Cypripedium acaule* (Orchidaceae). Lindleyana. **7:** 22-27.

Stenberg, M.L. and Kane, M.E. 1998. *In vitro* seed germination and green house cultivation of *Encyclia boothiana* var. *erythronioides*, an endangered Florida orchid. Lindleyana. **13:** 101-112.

Stephen, R.C. and Fung, K.K. 1971. Vitamin requirements of the fungal endophytes of *Arundina chinensis*. Can. J. Bot. **49:** 411.

Stewart, J. and Griffiths, M. 1995. Manual of orchids. Timber Press, Portland, Oregon.

Stewart, S.L. and Kane, M.E. 2006. Asymbiotic seed germination and *in vitro* seedling development of *Habenaria macroceratitis* (Orchidaceae), arare Florida terrestrial orchid. Plant Cell Tiss. Org. Cult. **86:** 147-158.

Stoutamire, W.P. 1963. Terrestrial orchid seedlings. Aust Plants **2:** 119–122.

Stoutamire, W.P. 1964. Seeds and seedlings of native orchids. Michigan Bot. **3:** 107-119.

Stoutamire, W.P. 1974. Terrestrial orchid seedlings. In: D.L.Withner (eds.). The orchids: scientific studies. Wiley& Sons, New York, USA.

Stoutamire, W.P. 1990. Eastern American *Cypripedium* species and the biology of *Cypripedium candidum*. In: C.E. Sawyers (eds.). North American native terrestrial orchid propagation and production. Chadds Ford Press, Pennsylvania, USA.

Sutter, E.G. 1996. General laboratory requirements, media and sterilization methods. In: Trigiano, R.N. and Gray, D.J. (eds.). Plant tissue culture concepts and laboratory exercises. CRC Press, Inc., New York, pp. 11-25.

Temjensangba, I. and Deb, C.R. 2005. Regeneration and mass multiplication of *Arachnis labrosa* (Lindl. ex Paxt.) Reichb: A rare and threatened orchid. Curr. Sci. **88(12):** 1966-1969

Thomale, H. 1957. Die Orchideen, 2nd ed. Verleg, Eugen Ulmer Stuttgart.

Thomas, T.D. and Michael, A. 2007. High- frequency plantlet regeneration and multiple shoot induction from cultured immature seeds of *Rhyncostylis retusa* Blume, an exquisite orchid. Plant Biotech. Rep.1: 243-249.

Tienken, H.G. 1947. Nutrient solution. Amer. Orch. Soc.Bull. **16:** 649.

Tomita, M. 2002. The cytokinin preference for immature embryo culture of some terrestrial orchids. Comb. Proc. Int. Plant Prop. Soc. **52:** 331-334.

Tonnier, J.P. 1954. Mise au point d'um milieu de semis pour les grains d'une espéce de Vanillier don't on n'obtient pas la germination sur les milieu de culture classiques. VIII Congr. Int Bot., Sec. 11-12, 411-412.

Tsuchiya, I. 1954. Possibility of germination of orchid seeds from immature fruits. Na Pua Okika o Hawaii Nei **4(1):** 11-16.

Tsukamoto, Y., Kano, K. and Katsuura, T. 1963. Instant media for orchid seed germination. Amer. Orch. Soc. Bull. **32(5):** 354-355.

Ulrich, J.M. 1960. Auxin production by mycorrhizal fungi. Physiol. Plant. **13:** 429.

Uniyal, M.R. 1975. *Astavarga. Sandigdha Vanaushadhi.* Dhanwantri Partrika. Sri Jwala Ayurevd Bhawan Aligarh, India.

Vacin, E.F. and Went, F.W. 1949. Some pH changes in nutrient solutions. Bot. Gaz.**110 (4):** 605-613.

Vacin, E.F. 1950a. Some problems of germinating *Cymbidium* seeds and growing seedlings. Cymbidium Soc. News **5(2):** 8.

Vacin, E.F. 1950b. Behaviour of nutrient solutions used in asymbiotic germination of orchid seeds. Bull. Pac. Orch. Soc. Bull **39:**907-910.

Van der Kinderen, G. 1987. Abscisic acid in terrestrial orchid seeds: a possible impact on their germination. Lindleyana 2:84–87

Van Overbeek, J., Conklin, M.E. and Blakeslee, A.F. 1941. Factors in coconut milk essential for growth and development of very young *Datura* embryos. Science.**94:** 350-351.

Van Overbeek, J. Conklin, M.E. and Blakeslee, A.F. 1942. Cultivation *in vitro* of small *Datura* embryos. Amer. J. Bot. **29:** 472-477.

Van Waes, J.M. and Debergh, P.C. 1986a. Adaptation of the tetrazolium method for testing the seed viability, and scanning electron microscopy study of some western European orchids. Physiol. Plant. **66:** 435–442.

Van Waes, J.M. and Debergh, P.C. 1986b. *In vitro* germination of some western European orchids. Physiol. Plant. **67:**253–261.

Van Waes, J. 1984. *In vitro* studies van de kiemingsfysiologie van Westeuropese Orchideeen. Thesis. Gent, Belgium: Rijkuniversiteit Gent.

Van Waes, J. 1987. Effect of activated charcoal on *in vitro* propagation of western European orchids.Acta Hort. **212:** 131–138.

Vasudevan, R. and Staden, J.V. 2010. Fruit harvesting time and corresponding morphological changes of seed integuments influence *in vitro* seed germination of *Dendrobium nobile* Lindl. Plant Growth Regul. **60(3):** 237- 246.

Vermeulen, X.P. 1947. Studies on *Dactylorclzis*. Ph.D, diss., University of Amsterdam.

Vogelpoel, L. 1980. *Disa uniflora*—its propagation and culture. Am. Orchid Soc. Bull. **49:** 961–974.

Vogelpoel, L. 1987. New horizons in *Disa* breeding. The parent species and their culture I. Orchid Rev. **95:**176–181.

Walter, K.S. and Gillett, H.J. 1998. 1997 IUCN Red List of Threatened Plants. World Conservation Monitoring Centre, IUCN- The world conservation union, Gland, Switzerland and Cambridge, UK.

Warcup, J.H. 1975. Factors affecting symbiotic germination of orchid seed. In: Sanders, F.E. Mose, B. and Tinker, P.B. (eds.). Endomycorrhizas. Academic Press, London, pp. 78-104.

Whigham, D.F., O'Neilla, J.P., Rasmussen, H.N., Caldwell, B.A. and McCormick, M.K. 2006. Seed longevity in terrestrial orchids- Potential for persistent *in situ* seed banks. Biol. Cons. **129:** 24-30.

White, P.R. 1934. Potentially unlimited growth of excised tomato root tips in a liquid medium. Plant Physiol. **9:** 585-600.

Wilkinson, K.G., Dixon, K.W., Sivasithamparam, K. and Ghisalberti, E.L. 1994. Effects of IAA on symbiotic germination of an Australian orchid and its Production By Orchid-Associated Bacteria. Plant & Soil 159: 291-295.

Withner, C.L. 1942. Nutrition experiments with orchid seedlings. Amer. Orch. Soc. Bull. **11(4):** 112-114.

Withner, C.L. 1943. Ovule culture: a new method for starting orchid seedlings. Amer Orch. Soc. Bull. **11:** 261-263.

Withner, C.L. 1947. Making orchid cultures. Amer. Orch. Soc. Bull. **16:** 98-104.

Withner, C.L. 1951. Effects of plant hormones and other compounds on the growth of orchids. Amer. Orch. Soc. Bull. **20(5):** 276-278.

Withner, C.L. 1955. Ovule culture and growth of *Vanilla* seedlings. Amer. Orch. Soc. Bull. **24(6):** 380-392.

Withner, C.L. 1959a. Orchid Physiology. In: Withner, C.L. (ed.), The Orchids. A Scientific Survey, the Ronald Press Co., NY.

Withner, C.L. 1959b. Orchid culture media and nutrient solutions. In: Withner, C.L. (ed.), The Orchids. A Scientific Survey, the Ronald Press Co., NY: 589-599.

Wodrich, K. 1997. Growing South African indigenous orchids. AA Balkema, Rotterdam, the Netherlands.

Wotavová-Novotná, K. Vejsadová, H. and Kindlmann, P. 2007. Effects of sugars and growth regulators on *in vitro* growth of *Dactylorhiza* species. Biol. Plant. **51(1):**198-200.

Wynd, F.L. 1933a. The sensitivity of orchid seedlings to nutritional ions. Ann. Miss. Bot. Gard. **20:** 223-237.

Wynd, F.L. 1933b. Sources of carbohydrates for germination and growth of orchid seedlings. Ann. Miss. Bot. Gard. **20:** 569-581.

Yam, T.W. and Arditti, J. 2009. History of orchid propagation: a mirror of the history of biotechnology. Plant Biotechnol. Rep.3: 1-56.

Yamada, M. 1963. A simplified method of orchid seed germination. Hawaiian Orch. Soc. Inc., Honolulu, Mimeograph.

Yanetti, R.A. 1990. *Arethusa bulbosa:* germination and culture. In: C.E. Sawyer (ed.) North American native terrestrial orchid propagation and production. Chadds Ford Press, Pennsylvania, USA.

Yates, R.C. and Curtis, J.T. 1949. The effects of sucrose and other factors on the shoot-root ratio of orchid seedlings. Amer. J..Bot. **36**: 390-396.

Zeigler, A.R. Sheehan, J.J. and Pool, R.T. 1967. Influence of various media and photoperiod on growth and amino acid content of orchid seedlings. Amer. Orch. Soc. Bull. **36**: 195-202.

Zettler, L.W. 1997. Orchid–fungal symbiosis and its value in conservation. McIlvainea 13:40–45.

Zettler, L.W. and McInnis, T.M. Jr. 1992. Propagation of *Platanthera intergrilabia* (Correll) Luer, an endangered terrestrial orchid, through symbiotic seed germination. Lindleyana **7**: 154–161.

Zettler, L.W. and Hofer, C.J. 1998. Propagation of the little club-spur orchid (*Platanthera clavellata*) by symbiotic seed germination and its ecological implications. Environmental and Experimental Botany **39**: 189-195.

Ziegenspeck, H. 1936. Orchidaceae. In: O.Von Krishner, E. Loew and C.Schrter (eds.), Lebensgeschichte der Biûtenflazen Mitteleuropas, vol.1. part 4. Stuttgart, Germany: Eugen Ulmer Verlag.

Chapter 5

In vitro Approaches in Medicinal Plants–A Viable Strategy to Strengthen the Resource Base of Plant Based Systems of Medicines

A.A. Waman, P. Bohra, B.N. Sathyanarayana and
B.G. Hanumantharaya

Introduction

Biotechnology has many practical implications in the medicinal plants sector. Tissue culture, though quite old technology has a vast potential in order to provide large number of uniform, elite and disease-free planting materials. This ensures restricted dependence on the natural populations and thereby restoration of forests. Apart from this, the potential of this technique in production of secondary metabolites and crop improvement through various means is worth exploring. This report mainly concerned the potential of plant tissue culture as an aid in the multiplication of medicinal plant species. Suitable representative examples from authors' experiences have been provided to improve the understanding of the article.

After centuries of painstaking efforts taken by our great sages and the contributions of the earlier workers, the Indian System of Medicines was born, mainly based upon the usage of a plant, combination of plants or parts thereof for treating various human ailments. During that period plant based medicines were the only scientific (though not proven, then) source to cure most of the known ailments, but majority of people were unaware about the miraculous effects of natural products, and they were taking shelter of 'superstitious healing techniques'. Gradually, people started realizing the potentials of plants and began using the 'nature gifted therapy'. As most of the population in our country was illiterate during that period, the traditional therapeutic knowledge was restricted with some sages and a few local *vaidyas*. Before the system could develop the wings to fly across the boundaries, the modern system of medicines developed with a mind boggling velocity. The modern medicines, though gradually, did spread its roots in many countries including India and became popular due to their quick relief actions. But, a large number of discoveries have been made in recent past which suggests that the repeated use (or we can say abuse) of the modern medicines can cause many health complexities. Along with curing of the ailment at a rapid speed, they are known to cause many side effects including major health problems.

Almost each and every plant on this earth is known to synthesize a variety of compounds/ products, which are rather difficult to imitate/ produce under the artificial condition using hundreds of chemical factories (Waman and Karanjalker, 2010). Most of the plant based systems of medicine use drugs/ extracts obtained from one or more plant/s parts. These extracts contain a mixture of compounds, which are probably responsible for cushioning our body from any damage caused due to the major active ingredient acting against the ailment. That means, the natural drugs are free from side effects and thus are getting popular not only in India but also in many developed countries.

Utilities of *in vitro* strategies in medicinal plants

As stated earlier, almost all the plants on the earth possesses some or the other medicinal property, though most of the properties are yet to be explored. There are some plants of which fruits/ parts are used for edible purpose and are known to have one or the other medicinal properties. That is the reason why fruits and vegetables are popularly called as the 'protective foods'. For instance, all yellow coloured fruits are rich in carotenoids, which are known to improve the eye sight. The recipe prepared from the core of banana pseudostem is popularly used for treating kidney stones in southern states of our country. Spices and condiments, which are indispensible components of all Indian cuisines, are known to have antioxidant, antimicrobial, anti-viral and many other properties. All these groups of plants are medicinal and are useful in preventing many disorders in human body, though they are not considered as the major drug sources.There is another group of plants which are mainly used for curing various ailments, normalizing body functions. A large number of scientific references are available which justifies the use of the 'typical medicinal plants (those yielding secondary metabolites and used mainly for therapeutic purpose)' mentioned in plant based systems of medicines for curing different ailments. These plants could be large sized trees, climbers, small shrubs, or otherwise tiny herbs. Natural/ conventional propagation of these species have their own advantages and disadvantages. Micropropagation could be an efficient strategy for overcoming most of the problems occurring in conventional propagation. National Medicinal Plant Board (NMPB) under the aegis of Government of India has prioritized 32 different medicinal plants based upon their importance in the drug industry and in national economy. A large number of reports are available which describes different protocols to multiply these plants under *in vitro* condition. Table 5.1 represents a glimpse of studies made on these medicinal plant species. These species, being prioritized, are of immense practical value in domestic as well as international markets. *In vitro* approaches could be of great help in supplying the quality propagules in large numbers as well as production of pharmaceutical macro-molecules in the purest forms. This will not only ensure a regular supply of disease free/ elite planting material to the growers, but also help in restoration of many threatened plant species in wild due to the reduced burden on forests. Apart from the application of plant tissue culture technology in the multiplication of medicinal plants, there are many other applications such as *in vitro* production of secondary metabolites, crop improvement through *in vitro* mutagenesis, protoplast fusion and somaclonal variations, genetic transformation etc. The present review concerned the application of biotechnology in production of quality planting materials of medicinal plants, thereby improving the resource base of the pharmaceutical industries.

(a) Utilities in medicinal tree species

A large number of plants used on regular basis in pharmaceutical industries are woody/ tree in nature. In order to understand the need of using *in vitro* strategies for multiplication of tree medicinal plant species and woody climbers, let us take an example of Ashoka (*Saraca indica* L. Syn. *S. asoca* De Wilde), an important plant mentioned in Ayurveda. It is a handsome ornamental cum medicinal plant, commonly known as *Sita Ashoka*, belonging to the family *Caesalpiniaceae*. This plant of Indian origin is extensively found in many tropical regions of the world including the

Andaman Islands, Malayan Peninsula, Sri Lanka, Myanmar and Bangladesh. The natural populations of Ashoka are commonly found in Khasi hills of Assam, hilly areas of West Bengal, Western Ghats of Maharashtra, North-eastern area and all Southern States of India (Purohit and Vyas, 2007). Dark bark of the tree contains proanthocyanidins- epicatechin and procyanidin B_2, n- alkanes, esters, tannins, ketosterol, catechol and other organic calcium compounds (Rastogi and Mehrotra, 1998). It is an excellent medicine for most of the female related problems and hence is rightly called as Ashoka, the sorrow less tree. Drug obtained from the bark imparts healthy tone and strength to the uterus mainly by its action on the endometrium and ovarian tissues. In Ayurveda, it is extensively useful in gynecological problems *viz.* leucorrhea, internal bleeding, haemorrhoids, haemorrhagic dysentery and uterine bleeding associated with fibroids (Sayed and Mukundan, 2005).

DR: Direct regeneration; IR: Indirect regeneration; SE: Somatic embryogenesis.

Table 5.1. Protocols developed for *in vitro* multiplication in 32 prioritized plants of India.

Plant species	Pathway	Explants	Reference
Amla (Emblica officinalis)	DR	Node	Divya and Seema, 2008
Ashoka (Saraca indica)	DR	Node, shoot tip	Subbu et al., 2008
Ashwagandha (Withania somnifera)	IR	Seedling	Rani and Grover, 1999
Atis (Aconitum heterophyllum)	IR	Nodes	Neelofar et al., 2006
Bael (Aegle marmelos)	IR	Nucellar tissue	Hossain and Kareem, 1993
Bhuamalaki (Phyllanthus niruri)	DR	Nodes	Ong and Chan, 2006
Brahmi (Bacopa monnieri)	DR	Nodes	Sharma et al., 2010
Chandan (Santalum album)	DR	Leaves	Abdul, 2005
Chirata (Swertia chirata)	SE	Leaves	Wang et al., 2009
Daru Haridra (Berberis aristata)	-	-	-
Giloe (Tinospora cordifolia)	DR	Nodes	Afshan and Nag, 2008
Gudmar (Gymnema sylvestre)	SE	Leaves	Kumar et al., 2002
Guggal (Commiphora wightii)	DR	Nodes	Tejovathi et al., 2011
Isabgol (Plantago ovata)	SE	Leaf	Jasrai et al., 1993
Jatamansi (Nardostachys jatamansi)	DR	Roots	Mathur, 1992
Kalihari (Gloriosa superba)	DR	Apical and axillary buds	Hassan and Roy, 2005

Table 5.1 (contd...)

Plant species	Pathway	Explants	Reference
Kalmegh (Andrographis paniculata)	SE	Leaf, internodes	Martin et al., 2004
Kesar (Crocus sativus)	SE	Meristem	George et al., 1992
Kokum (Garcinia indica)	DR/SE	Immature seeds	Thengane et al., 2006
Kuth (Sausurrea lappa)	DR	Shoot tip	Arora and Bhojwani, 1989
Kutki (Picrorhiza kurroa)	DR	Nodes	Chandra et al., 2006
Makoi (Solanum nigrum)	DR	Nodes	Sridhar and Naidu, 2011
Mulethi (Glycyrrhiza glabra)	DR	Shoot tip and nodes	Thengane et al., 1998
Patharchur (Coleus forskhlii)	DR	Axillary buds	Sharma et al., 1991
Pippali (Piper longum)	DR	Shoot tips	Soniya and Das, 2002
Safed musli (Chlorophytum borivilianum)	DR	Shoot tips	Dave et al., 2003
Sarpagandha (Rauwolfia serpentina)	DR	Nodes	Goel et al., 2007
Senna (Cassia angustifolia)	DR	Nodes	Siddiqui and Anis, 2007
Shatavari (Asparagus racemosus)	DR	Nodes	Bopana and Saxena, 2008
Tulsi (Ocimum spp.)	IR	Axillary buds	Gogoi and Kumaria, 2011
Vatsanabh (Aconitum ferox)	SE	Leaf and petiole	Giri et al., 1993
Vai vidang (Embelia ribes)	DR	Hypocotyls	Annapurna and Rathore, 2010

The bark, seeds, leaves comprises the raw drug and is in a great demand from the pharmaceutical industry. Being tree species, no systematic farming/ plantations of Ashoka exist and forests are thus major source of supply (Waman and Bohra, 2013). Conventionally, the tribes harvest the bark in a non scientific way, without paying attention towards the health of plant. Further, the injudicious method of collection leads to severe injury to the plants and the plants succumb to fungal and bacterial infections (Murthy *et al.*,2008) which could cause death of the plant. As a result of this, the species has been included in the red list of medicinal plant species in South India under EN/R (endangered/ regionally category) (Rajasekharan and Ganeshan, 2002). Under natural conditions, the ripe seeds of Ashoka fall in the forests and get germinated or are sometimes collected by the tribes which favour dispersal into new areas. But, the large scale multiplication through sexual means is difficult as the seed set is poor in its major areas such as the forests of Western Ghats. Further, the seeds have a great predation by insects, squirrels and rats, making its availability rare. Adding to this, the leftover seeds remaining after all these problems exhibit very low viability (Murthy *et al.*, 2008). All these problems observed in the multiplication of Ashoka are more or less common in most of the medicinal trees. Micropropagation of tree species offers a rapid means of producing clonal planting stock for aforestation, woody biomass production and conservation of the elite and rare germplasm (Bonga and Durzan, 1982).

In spite of numerous advantages, such woody taxa are generally not amenable to regenerate under *in vitro* conditions. In order to standardize a reliable protocol for mass multiplication of this plant, studies were initiated in author's institute (Waman *et al.*, 2010; Waman and Bohra, 2013) and conspicuous problems were observed during culture. For obtaining different explants, seeds of Ashoka were soaked for 24 h in water followed by sowing in seed pans. Previous report by Purohit and Vyas (2007) suggested a mean germination period of three weeks. In our studies, seed germination commenced after 33 days of sowing, and about 70 % of the seeds got germinated after a long period of eight weeks after sowing. Though, the seeds were collected from freshly fallen pods of healthy trees, the germination percentage was not observed to be impressive. This delay in germination could be attributed to the cooler climatic conditions as the experiment was conducted during rainy season under Bengaluru conditions. The other promising method to obtain culturable explants, coppicing, was also tried and the young sprouts of appropriate age were obtained after one and half months of coppicing. Thus, it was clear from both instances that, Ashoka like many other medicinal trees and climbers exhibit slower growth response, thereby making the protocol lengthier to follow. In most of the tree species, improved clones are not available for multiplication and the sexual means of propagation may lead to variability in the daughter plants. Recent report suggested the occurrence of polyembryony in the seeds of Ashoka (Waman and Bohra, 2013). This phenomenon is of great practical utility if plus trees showing polyembryony are identified. This will eliminate the disadvantages of sexual propagation, making the progeny more uniform. The same logic holds good for other tree species and climbers also.

Contamination is the most devastating problem in the tissue culture of woody plants, though it impacts the other species too. If a reliable protocol is standardized for aseptic establishment of cultures, tissue culture could be a boon for the growers and those involved in raw drug industries. During culture of explants under *in vitro* conditions, decontamination of node, internode and shoot tip was not possible. Various concentrations and combinations of sterilants, fungicides, and bactericides could not help in obtaining aseptic cultures in these explants. Antibiotics have been considered to be highly effective, provided they are used judiciously and their phyto-toxocity on cultures is assessed. Use of chloramphenicol for controlling contamination in above mentioned explants of Ashoka was reported by Subbu *et al.* (2008). The antibiotic, however, was ineffective in eliminating the contaminants from the cultures, probable because of variations in the place of collection of explants, season of collection, rainfall pattern of the area and other weather parameters which determines the kind of microflora present in a locality (Waman and Bohra, 2013). Further, phenolic exudation limited the establishment of cultures of Ashoka under *in vitro* condition. Non responsiveness of the explants to the inputs in the cultures has been noticed in many plant species. Leaf explants, obtained from both young and mature leaves adhered to this phenomenon during our studies also (Waman *et al.*, 2010). Various combinations of 6- benzylaminopurine and Kinetin were tried for induction of multiple shoots. Of the two explants used, the young explants exhibited curling of the surface shortly after (2-4 days) inoculation, but no subsequent growth was observed in the cultures and the explants got dried up. On the contrary, mature leaf plants remained green without any response towards caulogenesis or morphogenesis. Callus obtained from epicotyls and hypocotyls failed to regenerate.

Most of the woody plant species are known to pose these kinds of difficulties under *in vitro* conditions. But, a multi-pronged strategy could be of great use to solve these problems. For example, proper maintenance of mother plants, multi-step sterilization of explants and incorporation of antibiotics in the medium could help to reduce the chances of contamination. *In vitro* seed germination also provides an opportunity to use stronger sterilants with extended sterilization duration to obtain aseptic explants. Identification of polyembryonic lines could help in maintaining uniformity in the progeny. Antioxidants such as ascorbic and citric acid could also be incorporated to reduce the problems of phenolics exudation (Divya and Seema, 2008). Frequent subculture of the explants on fresh medium helps in reduction of the browning to a greater extent. A combination of cytokinins or use of stronger cytokinin such as thiadiazuron could induce caulogenesis and organogenesis. All these techniques are quite promising and have been tried in many other medicinal tree species. Thus, it is clear that tissue culture is a promising technique in medicinal trees not only for multiplying them in large number but also for conserving the natural population and reducing the burden on the wild stock.

(b) Utilities in medicinal shrubs and herbs

Most of the medicinal shrubs are propagated through seeds or vegetative methods such as cuttings, grafting, budding etc. In most of the species, almost all the plant parts are used in the pharmaceutical industries. Many of the species are annual/ biennial in nature and are in good demand, so the cultivation of these species is becoming popular amongst the growers. For establishing commercial scale plantations in order to supply the drugs to industries, farmers need large quantity of uniform planting material. Meeting this demand through traditional techniques is quite difficult as there are very few nurseries which supply the planting materials on larger scale. The Medicinal and Aromatic Plants Section of the Department of Horticulture of author's University, State Departments of Horticulture and Foundation for Revitalization of Local Health and Traditions (FRLHT) Bengaluru are few of the authentic suppliers of medicinal plants, apart from a number of State Agricultural Universities and Forest departments. But, the demand for a few commercially important plant species is huge and could be effectively met using tissue culture technology.

Let us consider an example of Ashwagandha (*Withania somnifera* L.), an important medicinal member of the Solanaceae family. Tuberous roots of Ashwagandha (also called Indian ginseng and winter cherry) contain several pyrazol alkaloids, withasomnine, steroidal lactones, withaferine-A, and withanolides, owing to which the roots are being used as sedative, astringent, stimulant, aphrodisiac, diuretic, and tonic in Ayurvedic medicines (Kapoor, 1990; Prajapati *et al.*, 2003). Recently, a new dimeric withanolide isolated from roots was found to be useful against human gastric, breast, central nervous system, colon, and lung cancer lines (Subbaraju *et al.*, 2006). This species is commercially propagated through seeds, as the plant lacks the natural ability to multiply though vegetative means (Sen and Sharma, 1991). Also, limited seed viability (Rani and Grover, 1999) and nursery diseases *viz.* seed rot and blight are known to affect its commercial cultivation (Farooqi and Sreeramu, 2004). All these factors make tissue culture a reliable and effective method for multiplication of ashwagandha and alike species. Further, in crops where aseptic cultures are not easily obtained from the mature explants, *in vitro* seed germination could be a more convenient option. Especially in crops which mostly breed true to type, even when propagated through sexual means, this technique is of merit. *In vitro* germination of seeds provides an opportunity to use a variety of explants for better comparison of succeeding treatments such as standardization of explant source, growth regulators etc. (Waman *et al.*, 2011b). Herbal medicinal plants such as basil, brahmi, mandukparni, mints *etc.*, on the other hand, could be propagated much easily through seeds or cuttings. So, tissue culture techniques would not be always economically feasible unless low cost alternatives are used to cut down the production cost. The problems of contaminations, culture browning and drying, poor culture response though present to some extent, are less severe in the medicinal shrubs and the herbal plants. This is of great advantage to get maximum output per culture. Thus, tissue culture techniques help in rapid multiplication of medicinal plants species to meet demands from pharma industries.

Table 5.2. Select protocols developed for multiplication of commercially important medicinal plants at authors' Laboratory

Plant species	Regeneration pathway	Explant	Reference
Bacopa monnieri	Somatic embryogenesis	Leaves	Sharath et al., 2007
Centella asiatica	Indirect regeneration	Leaves and Petiole	Hanumantharaya et al., 2009
Centella asiatica	Direct regeneration	Nodes	Hanumantharaya et al., 2010
Centella asiatica	Somatic embryogenesis	Nodes	Hanumantharaya et al., 2011b
Chlorophytum borivilianum	Direct regeneration	Tubers with stem disc	Maruthi Prasad et al., 2007a
Chlorophytum borivilianum	Somatic embryogenesis	Tubers with stem disc	Maruthi Prasad et al., 2007b
Gloriosa superba	Somatic embryogenesis	Nodes and leaves	Kalpana et al., 2002
Murraya koenigii	Direct regeneration	Leaves	Mathew and Prasad, 2007
Pogostemon patchouli	Direct and indirect regeneration	Shoot tips and leaves	Babu et al., 2010
Saraca indica	Indirect regeneration	Leaves, Epicotyls and Hypocotyls	Waman et al., 2010
Withania somnifera	Indirect regeneration	Seedling explants	Waman et al., 2010a
Withania somnifera	Indirect regeneration	Seedling explants	Waman et al., 2011c

Table 5.2 describes the details of protocols developed for the commercially important medicinal plants at the author's institute. As

described earlier, tissue culture might not be affordable in few medicinal plants in which the demand is quite high but the produce is sold at the lower price. In order to make this cost effective, a good number of low cost substitutes are available for tissue culture now. These alternatives have been widely tried in different plant species. The cost of plant production could be greatly reduced by reducing the input costs e.g., using cheaper substitute to sucrose, agar, water source, hardening, container *etc.* So far, these techniques have not been commercially used in multiplication of medicinal plants, but a few attempts have been made as described in Table 3. Thus, there is vast potential of these techniques in the context of medicinal plants' multiplication.

Table 5.3: Promising low cost alternatives used for *in vitro* multiplication of important medicinal plants

Regular component	Cost effective alternative	Plant species	Reference
Agar gelled medium	Liquid medium with coir as substrate Liquid medium Liquid medium Isabgol Liquid medium with glass beads	*Andrographis paniculata* *Chlorophytum borivilianum* *Centella asiatica* *Curcuma longa* *Rauwolfia serpentina*	Gangopadhyay *et al.*, 2002 Rizvi *et al.*, 2007 Raghu *et al.*, 2007 Tyagi *et al.*, 2007 Goel *et al.*, 2007
In vitro rooting and *ex vitro* hardening	Single step *ex vitro* rooting and hardening	*Withania somnifera* *Centella asiatica* *Azadirachta indica*	Waman *et al.*, 2011a Raghu *et al.*, 2007 Lavanya *et al.*, 2009
Sucrose (3%)	Market grade sugar (1%) Market grade sugar (3%) Market grade sugar (3%) Market grade and Darula sugar	*Withania somnifera* *Centella asiatica* *Curcuma longa* *Rauwolfia serpentina*	Waman *et al.*, 2011b Hanumantharaya *et al.*, 2011a Tyagi *et al.*, 2007 Goel *et al.*, 2007
Full strength MS medium	Quarter and half strength MS medium Quarter and half strength MS medium Half strength MS medium Half strength MS medium	*Centella asiatica* *Chlorophytum borivilianum* *Centella asiatica* *Saraca indica*	Hanumantharaya *et al.*, 2011a Maruthi Prasad *et al.*, 2007a Raghu *et al.*, 2007 Waman *et al.*, 2010
Baby jar bottles	Bioreactor	*Stevia rebaudiana* *Digitalis lanata* *Artemisa annua* *Asparagus officinalis* *Atropa belladonna*	Akita *et al.*, 1994 Greidziak *et al.*, 1990 Park *et al.*, 1989 Takayama, 1991 Takayama, 1991
Double distilled water	Tap water	*Centella asiatica*	Raghu *et al.*, 2007

References

Abdul, M. 2005, *In vitro* regeneration of Sandal (*Santalum album* L.) from leaves. Turk. J. Bot. **29**: 63-67.

Afshan, T. and Nag, K.K. 2008, *In vitro* regeneration of *Tinospora cordifolia*, a medicinal climbing shrub. Vegetos. **21(2)**: 125-128.

Akita, M., Shigeoka, T., Koizumi, Y. and Kawamura, M. 1994. Mass propagation of shoots of *Stevia rebaudiana* using a large-scale bioreactor. Plant Cell Rep. **13**: 180-183.

Annapurna, D. and Rathore, T.S. 2010. Direct adventitious shoot induction and plant regeneration of *Embelia ribes* Burm. f. Plant Cell Tiss. Organ Cult. **101(3)**: 269-277.

Arora, R. and Bhojwani, S.S. 1989. *In vitro* propagation and low temperature storage of *Saussurea lappa*- an endangered medicinal plant. Plant Cell Rep. **8**: 44-49.

Babu, P., Ramachandra, K.M. and Sathyanarayana, B.N. 2010. *In vitro* regeneration potential of patchouli shoot tip and leaf explants. Plant Cell Biotechnol. Mol. Biol. **11(1-4)**: 43-50.

Bonga, J.M. and Durzan, D.J. 1982. Cell and Tissue Culture in Forestry. Martinus Nijhoff, The Hague.

Bopana, N. and Saxena, S. 2008. *In vitro* propagation of a high value medicinal plant: *Asparagus racemosus* Willd. *In vitro* Cell. Dev. Biol.– Plant. **44(6)**: 525-532.

Chandra, B., Palni, L.M.S. and Nandi, S.K. 2006. Propagation and conservation of *Picrorrhiza kurroa* Royle ex Benth.: an endangered Himalayan medicinal herb of high commercial value. Biodiver. Conser. **15**: 2325.

Dave, A., Bilochi, G. and Purohit, S.D. 2003. Scaling up production and field performance of micropropagated medicinal herb Safed musli (*Chlorophytum borivilianum*). *In vitro* Cell. Dev. Biol. **39**: 419.

Divya, B. and Seema, G. 2008. *In vitro* shoot proliferation in *Emblica officinalis* var. Balwant from nodal explants. Indian J. Biotechnol. **7**: 394-397.

Farooqi, A.A. and Sreeramu, B.S. 2004. Cultivation of medicinal and aromatic plants. University Press, Hyderabad, India.

Gangopadhyay, G., Das, S., Mitra, S. K., Poddar, R., Modak, B. K. and Mukherjee, K. K. 2002. Enhanced rate of multiplication and rooting through the use of coir in aseptic liquid culture media. Plant Cell Tiss. Org. Cult. **68**: 301–310.

George, P.S., Sujata, V., Ravishankar, G.A. and Venkataraman, L.V. 1992. Tissue culture of saffron (*Crocus sativus* L.): Somatic embryogenesis and shoot regeneration. Fd. Biotechnol. **6(3)**: 217-223.

Giri, A., Ahuja, P.S. and Ajay, K.P.V. 1993. Somatic embryogenesis and plant regeneration from callus cultures of *Aconitum heterophyllum* Wall. Plant Cell, Tiss. Organ Cult. **32(2)**: 213-218.

Goel, M.K., Kukreja, A.K. and Khanuja, S.P.S. 2007. Cost-effective approaches for *in vitro* mass propagation of *Rauwolfia serpentina* Benth. ex Kurz. Asian J. Plant Sci. **6**: 957–961.

Gogoi, K. and Kumaria, S. 2011. Callus - mediated plantlet regeneration of *Ocimum tenuiflorum* L . using axillary buds as explants. Intl. Res. J. Plant Sci. **2(1)**: 1-5.

Greidziak, N., Diettrich, B. and Luckner, M. 1990. Batch cultures of somatic embryos of *Digitalis lanata* in gas lift fermenters, development and cardenolide accumulation. Plant Med. **6**: 175-178.

Hanumantharaya, B.G., Sathyanarayana, B.N. and Waman, A.A. 2010. An efficient *in vitro* regeneration protocol to conserve wild population of *Centella asiatica* L. Plant Cell Biotechnol. Mol. Biol.**11(1-4)**: 51-58.

Hanumantharaya, B.G., Sathyanarayana, B.N. and Waman, A.A. 2011a. Reduced media salt concentration improves *in vitro* rooting in Indian Pennywort. J. Cell Tiss. Res. **11(2)**: 2771-2774.

Hanumantharaya, B.G., Sathyanarayana, B.N., Waman, A.A. and Bohra, P. 2009. Callus culture and high frequency plantlet regeneration technique – an efficient tool to replenish *Centella asiatica* (L.) population in wild. J. Plant Biol. **36(3)**: 89–93.

Hanumantharaya, B.G., Sathyanarayana, B.N., Waman, A.A. and Guruprakash, R.G. 2011b. High frequency Somatic embryogenesis in Indian pennywort. J. Med. Arom. Plant Sci. **33(4)**: 451-456.

Hassan, A.K.M. S. and Roy, S.K. 2005. Micropropagation of *Gloriosa superba* L. through high frequency shoot proliferation Plant Tiss. Cult. **15(1)**: 67-74.

Hossain, M. and Karim, M.R. 1993. Plant regeneration from nucellar tissues of *Aegle marmelos* through organogenesis. Plant Cell Tiss. Organ Cult. **34 (2)**: 199-203.

Jasrai, Y.T, Yadava N. and Mehta, A. R. 1993. Somatic embryogenesis from leaf induced cell cultures of *Plantago ovata* Forsk. J. Herbs Spices Med. Plants. **1(4)**: 11-16.

Kalpana, S., Prakash, H.S., Prasad, T.G. and Sathyanarayana, B.N. 2002. Somatic embryogenesis and plantlet regeneration from node and leaf callus cultures of *Gloriosa superba* L. J. Plant Biol. **29**: 179-189.

Kapoor, L.D. 1990. Handbook of Medicinal Plants. CRC Press, LLC, Boca Raton, Florida, pp. 337-338.

Kumar, H.G.A., Murthy, H.N. and Paek, K.Y. 2002. Somatic embryogenesis and plant regeneration in *Gymnema sylvestre*. Plant Cell Tiss. Organ Cult. **71**:85–88.

Lavanya, M., Venkateshwarlu, B. and Poornasri D. 2009. Acclimatization of neem microshoots adaptable to semi- sterile conditions. Indian J. Biotechnol. **8**: 218-222.

Martin, K.P. 2004. Plant regeneration protocol of medicinally important *Andrographis paniculata* (Burm. f.) Wallich ex Nees via somatic embryogenesis. *In vitro* Cell. Dev. Biol.—Plant. **40**:204-209.

Maruthi Prasad, B.N., Sathyanarayana, B.N. Gowda B. and Sharath, R. 2007a. *In vitro* regeneration of drug-yielding tuber crop *Chlorophytum borivilianum*. Med. Arom. Plant Sci. Biotechnol. **1(1)**: 124-127.

Maruthi Prasad, B.N., Sathyanarayana, B.N., Jaime A.T.S., Sharath, R. and Gowda B. 2007b. Regeneration in *Chlorophytum borivilianum* through somatic embryogenesis. Med. Arom. Plant Sci. Biotechnol. **1(1)**: 128-132.

Mathew, D. and Prasad, M.C. 2007. Multiple shoot and plant regeneration from immature leaflets of *in vitro* origin in Curry leaf (*Murraya koenigii* Spreng). Indian J. Plant Physiol.**12(1)(N.S.)**: 18-22.

Mathur, J. 1992. *In vitro* morphogenesis in *Nardostachys jatamansi* DC: shoot regeneration from callus derived roots. Ann. Bot. **70(5)**: 419-422.

Murthy, S.M., Mamatha, B. and Shivananda, T.N. 2008. *Saraca asoca-* an endangered plant. Biomed. **3(3 & 4)**: 224-228.

Neelofer, J., Shawl, A.S., Dar, G.H., Arif, J. and Phalisteen, S. 2006. Callus induction and organogenesis from explants of *Aconitum heterophyllum-* medicinal plant. Biotechnol. **5(3)**: 287-291.

Ong, P.L. and Chan, L.K. 2006. *In vitro* propagation, flowering and fruiting of *Phyllanthus niruri* L. Intl. J. Bot.**2(4)**: 409-414.

Park, J.M., Hu, W.S. and Staba, E.J. 1989. Cultivation of *Artemisia annua* L. plantlets in a bioreactor containing a single carbon and a single nitrogen source. Biotechnol. Bioeng. **34**: 1209-1213.

Prajapati, N.D., Purohit, S.S., Sharma, A.K. and Kumar, T. 2003. A Handbook of Medicinal Plants: a Complete Source Book. Agribios Publications, Jodhpur, India.

Purohit, S.S. and Vyas, S.P. 2007. Medicinal Plants Cultivation: A Scientific Approach.Agribios Publications, Jodhpur, India.

Raghu, A.V., Martin, G., Priya, V., Geetha, S.P. and Balachandran, I. 2007. Low cost alternatives for the micropropagation of *Centella asiatica*. J. Plant Sci. **2(6)**: 592-599.

Rajasekharan, P.E. and Ganeshan, S. 2002. Conservation of medicinal plants biodiversity- and Indian perspective. J. Med. Arom. Plant Sci.**24**: 132-147.

Rani, G. and Grover, I.S. 1999. *In vitro* callus induction and regeneration studies in *Withania somnifera*. Plant Cell Tiss. Organ Cult. **57**: 23-27.

Rastogi, R.P. and Mehrotra, B.N. 1998. Compedium of Indian Medicinal Plants. Vol. V. National Institute of Science Communication, New Delhi, India.

Rizvi, M.Z., Kukreja, A.K., and Khanuja, S.P.S. 2007. *In vitro* culture of *Chlorophytum borivilianum* Sant. et Fernand. in liquid culture medium as a cost-effective measure. Curr. Sci. **92(1)**: 87-90.

Sayed, N.Z. and Mukundan, U. 2005. Medicinal and Aromatic Plants of India, Part – I. Ukaaz Publication, Hyderabad, p.35.

Sen, J. and Sharma, A.K. 1991. Micropropagation of *Withania somnifera* from germinating seeds and shoot tips. Plant Cell Tiss. Organ Cult. **26**: 71-73.

Sharath, R., Krishna, V., Sathyanarayana, B.N., Maruthi Prasad, B.N. and Harish, B.G. 2007. High frequency regeneration through somatic embryogenesis in *Bacopa monnieri* (L.) Wettest, an important medicinal plant. Med. Arom. Plant Sci. Biotechnol. **1(1)**: 138-141.

Sharma, N., Chandel, K.P.S. and Srivastava, V.K. 1991. *In vitro* propagation of *Coleus forskohlii,* a threatened medicinal plant. Plant Cell Rep.**10**: 67-71.

Sharma, S., Barkha, K., Neelima, R., Sudhir, C., Vikas, J., Neha, V., Ashok, G. and Sarita A. 2010. *In vitro* rapid and mass multiplication of highly valuable medicinal plant *Bacopa monnieri* (L.) Wettest. African J. Biotechnol. **9(49)**: 8318-8322.

Siddique, I. and Anis, M. 2007. *In vitro* shoot multiplication and plantlet regeneration from nodal explants of *Cassia angustifolia* (Vahl.): a medicinal plant. Acta Physiol. Planta. **29(3)**: 233-238.

Soniya, E.V. and Das, M.R. 2002. *In vitro* micropropagation of *Piper longum* – an important medicinal plant. Plant Cell Tiss. Organ Cult. **70(3)**: 325-327.

Sridhar, T.M. and Naidu, C.V. 2011. Effect of different carbon sources on *in vitro* shoot regeneration of *Solanum nigrum* (Linn.) - an important antiulcer medicinal plant. J. Phytol.**3(2)**:78-82.

Subbaraju, G.V., Vanisree, M., Rao, C.V., Sivarama, K.C. Sridhar, P., Jayaprakasam, B. and Nair, M.G. 2006. Ashwagandhanolide, a bioactive dimeric thiowithanolide isolated from the roots of *Withania somnifera* (L.). J. Nat. Prod. **69(12)**: 1790-1792.

Subbu, R.R., Chandraprabha, A, and Sevugaperumal. 2008. *In vitro* clonal propagation of vulnerable medicinal plant, *Saraca asoca* (Roxb.) De Wilde. Nat. Prod. Radiance. **7(4)**: 338-341.

Takayama, S. 1991. Mass propagation of plants through shake and bioreactor culture techniques. In: Y.P.S. Bajaj (ed.). Hightech Micropropgation and Biotechnology in Agriculture and Forestry Vol. 17, pp. 1-46.

Tejovathi, G., Harisharan, G. and Rekha, B. 2011. *In vitro* propagation of endangered medicinal plant–*Commiphora wightii*. Indian J. Sci. Technol. **4(11)**: 1537-1541.

Thengane, S.R., Deodhar, S.R., Bhosle, S.V. and Rawal, S.K. 2006. Direct somatic embryogenesis and plant regeneration in *Garcinia indica* Choiss. Curr. Sci.**91(8)**: 1074-1078.

Thengane, S.R., Kulkarni, D.K. and Krishnamurthy, K.V. 1998. Micropropagation of licorice (*Glycyrrhiza glabra*) through shoot tip and nodal culture. *In Vitro* Cell. Dev. Biol. Plant **34**: 331.

Tyagi, R.K., Agrawal, A., Mahalakshmi, C., Zakir, H. and Tyagi, H. 2007. Low-cost media for *in vitro* conservation of turmeric (*Curcuma longa* L.) and genetic stability assessment using RAPD markers. *In vitro* Cell. Dev. Biol. Plant. **43**:51–58.

Waman, A.A. and Bohra, P. 2013. Choice of explants – a determining factor in tissue culture of Ashoka (*Saraca indica* L.). Intl. J. For. Usufructs Manag. **14(1)**: 10–17.

Waman, A.A. and Karanjalker, G.R. 2010. Diversifying the income avenues through herbal spices. Indian J. Arecanut Spices Med. Plants. **12 (1)**: 18-21.

Waman, A.A., Sathyanarayana, B.N., Umesha, K., Gowda, B., Ashok, T.H., Rajesh, A.M. and Guruprakash, R.G. 2011a. Optimization of growth regulators and explant source for micropropagation and cost effective *ex vitro* rooting in 'Poshita' Winter Cherry (*Withania somnifera* L.). J. Appl. Hort. **11(2)**: 150-153.

Waman, A.A., Umesha, K. and Sathyanarayana, B.N. 2010. First report on callus induction in Ashoka (*Saraca indica* L.): An important medicinal plant. Acta Hortic. **865**: 383-386.

Waman, A.A., Umesha, K., Sathyanarayana, B.N., Ashok T.H. and Gowda, B. 2011c. Callus culture and plant regeneration from seedling explants in 'Poshita' Indian Ginseng. Hort. Environ. Biotechnol. **52(1)**: 83-88.

Waman, A.A., Umesha, K., Sathyanarayana, B.N., Rajesh, A.M. and Guruprakash, R.G. 2011b. Cost effective *in vitro* seed germination in *Withania somnifera* L. Cv. 'Poshita' as affected by different chemicals. Crop Res. **42(1, 2 & 3)**: 163-165.

Wang, L., Lizhe, A., Yanping, H., Lixin, W. and Yi, L. 2009. Influence of phytohormones and medium on the shoot regeneration from leaf of *Swertia chirata* Buch.-Ham. ex Wall. *in vitro*. African J. Biotechnol. **8(11)**: 2513-2517.

Chapter 6

Somaclonal Variation in Plant Tissue Culture and its Role in Crop Improvement

P. Chandramati Shankar and H. Fathima Nazneen

Introduction

Somaclonal variation is a phenomenon that results in the phenotypic variation of plants regenerated from cell culture. When plant tissues are passed through *in vitro* culture, many regenerated plants appear to be no longer clones to their donor genotype. These variations are termed as somaclonal variations. It is seen among clonally propagated plants of a single donor clone (Larkin and Scowcroft, 1981). The occurrence of somaclonal variation is a matter of great concern for any micropropagation system. Somaclonal variation has relevance in the clonal propagation of valuable or endangered plant germplasm, and in the production of transgenic plants. It may also be an effective means of generating useful mutants. In order to evaluate its presence several strategies were used to detect somaclonal variants, based on one or more determinants from among morphological traits, cytogenetic analysis (numerical and structural variation in the chromosomes), and molecular and biochemical markers. In addition, studies on somaclonal variation are important for its control and possible suppression with the aim of producing genetically identical plants, and for its use as a tool to produce genetic variability, which will enable breeders the genetic improvement. Somaclonal variation has been studied extensively in herbaceous plants, whereas few studies have focused on temperate perennial fruit crops (Leva *et al.*, 2012).

Evans and Sharp (1986) reported four critical variables for somaclonal variation: explant origin, genotype, cultivation period and the cultural condition in which the culture is made. Identification of possible somaclonal variants at an early stage of development is considered to be very useful for quality control in plant tissue culture, transgenic plant production and in the introduction of variants. The variations can be either genotypic or phenotypic. The phenotypic variations can be either genetic or epigenetic (non inheritable) in origin. Typical genetic alterations are: changes in chromosome numbers (polyploidy and aneuploidy), chromosome structure (translocations, deletions, insertions and duplications) and DNA sequence (base mutations) insertion of transposable elements. Typical epigenetic related events are: gene amplification and gene methylation.

Origin and Causes

Although somaclonal variation has been studied extensively, the mechanisms by which it occurs remain largely unknown (Skirvin *et al.*, 1993, 1994). A variety of factors may contribute to the phenomenon. The method by which the regeneration of plantlets is induced, type of tissue, explant source, media components and the duration of the culture cycle are some of the factors that are involved in inducing variation during *in vitro* culture (Pierik,1987).

Explant Source

Genetic fidelity largely depends on explant source (Krikorian *et al.*, 1993). The explant tissue can affect the frequency and nature of somaclonal variation. The use of meristematic tissues, as starting materials for tissue culture reduces the possibility of variation. In contrast, highly differentiated tissues, such as roots, leaves, and stems, generally produce more variants, probably due to the callus-phase, than explants that have pre-existing meristems (Sharma *et al.*, 2007).

Medium Components

The hormonal components of the culture medium are powerful agents of variation. Unbalanced concentrations of auxins and cytokinins may induce polyploidy, whereas under a low concentration or total absence of growth regulators the cells show normal ploidy (Swartz, 1991). In addition, rapid disorganised growth can induce somaclonal variation. Auxins added to cultures of unorganised calli or cell suspensions increase genetic variation by increasing the DNA methylation rate (LoSchiavo *et al.*, 1989).

Duration and number of culture cycles

The frequency of somaclonal variation increases as the number of subcultures and their duration increases, especially in cell suspensions and callus cultures. Moreover, the rapid multiplication of a tissue or long-term cultures may affect genetic stability and thus lead to somaclonal variation (Reuveni and Israeli, 1990; Rodrigues *et al.*, 1998, Bairu *et al.*, 2006). A statistical model has been proposed for predicting the theoretical mutation rate, primarily on the basis of the number of multiplication cycles (Cote *et al.*, 2001). However, the model has limited application, due to the complexity of biological systems.

Effect of genotype

Conditions of culture *in vitro* can be extremely stressful for plant cells and may initiate highly mutagenic processes (Kaeppler *et al.*, 1993). However, different genomes respond differently to the stress-induced variation, which indicates that somaclonal variation also has genotypic components. The differences in genetic stability are related to differences in genetic make-up, because some components of the plant genome may become unstable during the culture process, for example the repetitive DNA sequences, which can differ in quality and quantity between plant species (Lee and Phillips, 1988). In banana tissue culture the most important factor that influenced dwarf off-type production was found to be the inherent instability of the cultivars; for example, the cultivar 'New Guinea Cavendish' showed a higher level of instability *in vitro* than 'Williams'. The dwarf off-types remained stable during *in vitro* culture, and the conditions under which tissue was cultured that induced dwarfism did not induce reversion of the dwarf off-type trait (Damasco *et al.*, 1998). In *Musa* species, the type and rate of variation was specific and depended on genotype (Stover, 1987; Israel *et al.*, 1991) and genome composition. An interaction between genotype and the tissue culture environment is also reported (Martin *et al.*, 2006).

Induction, Isolation, Screening and characterization of Somaclonal Variation

Induction of somaclonal variation

Somaclonal variation is induced in callus (a mass of undifferentiated cells), which are obtained by placing a suitable explants on a appropriate medium. When the callus grows; they are cut into small pieces and sub cultured to fresh medium. Somaclonal variations can be also induced in cell suspension cultures, which are established by transferring actively growing callus cells to constantly agitated liquid medium and multiplied through periodic subculturing onto the same fresh medium. The selection of variants *in vitro* level can be carried out for some traits by adding supplements in the medium for various biotic and abiotic factors, only the variant cells will survive (Fig 1.).

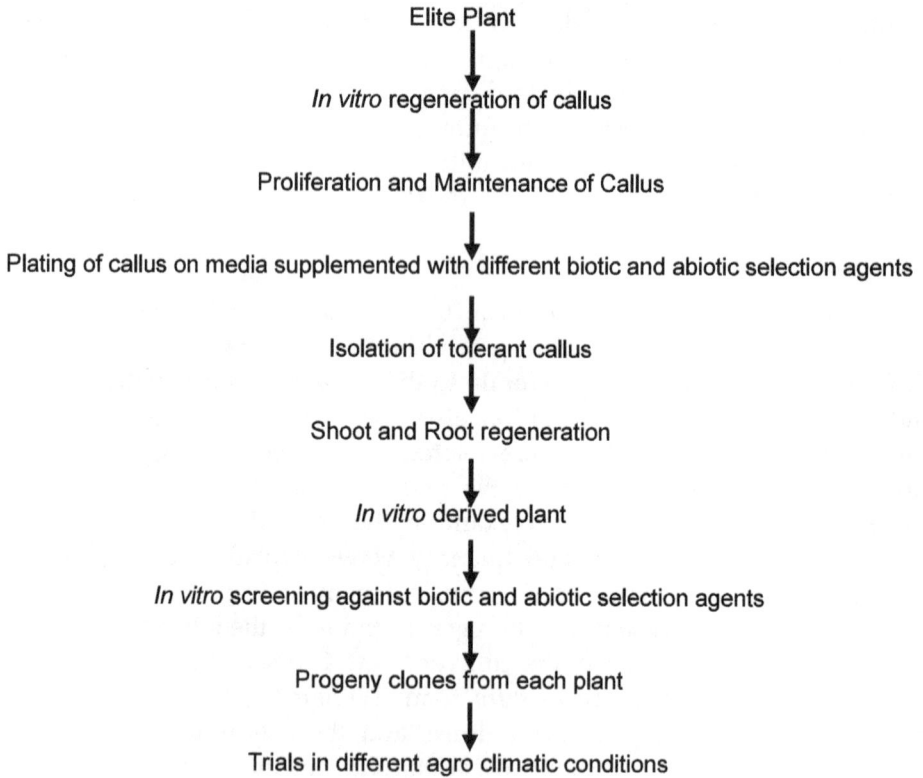

Elite Plant

↓

In vitro regeneration of callus

↓

Proliferation and Maintenance of Callus

↓

Plating of callus on media supplemented with different biotic and abiotic selection agents

↓

Isolation of tolerant callus

↓

Shoot and Root regeneration

↓

In vitro derived plant

↓

In vitro screening against biotic and abiotic selection agents

↓

Progeny clones from each plant

↓

Trials in different agro climatic conditions

Fig. 6.1. Schematic diagram of generation of somaclonal variants under *in vitro* conditions

Isolation of Somaclonal Variants

Variants for several traits can be far more easily isolated from cell cultures than from the whole plant. This is because the single cells, can be easily and effectively screened and monitored for mutant traits. The isolation of somaclonal variants can be broadly categorized into two groups: (i) screening of cells and (ii) cell selection.

Screening of cells. In this method a large number of cells are screened for the detection of variant individuals. This is a suitable method for the isolation of mutants for yield and yield traits. Screening has been widely employed for the isolation of cell clones that produce higher quantities of certain biochemical. The variant are screened in R1 progeny (progeny of regenerated, Ro, plants) and their R2 progeny lines are evaluated for confirmation.

Cell Selection: In this method a suitable selection pressure is applied which permits the preferential survival and growth of variant cells only. Selecting agents such as toxins, herbicides, pesticides, salts etc are incorporated into the culture medium at various concentration. The selection pressure allows only the mutant cells to survive or divide, it is called positive selection. In the case of negative selection, the wild type cells divide normally and therefore are killed by a counter selection agent. The mutant cells are unable to divide as a result of which they escape the counter selection agent. These cells are later rescued by removal of the counter selection agent. Negative selection approach is utilized for the isolation of auxotrophic mutants.

Characterization of Variants

Somaclonal variants isolated through cell selection are often unstable. The variants fail to exhibit their resistance during further screening or selection. Several clones lose their resistance to the selection agent after a period of growth in the absence of selection pressure. Such clones are called unstable variants and may result from changes in gene expression and from gene amplification.

Some variant phenotypes are quite stable during the cell culture phase, but they disappear when plants are regenerated from the variant cultures, or when the regenerated plants reproduce sexually. But sometimes the phenotypes are expressed in the regenerated plants due to stable changes in genes such changes are known as epigenetic changes and are attributed to stable changes in gene. These kind of somatic variants are important in crop improvement and plant breeding.

Molecular Basis of Somaclonal Variation

Somaclonal variation may arise due to any of the following events at molecular level such as changes in chromosome number or structure, mitotic crossing over, gene mutation, alteration in gene expression, gene amplification, transposable element activation, and rearrangements in cytoplasmic genes.

The molecular basis of somaclonal variation is complex. Epigenetically it involves modifications of the activation sites of certain genes, but not the basic structure of DNA.

Quantification of the tissue-culture induced variation

Different methods can be used to analyze plant genetic structure in tissue cultured plant clones, e.g., cytogenetic analysis, isoenzyme markers and different DNA molecular markers etc., many of these methods have limitations e.g., Karyology analysis cannot show structural changes in specific genes or in small chromosome arrangements. Molecular markers that can be used to detect such genetic changes are i) Isoenzyme: This is a good method to detect genetic changes; the disadvantage of these markers are that they are susceptible to ontogenetic variation and are limited in number, and only DNA segments coding for soluble proteins can be sampled. ii) RFLP (Restriction Fragment Length Polymorphism) markers are unlimited in number and can be used for sampling various genome regions. The limitations of this technique is; it is slow, expensive and requires large quantities of tissue. iii) RAPD (Random Amplified Polymorphic DNA) analysis is a PCR based marker using short primers of arbitrary sequence and it has been demonstrated to be sensitive in detecting variation among individuals. The advantages of this technique are: a) a large number of samples can be quickly analyzed using only micro-quantities of DNA b) the DNA amplicons are independent from the ontogenetic expression; and c) many genomic regions can be sampled with a potentially unlimited number of markers. RAPD markers have been used widely in studying the genetic diversity of somaclonal variations in various plant species (Rani *et al.*, 1995; Soniya *et al.*, 2001). The identification of variants at an early stage of tissue culture is very useful for quality control in transgenic plant production and in the introduction of variants.

Somaclonal Variation; Role in Improvement of crop plants its Benefits and disadvantages

Role in Improvement of crop Plant: Somaclonal variation and *in vitro* selection for obtaining plant genotype with improved tolerance to the biotic or abiotic stress, such as drought, high salinity, heavy metal stress, acid soil, and disease tolerance (Ahmed *et al.*, 1996; Yusnita *et al.*, 2005) have been obtained. In addition, the plants have shown some desired characters such as having bigger fruit size, more interesting flower texture, more delicious taste and higher production (Pedrieri, 2001; Ahloowalia and Maluszynski, 2001; Witjaksono, 2003).

Anwar *et al.* (2010) reported saline resistant sweet potato which was produced by passing the cultures through an *in vitro* salinity screen system where media were supplemented with 0, 75, 150 and 200 mM of NaCl. The data for parameters (number of roots, length of roots, leaf and root condition) suggested a significant variation in salinity tolerance among regenerated and control plants that proved the occurrence of somaclonal variation in regenerated plants In 1984, Taiwan Banana Research Institute started a selection programme for resistance to Fusarium wilt in Cavendish bananas using somaclonal variation (Hwang and Ko, 1987). In 1992, a resistant cultivar, 'Tai Chiao No. 1', was successfully selected and released to banana growers in Taiwan (Hwang *et al.*, 1992). Submergence tolerance is an important agronomic trait for rice grown in eastern India; where flash flooding occurs frequently and unpredictably during the monsoons. Generation of somaclones for the two submergence tolerant rice cultivar FR13A and FR43B through gamma irradiation were reported (Joshi and Rao, 2009).

Table 6.1. Several disease tolerant plants resulted from somaclonal and *in vitro* selection (Yusnita *et al.*, 2005)

Plant	*In vitro* culture system for	Selecting agent	Resistant to
Potato	Callus culture	Fungi Filtrate culture	*Phytophthora*
Tomato	Callus from leaf explants	-	*Fusarium oxysporum* f.sp. *infestan lycopersici* ras 2
Papaya	Shoot culture	-	*Phytophthora palmivora*
Soybean	Embryonic culture	-	*Fusarium oxysporum* f.sp *cubense* ras 4
Banana	Multiple bud clumps	Fusaric acid	*Fusarium* sp. ras I
Mango	Somatic embryo culture	Fungi Filtrate culture	*Colletotrichum gloesporoides*
Strawberry	Morphogenetic callus	Fungi Filtrate culture	*Rhizoctonia fragariae* and *Botrytis cemerea*
Apple	Shoot culture	Fungi Filtrate culture	*Phytophthora cactorum*
Wheat	Morphogenic callus	Fungi Filtrate culture	*Fusarium graminearum* and *Fusarium culmorum*

Benefits of somaclonal variation

Somaclonal variation represents a new source of genetic variability; therefore, it constitutes another tool for plant breeders (Larkin and Scowcraft, 1981). Exposure of *in vitro* cultures to mutagenic agents or to stress conditions can increase the number of somaclonal variants. Using this system, successful selection of mutants with various desired characteristics has been reported in some plant species, such as increased herbicide tolerance in hybrid poplar and other species (Michler and Haissig, 1988; Chaleff, 1986), disease resistance in *Larix* and other species (Diner, 1991; Sachristan, 1986), and heavy metal and salt tolerances in several species (Misra and Gedamu,

1989) as well as multi-gene agronomic characters (Mohammed, 1991). Serres *et al.* (1991) reported somaclonal hybrid *Populus* variants that were dwarf but fast-growing and with color-changed leaves. Thus, selection of somaclonal variants is particularly useful in the creation of new ornamental characteristics. The problems with use of this technique are: 1) there is a low frequency of variants, so a large-scale experiment is needed, 2) most variants are not desirable (useful), and 3) many of the variants may be epigenetic in nature, and not true variants. In some cases, the resultant variants are due to stable mutations with sexual transmission of the traits to progeny (Chaleff, 1986; Serres *et al.*, 1991,) and thus a useful source of variation.

One of the major potential benefits of somaclonal variation is the creation of additional genetic variability in co adapted, agronomically useful cultivars, without the need to resort to hybridization (Larkin , 1987).

Methodology of introducing somaclonal variations is simpler and easier as compared to recombinant DNA technology. Development and production of plants with disease resistance e.g., rice, wheat, apple, tomato etc. Develop biochemical mutants with abiotic stress resistance e.g., aluminium tolerance in carrot, salt tolerance in tobacco and maize. Development of somaclonal variants with herbicide resistance e.g., tobacco resistant to sulfonylurea. Development of seeds with improved quality e.g., a new variety of *Lathyrus sativa* seeds (Lathyrus Bio L 212) with low content of neurotoxin. Bio-13 – A somaclonal variant of Citronella java (with 37% more oil and 39% more citronellon), a medicinal plant has been released as Bio-13 for commercial cultivation by Central Institute for Medicinal and Aromatic Plants (CIMAP), Lucknow, India. Super tomatoes- Heinz Co. and DNA plant Technology Laboratories (USA) developed Super tomatoes with high solid component by screening somaclones which helped in reducing the shipping and processing costs. Micropropagation can be carried out throughout the year independent of the seasons.

Disadvantages

Tissue culture plays an important role in horticulture and forestry industries where clonal uniformity is required during rapid propagation of elite genotypes. In such cases occurrence of somaclonal variation is a great disadvantage.

The problems with use of this technique are: 1) there is a low frequency of variants, so a large-scale experiment is needed, 2) most variants are not desirable, and 3) many of the variants may be epigenetic in nature, and not true variants. In some cases, the resultant variants are due to stable mutations with sexual transmission of the traits to progeny (Chaleff, 1986; Serres *et al.*, 1991) and thus a useful source of variation.

Conclusion

The development of plant cell and tissue culture over the last 20 years has made it possible to transfer part of the breeding work from field to laboratory conditions. Clonal propagation via tissue cultures is in use today by many commercial growers around the world and is playing an increasing role in forest nurseries and the horticultural industry, yielding relatively high economic return. The main challenge for mass production of plants by tissue culture is the production of plants that show genetic fidelity to the donor plant. Long-term cultures does not guarantee genetic fidelity and leads to somaclonal variation among the regenerated plants.

Somaclonal variation in tissue culture is a complex problem that needs several approaches to be appreciated correctly. The use of molecular markers, such as RAPDs, to assess the genetic stability of an *in vitro* production system may be inadequate, and an approach that focuses on morphological traits appears to be a valuable complementary tool.

References

Ahloowalia, B.S. and Maluszynski, M. 2001. Induced mutations-A new paradigm in plant breeding. Euphytica **118**: 167-173

Ahmed, K.Z., Mesterhazy, A., Bartok, T. and Sagi, F. 1996. *In vitro* techniques for selecting wheat (*Triticum aestivum*.L) for *Fusarium resistance* II. Culture filtrate technique and inheritance of *Fusarium I* resistance in the somaclones. Euphytica **91**: 341-34

Anwar, N., Kikuchi, A. and Watanabe, K.N. 2010. Assessment of somaclonal variation for salinity tolerance in sweet potato regenerated plants. African J. Biotech. **9(43):** 7256-7265.

Bairu, M.W., Fennell, C.W. and van Staden, J. 2006. The effect of plant growth regulators on somaclonal variation in Cavendish banana (*Musa* AAA cv. 'Zelig'). Scientia Horticulturae **108**: 347-351.

Chaleff, R.S. 1986. Isolation and characterization of mutant cell lines and plants: herbicide-resistant mutants, In I.K. Vasil (ed.) Cell Culture and Somatic Cell Genetics of Plants, Vol. 3, Plant Regeneration and GeneticVariability. Academic Press, New York. pp.499-512.

Cote, F., Teisson, C. and Perrier, X. 2001. Somaclonal variation rate evolution in plant tissue culture: contribution to understanding through a statistical approach. In Vitro Cell. Develop. Biol. Plant **37**: 539-542.

Damasco, O.P., Smith, M.K., Adkins, S.W., Hetherington, S.E., and Godwin, I.D. 1998. Identification and characterisation of dwarf off-types from micropropagated 'Cavendish' bananas. Acta Horticulturae **490**: 79-84.

Diner, A.M. 1991. *In vitro* disease resistance for expression of somaclonal variation in *Larix*. In M.R. Ahuja (ed.) Woody Plant Biotechnology. Plenum Press, New York. pp.63-65.

Evans D. A. and Sharp W. R. 1986. Somaclonal and Gametoclonal Variation. In: Evans, D.A., Sharp W.R. and Ammirato P.V. (Eds.) "Handbook of plant cell culture". New York: MacMillan. Publishing Company, 1988. v.4, p.97-132.

Hwang, S.C. and Ko, W.H. 1987. Somaclonal variation of banana and screening for resistance of Fusarium wilt. p.157-160. In: G.J. Persley and E.A. De Langhe (eds.), Banana and Plantain Breeding Strategies. ACIAR Proceedings No. 21, ACIAR, Canberra, Australia.

Hwang, S.C., Ko, W.H. and Chao, C.P. 1992. Control of fusarial wilt of Cavendish banana by planting a resistant clone derived from breeding. p.259-280. In: Proceedings of the Symposium on the Non-chemical Control of Crop Diseases and Pests, Plant Prot. Bull. (Taiwan) Special Publication, Taiwan.

Israeli, Y., Lahav, E. and Reuveni, O. 1995. *In vitro* culture of bananas. In: Gowen, S. (ed.) Bananas and plantians. Chapman and Hall, London,.p 147-178.

Israeli, Y., Reuveni, O. and Lahav, E. 1991. Qualitative aspects of somaclonal variations in banana propagated by *in vitro* techniques. Scientia Horticulturae **48:** 71-88.

Joshi, R.K. and Rao, G. J. N. 2009. Somaclonal variation in submergence tolerant rice cultivars and induced diversity evaluation by PCR markers. Intern. J. Genetics Mol. Biol. **1 (5):** 80-88.

Kaeppler, S. and Phillips, R. 1993. DNA methylation and tissue culture induced variation in plants. *In Vitro* Cell. Develop. Biol. Plant **29:** 125-130.

Krikorian, A.D., Irizarry, H., Cronauer-Mitra, S.S. and Rivera, E. 1993. Clonal fidelity and variation in plantain (*Musa* AAB) regenerated from vegetative stem and floral axis tips *in vitro*. Annals Bot. **71:** 519-535.

Larkin, P.J. 1987. Somaclonal variation: history, method, and meaning. Iowa State J. **61:** 393–434.

Larkin, P.J. and W.R. Scowcroft. 1981. Somaclonal variation - a new source of variability from cell culture for plant improvement. Theor. App. Genet. **60:** 197-214.

Lee, M. and Phillips, R.L. 1988. The Chromosomal basis of somaclonal variation. Annual Review Plant Physiol. Plant Mol. Biol. **39:** 413-437.

Leva, A.R., Petruccelli, R. and Rinaldi, L.M.R. 2012. Somaclonal variation in tissue culture: A case study with Olive (http://creativecommons.org/licenses/by/ 3.0), http://dx.doi.org/10.5772/50367

LoSchiavo, F., Pitto, L., Giuliano, G., Torti, G., Nuti-Ronchi, V., Marazziti, D., Vergara, R., Orselli, S. and Terzi, M. 1989. DNA methylation of embryogenic carrot cell cultures and its variations as caused by mutation, differentiation, hormones and hypo methylating drugs. Theor. App. Genet. **77:** 325-331.

Martin, K., Pachathundikandi, S., Zhang, C., Slater, A. and Madassery, J. 2006. RAPD analysis of a variant of banana (*Musa sp.*) cv. grande naine and its propagation via shoot tip culture. Plant **42:** 188-192.

Michler, C.H. and Haissig, B.E. 1988. Increased herbicide tolerance of *in vitro* selected hybrid poplar, pp.183-189. In: M.R. Ahuja (ed.) Somatic Cell Genetics of Woody Plants. Kluwer Acad. Publishers, Boston.

Misra, S. and Gedamu, L. 1989. Heavy metal tolerant transgenic *Brassica napus* L. plants. Theor. App. Genet.**78**: 161-168.

Mohmand, A.S. 1991. Somaclonal variation in some agronomic character in wheat. Acta Horti. **289**: 247-250.

Pedrieri. S. 2001. Mutation induction and tissue culture in improving fruits. Plant Cell, Tiss. Organ Cult. **64**: 185-210.

Pierik, R.L.M. 1987. *In vitro* culture of higher plants. Kluwer Academic Publishers, Dordrecht

Rani, V., Parida, A. and Raina, S.N. 1995. Random amplified polymorphic DNA(RAPD) markers for genetic analysis in micropropagated plants of *Populus deltoids* Marsh. Plant Cell Report **14**: 459-462

Reuveni, O. and Israeli, Y. 1990. Measures to reduce somaclonal variation in *in vitro* propagated bananas. Acta Horticulturae **275**: 307-313.

Rodrigues, P.H.V., Tulmann Neto, A., Cassieri Neto, P. and Mendes, B.M.J. 1998. Influence of the number of subcultures on somoclonal variation in micropropagated Nanico (*Musa spp.*,AAA group). Acta Horticulturae **490**: 469-473.

Sacristan, M.D. 1986. Isolation and characterization of mutant cell lines and plants: disease-resistant mutants. In I.K. Vasil (ed.) Cell Culture and Somatic Cell Genetics of Plants, Vol. 3, Plant Regeneration and Genetic Variability. Academic Press, New York. pp.513-525.

Serres, R., Ostry, M., McCown B. and Skilling, D. 1991. Somaclonal variation in *Populus* hybrids regenerated from protoplast culture, In M.R. Ahuja (ed.) Woody Plant Biotechnology. Plenum Press, New York. pp.59-61.

Sharma, S., Bryan, G., Winfield, M. and Millam, S. 2007. Stability of potato (*Solanum tuberosum* L.) plants regenerated via somatic embryos, axillary bud proliferated shoots, microtubers and true potato seeds: a comparative phenotypic, cytogenetic and molecular assessment. Planta **226**: 1449-1458.

Skirvin, R.M., Norton, M. and McPheeters, K.D. 1993. Somaclonal variation: has it proved useful for plant improvement? Acta Horticulturae **(336)**: 333-340.

Skirvin, R.M., McPheeters, K.D., Norton, M. 1994. Sources and frequency of somaclonal variation. HortScience **29**: 1232-1237.

Soniya, E.V., Banerjee, N.S., Das, M.R. 2001. Genetic analysis of somaclonal variation among callus-derived plants of tomato. Cur Sci. **80**: 1213-1215

Stover, R.H. 1987. Somaclonal variation in Grande Naine and Saba bananas in the nursery and in the field. In: Persley GJ, De Langhe E (eds.) ACIAR proceeding no. 21, Canberra.

Swartz, H.J. 1991. Post culture behaviour, genetic and epigenetic effects and related problems. In: Debergh, P.C. and Zimmerman, R.H. (eds.) Micropropagation: technology and application. Dodrecht : Kluwer Academic Publishers;.p 95-122.

Witjaksono. 2003. Peran bioteknologi dalam pemuliaan tanaman buah tropika. Seminar Nasional Peran Bioteknologi dalam PengembanganBuah Tropika. Kementerian Riset dan Teknologi RI & Pusat Kajian Buah Buahan Tropika, IPB. Bogor, 9 Mei 2003.

Yusnita, Widodo, and Sudarsono. 2005. *In vivo* selection of peanut somatic embryos on medium containing culture filtrates of *Sclerotium rolfsii* and plantlet regeneration. Hayati **12 (2):** 50-56.

Chapter 7

In vitro Multiplication of *Pronephrium triphyllum* (Sw.) Holttum - An Endangered Fern

M. Johnson and V.S. Manickam

Introduction

The present study was aimed to produce a reproducible protocol for the large scale multiplication of an endangered fern *Pronephrium triphyllum* (Sw.) Holttum via *in vitro* organogenesis. The spore derived young croziers were harvested and used as explant source. The croziers were cut into small pieces and cultured on different media augmented with various concentrations and combinations of plant growth regulators. Murashige and Skoog's medium supplemented with 2, 4-D 1.0 mg/l induced maximum percentage (68.3 ± 1.43) of callus formation. Highest frequency (68.3 ± 1.43) of shoot formation was observed in Murashige and Skoog's medium augmented with Kinetin 1.0 mg/l and Naphthalene Acetic Acid 0.5 mg/l. Half strength Murashige and Skoog's Medium augmented with IBA 1.0 mg / l induced maximum number (8.1 ± 1.31) of roots formation with high frequency (71.3 ± 1.81), followed by IBA 2.0 mg/l. The micropropagated plants showed maximum percentage (78.6 ± 1.24) establishments during hardening and 76.8 ± 0.62 percentage establishments in the field at Kodaikannal Botanic Garden, at Kodaikannal, Tamil Nadu, and India.

The process of regenerating an entire plant from a single cell is known as totipotency. But for this regeneration ability of the plants cell, many exciting results in plant tissue culture and genetic engineering could not have been achieved. The emergence of modern biotechnology presents an important approach for establishing a link between conservation and sustainable utilization of genetic diversity. Plant tissue culture has been viewed as a key technology for enhancing the capability for the production of large quantities of good quality plants. The plant cell, tissue and organ culture techniques have emerged as an inseparable tool with possibilities for complementing and supplementing the conventional methods of plant breeding. Plant tissue culture has emerged as a powerful tool with the potential not only for rapid multiplication of plant species but also for conservation of rare and endangered ones. Different pathways of *in vitro* culture techniques have been developed not only to achieve faster propagation, but also to unravel intricacies of morphogenesis involved in these processes. The pteridophytes, offer a vast scope of morphogenetic studies as the experimental work can go well beyond the mere callus formation and differentiation and cover the important phenomena of induced apogamy and apospory vis a vis nutritive environment. Application of tissue culture methods not only can increase the sporophyte production but also can provide usefull insights into fern biology. Differentiation and morphogenesis directly depend on the nutrients and growth regulators in the nutrient milieu. Micropropagation of any plant material depends on the appropriateness of the nutrients supplied in the culture medium. Any change in the medium results in variation in response of the explants. Not only nutrients, but also various other factors affecting plant regeneration from excised cells or tissues/callus have been studied (George *et al.*, 1996). Formation and maintenance of callus cultures in vascular cryptogams have also been successfully achieved. Callusing has been induced from gametophytes in certain species (Mehra and Sulklyan, 1969; Kato, 1969; Vallinayagam, 2003; Johnson and Manickam, 2010).

Abbreviations: MS – Murashige and Skoog's medium, KC (M) N – Knudson C medium modified with Nitsch's trace elements, ½ MS – Half strength Murashige and Skoog medium, PGRs – Plant Growth Regulators, 2, 4 – D - 2, 4-Dichlorophenoxy acetic acid, 2, 4, 5 – T - 2, 4, 5 - Trichlorophenoxy acetic acid, 2 ip - 2 isopentenyl adenine, BAP - Benzyl - 6 - Amino Purine, CPA - Trichlorophenoxy acetic acid, Kin - Kinetin, NAA - Naphthalene Acetic Acid, IBA – Indole-3-Butyric Acid , IAA – Indole-3-Acetic acid.

Success of tissue culture is not based on the discovery of suitable explants and development of appropriate cultural technique but on with application. There are many reports available on callus induction from different explants such as gametophyte, young leaves, stolon tips, stipes, rhizome segments etc. A lot of research has been carried out on callus formation from rhizome (Kshirsagar and Mehra, 1978), roots (Mehra and Palta, 1971); runner segments and leaves (Cheema, 1983) of pteridophytes. The regeneration of shootlets from callus derived from gametophyte and sporophyte was reported by Cheema (1983), Cheema and Sharma (1994), Cheema and Kaur (1986), Padhya (1985, 1987), Byrene and Caponneti (1992) and Kwa *et al.* (1997). But there is no report on the callus production and regeneration of sporophyte from the crozier derived calli of *Pronephrium triphyllum* (Sw.) Holttum. To fuflill the lacuna the present investigation was aimed to produce a reporducible protocol for the calli mediated sporophyte formation from the croziers of *Pronephrium triphyllum* (Sw.) Holttum.

Material and Methods

The spore derived croziers of *Pronephrium triphyllum* (Sw.) Holttum were harvested and served as the source of explants. The young sporophytes were cut into small segments (1 cm in length) before implanted on medium aseptically. The croziers were cultured on 0.5% (w/v) agar gelled MS's medium and KC(M)N medium augmented with various concentrations (0.5 mg/l - 2.5 mg/l) and combinations of plant growth regulators viz., BAP, Kin, IAA, NAA, 2, 4-D, 2, 4, 5-T, and CPA. The cultures were incubated under 12h / photoperiod at 25° ± 2° C. For sporophyte formation, the crozier derived calli were sub - cultured in the same medium augmented with various concentrations and combination of PGRs. After few weeks the *in vitro* raised sporophytes were sub - cultured on to root inducing medium with different concentrations of auxins. The culture tubes containing tissue raised micropropagated plants of *P.triphyllum* were kept at room temperature (30-32°C) for a week before transplantations. For acclimatization, the plants with well developed roots (5-8 cms) were removed from culture tubes, washed in running tap water to remove the remnants of agar and each group was planted separately onto 10 cm diameter polycup filled with different potting mixtures - river sand, garden soil and farm yard manures (1 : 1 : 1) and sand and soil (2 : 1). The plants were kept in mist chamber with a relative humidity of 70%. Plants were irrigated at 8h intervals for 3-4 weeks and establishment rate was recorded. The plantlets established in community pots were transferred to shade net house for 3-4 weeks and then repotted in larger pots (20 cm diameter) with one plant in each pot then transferred to its native habitat and also to the natural forest segment at KBG, Kodaikannal. For cytological analysis, the *in vitro* raised *P. triphyllum* immature sporangia in fertile fronds and young croziers were squashed in acetocarmine after being fixed in 1 : 3 : 6 mixture of glacial acetic acid, chloroform and 100% ethyl alcohol for 24 hours and then preserved in 95% ethyl alcohol. Meiotic and mitotic chromosomes were observed in several cells for establishing the correct counts.

Results and Discussion

Influence of PGRs on *in vitro* raised sporophyte

The *in vitro* raised sporophytes were subcultured on KC (M) N and MS medium supplemented with 3% sucrose and various concentrations of PGRs (2, 4-D and 2, 4, 5-T) for callus induction. The MS medium supplemented with 2, 4-D 1.0 mg/l induced maximum amount of callus formation at high frequency (68.3 ± 1.43). The KC (M) N medium supplemented with 2, 4-D 1.0 mg/l also induced high percentage (52.1 ± 1.48) of callus formation compared to other PGRs (Table 7.1).

Table 7.1: Influence of PGRs on callus induction of *P. triphyllum in vitro* raised sporophyte cultured on KC (M) N and MS medium

KC (M) N + PGRs	MS + PGRs	% of Callus formation in MS medium	% of Callus formation in KC (M) N medium
Basal	-	-	-
2, 4-D 0.5	0.5	33.8 ± 1.21	-
1.0	1.0	**68.3 ± 1.43**	52.1 ± 1.48
1.5	1.5	62.1 ± 1.31	43.1 ± 1.21
2.0	2.0	48.1 ± 1.24	-
2.5	2.5	43.1 ± 1.38	-
2, 4, 5-T 0.5	0.5	-	-
1.0	1.0	38.3 ± 2.13	-
1.5	1.5	-	-
2.0	2.0	-	-

Influence of PGRs on sporophyte derived calli

The crozier derived calli were subcultured on MS medium supplemented with various concentrations and combinations of PGRs for shoot induction. The high frequency (68.3 ± 1.43) of sporophyte proliferation was observed in MS medium supplemented with Kin 1.0 mg/l and NAA 0.5 mg/l (Fig. 1 C-F). The MS medium supplemented with Kin 1.0 mg/l and NAA 0.5 mg/l induced maximum number of sporophytes also. In media with other hormones, BAP 1.0 mg/l, Kin 1.0 mg/l, 2iP 1.0 mg/l there was formation of gametophyte (Aposporous gametophyte) formation in which sporophyte formation was not observed (Table 7.2 ; Fig. 7.1 A, B, G-I).

Table 7.2: Influence of PGRs on shoot differentiation in sporophyte derived calli of *P. triphyllum* on MS medium

MS medium + PGR (mg/l)	% of Sporophyte medium ±S.D.	% of Gametophyte formation ±S.D.
Basal	-	-
BAP 0.5	-	61.3 ± 1.21
1.0	-	68.4 ± 0.93
1.5	-	60.3 ± 1.38
Kin 0.5	-	73.1 ± 1.21
1.0	58.3 ± 1.38	**71.3 ± 1.43**
1.5	43.8 ± 1.42	68.3 ± 1.38
2iP 0.5	-	59.3 ± 1.48
1.0	-	60.8 ± 2.13
Kin 1.0 & NAA 0.5	**68.3 ± 1.43**	-
Kin 1.0 & NAA 1.5	60.3 ± 1.21	-

Influence of PGRs on Root Formation

The *in vitro* raised sporophytes were subcultured on ½ MS medium and KC (M) N medium supplemented with auxins in various concentrations and combinations for root formation. Half strength MS medium supplemented with IBA 1.0 mg/l induced maximum number (8.1 ± 1.31) and highest percentage (71.3 ± 1.81) of root formation. NAA alone or in combination with IBA induced only basal callus formation without being followed by root formation (Table 7.3).

Table 7.3: Influence of PGRs on *in vitro* raised sporophyte of *P. triphyllum* for root formation

½ MS medium + Auxins (mg/l)	% of Root formation ± S.D.	Callus formation	Mean no. of roots ± S.D.
Basal	-	-	-
IBA 1.0	**71.3 ± 1.81**	-	8.1 ± 1.31
IBA 2.0	58.3 ± 1.26	-	3.4 ± 1.41
IAA 1.0	-	+++	-
IAA 2.0	-	+++	-
NAA 1.0	-	+++	-
NAA 2.0	-	+++	-

+ Sign indicates callus formation.

Hardening of micropropagated plants

After 15 days of rooting, the *in vitro* derived plantlets were washed thoroughly in running tap water to remove the pieces of agar adhered to the roots and implanted in the pots containing a mixture of (1:2:1) sterile soil : sand : farmyard manure irrigated with 10X diluted MS / KC liquid medium once in a week. The pots were covered with poly bags to maintain the humidity. The plantlets were kept in culture room for 15 days. After that, they were transferred to green house (R.H. 80%) under constant misting. After 3 weeks the plants were transferred to the field. The micropropagated plants showed 73.5 ±1.24% establishments during hardening and 72.3 ± 1.24% establishment in the field at KBG. Subsequently the micropropagated plants were distributed to various botanic gardens for *ex situ* conservation. Cytological studies of ten randomly selected plants established in KBG revealed the presence of 144 chromosomes in *P. triphyllum* and confirmed the genetic uniformity.

Fig 7.1. *In vitro* multiplication of *Pronephrium triphyllum* (Sw.) Holttum

A - Aposporous gametophyte proliferation from crozier derived calli; B - Sporophyte formation from crozier derived calli; C-F - Different stages of sporophyte formation from crozier derived calli; G - Aposporous gametophyte formation from crozier derived calli; H - Rhizoid formation from the crozier deirved calli; I - Sporophyte formation from aposporous gametophyte

The age of the explant greatly influenced its morphogenetic capacity (Padhya and Mehta, 1982). The youngest croziers exhibited highest ability for morphogenesis (Padhya, 1987) which is substantiated in the present study as well. The ability of the crozier derived calli to produce microplants free of abnormalities indicates the application of tissue culture technology for conservation of the rare ferns. So far only a few species of ferns have been successfully tested for conservation through *in vitro* means. Cheema and Kaur (1986) reported *in vitro* regeneration and clonal propagation of some aquatic ferns. Padhya (1987) reported mass clonal propagation of few species of *Nephrolepis* and *Thelypteris* through tissue culture. The reports available show callus induction from rhizome segments (Kshirsagar and Mehra, 1978), roots (Mehra and Palta, 1971) and leaves (Cheema, 1983). There are number of recent publications on the successful recovery of endangered plant taxa through the mediation of *in vitro* multiplication (Vallinayagam, 2003; Manickam *et al.*, 2003; Johnson and Manickam, 2006, 2010, 2011, 2012). Morphogenesis in cultured tissues and cells is regulated by plant hormones, especially auxins and cytokinins (Skoog and Miller, 1957; D'silva and D'souza, 1992; Wysosmika, 1993). In the present study, media supplemented with Kin 1.0 mg/l combined with NAA 0.5 mg/l induced shootlet formation which is similar to the findings of Hicks and Aderkas (1986); Harper, (1976); Beck and Caponetti (1983) and Johnson and Manickam (2011, 2012). Beck and Caponetti (1983) reported that kinetin in the medium promoted leafy shoot development. In the present investigation also similar observations were made in *P. triphyllum* callus. The media supplemented with kinetin alone promoted the gametophyte formation. Many studies have been conducted wherein morphological traits, karyotypic constitution and isozyme profiles are employed for studying and monitoring the variations arising in the tissue cultured plants (Maheswaran and Williams, 1987; Lakhanpaul *et al.*, 1991; Oh *et al.*, 1995, Nair, 2000; Johnson and Manickam, 2006; Johnson and Manickam, 2010; 2011; 2012). All the plants raised through tissue culture were morphologically uniform and showed uniform growth. The cytological stability and genetic uniformity of the micropropagated plants raised through tissue cultures were then tested through cytological analysis. Chromosome analysis in both micrpropagated and mother plants from the wild indicated the cytological uniformity of both with same chromosome numbers. However, there was no focus on conservation in these earlier studies. Since the species selected are important for conservation, the methods developed could now be employed for large scale multiplication for reintroduction and restoration. Murashige (1974) successfully propagated many varieties of ferns on mass scale employing tissue culture technology. Padhya (1987) successfully developed protocols for three ferns through tissue culture. The ability of the crozier derived calli to produce microplants free of abnormalities indicates the application of tissue culture technology for conservation of the rare ferns.

Acknowledgement

The authors sincerely acknowledge the Ministry of Environment and Forest, New Delhi for their financial support.

References

Beck, M.J. and Caponetti, J.D. 1983. The effects of Kinetin and Naphthalene acetic acid on *in vitro* shoot multiplication and rooting in the fishtail Fern. Amer. J. Bot. **70(1):** 1-7.

Byrne, T.E. and Caponetti, J.D. 1992. Morphogenesis in three cultivars of Botson fern II. Callus production from stolon tips and plantlet differentiation from callus. Amer. Fern Jl. **82(1):** 1-11.

Cheema, H.K. 1983. *In vitro* studies on callus induction and differentiation of gametophytes and sporophytes in ferns. In Verma, S.C. and Sare, T. S., (eds.) In National seminar on progress in Botanical Research. Bot. Dept. Punjab University. India, pp. 44-46.

Cheema, H.K. and Kaur, M. 1986. *In vitro* regeneration and effects of growth regulators on aquatic heterosporous fern. Res. Bull Punjab Uni. Sci. **36:** 35-37.

Cheema, H.K. and Sharma, M. B. 1994. Induction of multiple shoots from adventitious buds and leaf callus in *Ceratopteris thalictroides*. Ind. Fern J. **11:** 63-67.

D' Silva. and D' Souza, L. 1992. *In vitro* propagation of *Anacardium occidentale* L. Plant Cell Tiss. Org. Cult. **29:** 1-6.

George, E.F. 1996. Plant propagation by tissue culture - Part 2. Exegetics Ltd. Edington, England, pp. 1 -300.

Harper, K.L. 1976. Asexual multiplication of Leptosporangiate ferns through tissue culture, M.S. thesis, University of California.

Hicks, G. and Aderkas, P. V. 1986. A tissue culture of the ostrich fern *Matteuccia struthiopteris* (L.) Todaro. Plant Cell Tiss. Org. Cult. **5:** 199-204.

Johnson, M. and Manickam, V.S. 2006. *In vitro* studies on normal and abnormal life cycle of *Metathelypteris flaccida* (Bl.) Ching. Ethiopian J. Sci. Tech. **4(1):** 37-44.

Johnson, M. and Manickam, V.S. 2010. Influence of Plant Growth Regulators on *in vitro* raised gametophytes of *Cheilanthes viridis* (Forssk.) Swartz. J. Basic & Applied Biol. **4(4):** 18-23.

Johnson, M. and Manickam, V.S. 2011. Influence of media, sucrose and plant growth regulators on *in vitro* apospory induction and developmental changes in *Speharostephanos unitus*. J. Basic & Applied Biol. **5(1&2):** 344-351.

Johnson, M. and Manickam, V.S. 2012. *In vitro* organogenesis studies of *Cheilanthes viridis* – an endangered fern. Proceedings of the National conference on "Climate change, Biodiversity & Conservation". Edited by Rajkumar, S.D. Samuel, C. O. & Lal, J. K. IJBT Special Issue 2012. Gayathri Teknological Publication, Palayamkottai, India, **2**: 76-88.

Kato, Y. 1969. Physiological and morphogenetic studies of fern gametophytes by aseptic culture VII Experimental modification of dimensional growth in gametophytes of *Pteris vittata* L. Phytomorphology **19**: 114-121.

Kshirsagar, M.K. and Mehra, A.R. 1978. *In vitro* studies in ferns: growth and differentiation in rhizome callus of *Pteris vittata* L. Phytomorphology **28**: 50-58.

Kwa, S. H., Wee, Y. C. Lim, T.M. and Kumar, P. P. 1997. Morphogenetic plasticity of callus reinitiated from cell suspension cultures of the fern *Platycerium coronarium* (Koenig) Desv. Plant Cell Tiss. Org. Cult. **48**: 37-44.

Lakhanpaul, S., Mandal, B. B. and Chandel, K. P. S. 1991. Isozyme studies in the *in vitro* regenerated plants of *Ipomoea batatas* (L.) Lamarck. J. Root Crops. **17**: 305-310.

Maheswaran, G. and Williams, E. G. 1987. Uniformity of plants regenerated by direct somatic embryogenesis from zygotic embryos of *Trifolium repens*. Ann. Bot. **59**: 93-97.

Manickam, V.S., Vallinayagam, S. and Johnson, M. 2003. Micropropagation and conservation of rare and endangered ferns of the Southern Western Ghats through *in vitro* culture. Pteridology of New Millennium. Kluwer Academic Publishers, Netherlands, pp. 497 – 504.

Mehra, P. N. and Palta, H. K. 1971. *In vitro* controlled differentiation of the root callus of *Cyclosorus dentatus*. Phytomorphology **21**: 367-375.

Mehra, P. N. and Sulklyan, D. S. 1969. *In vitro* studies on apogamy, apospory and controlled differentiation of rhizome segments of the fern *Ampelopteris prolifera* (Retz.) Copel. Bot. J. Linn. Soc. **62**: 431-443.

Murashige, T. 1974. Plant propagation through tissue culture. Ann. Rev. Plant Physiol. **25**: 135-166.

Nair, L. G. 2000. Conservation through micropropagation restoration of selected woody medicinal plants. Ph.D. Thesis submitted Kerala University, Thiruvananthapuram, Kerala, India.

Oh, M. H., Choi, D. W., Kwon, Y. M. and Kim, S. G. 1995. An assessment of cytological stability in protoplast cultures of tetraploid. Plant Cell Tiss. Org. Cult. **41**: 243-248.

Padhya, M. A. 1995. Rapid propagation of *Adiantum trapeziformae* L. through tissue culture. Indian Fern J. **12**: 20-23.

Padhya, M. A. 1987. Mass propagation of ferns through tissue culture. Acta Hort. **212**: 645-648.

Padhya, M. A. and Mehta, A. R. 1982 Propagations of fern (*Nephrolepis*) through tissue culture. Plant cell Rep. **1**: 261-263.

Skoog, F. and Miller, C. O. 1957. Chemical regulation of growth and organ formation in plant tissues cultured *in vitro*. Sym. Soc. Exp. Biol. **11**: 118-131.

Vallinayagam, S. 2003. *In vitro* propagation of some rare and endangered ferns of Western Ghats, South India Ph. D., Thesis. Manonmaniam Sundaranar University Tirunelveli, Tamil Nadu, India.

Wysokinska, H. 1993. Micropropagation of *Penstemon serrulatus* and iridoid formation in regenerated plants. Plant Cell, Tiss. Org. Cult. **33:** 181-186.

Chapter 8

Effect of Plant Growth Regulators on *in vitro* Raised Gametophytes of *Phlebodium aureum* L.

M. Johnson and V.S. Manickam

Introduction

The present study was carried out to study the effect of plant growth regulators on gametophyte multiplication and calli induction, direct and calli mediated sporophyte formation on *Phlebodium aureum*. Highest percentage (75.8 ± 1.38) of gametophyte multiplciation was observed in Mitra medium fortified with 1.0 mg/l of Naphthalene acetic acid. Highest percentage (60.3 ± 1.3) of callus proliferation was obtained in Murashige and Skoog's medium augmented with 2, 4-Dichlorophenoxy acetic acid 1.0 mg/l. Highest percentage (73.8 ± 0.93) of sporophyte formation from *in vitro* gametophyte raised calli was observed in Murashige and Skoog medium supplemented with Kinetin 1.0 and Naphthalene Acetic Acid 0.5 mg/l. Highest percentage (73.1 ± 0.68) of rooting of the *in vitro* raised sporophytes was observed in half strength Murashige and Skoog medium fortified with 0.5 mg/l of Indole-3-Butyric Acid in combination with 0.5 mg/l of Naphthalene acetic acid.

Among the various biotechnological options, also reported in other agri-horticultural crops, micropropagation through tissue culture and *in vitro* spore germination are best applied and commercially exploited in fern species (Fay, 1994). Application of this technology (*in vitro* spore germination) for large-scale multiplication of certain species of ferns from the Western Ghats has been demonstrated (Sara *et al.*, 1998). That plant tissue culture as an effective tool to conserve plant genes and guarantee the survival of the endemic, endangered and over exploited genotypes is derived from the fact that, it makes use of small units (cells and tissues) without losing their mother plant, takes pressure off the warning wild populations and makes available large number of plants for reintroduction and commercial delivery. Tissue culture of fern species in India has been reviewed by Cheema (1997). Endangered ferns such as *Cheilanthes viridis, Pronephrium triphyllum, Sphaerostephanus unitus, Diplazium cognatum, Histiopteris incisa, Hypodematium crenatum, Thelypteris confluens, Athyrium nigripes, Pteris vittata* and *Cyathea crinita* have been multiplied *in vitro* for conservation through *in vitro* spore culture (Sara, 2001; Manickam *et al.*, 2003; Vallinayagam, 2003; Johnson and Manickam, 2011). Tissue culture technique can be used to induce shoot formation directly or by inducing callus formation and regeneration of shoots/roots from the callus. Tissue culture is a modern biotechnological innovation for rapid propagation of large number of uniform plants. Application of this method is the best option to overcome these problems and now it became a linch-pin to find many solutions to the stringent biological constrains and environmental uncertainties, wherein considerable efforts might lead to the production of high yielding improved plant types, production of secondary metabolites under *in vitro* conditions and practical for conservation of threatened and endangered species thereby guaranteeing the survival of remaining plants in the wild. The regeneration of shootlets from callus derived from gametophyte and sporophyte was reported by Cheema (1983), Cheema and Sharma (1994), Sara (2001), Cheema and Kaur (1986), Padhya (1985, 1987) and Kwa *et al.* (1991, 1995, 1997). Manickam *et al.*, (2003) studied the spore germination, gametophyte

---

Abbreviations: KC- Knudson C medium, KC (M) N – Knudson C medium modified with Nitsch's trace elements, KN – Knop's medium, Mi – Mitra medium, Mo – Moore's medium, MS – Murashige and Skoog's medium, ½ MS – Half strength Murashige and Skoog medium, PGRs – Plant Growth Regulators, 2ip - , 2, 4 – D - 2, 4-Dichlorophenoxy acetic acid, 2, 4, 5 – T - , BAP – Benzyl – 6 – Amino Purine , Kin - Kinetin, NAA - Naphthalene Acetic Acid, IBA – Indole-3-Butyric Acid , IAA – Indole-3-Acetic acid.

development and sporophyte formation of *Phlebodium aureum* L. and Johnson and Manickam (2010) studied the *in vitro* proliferation of secondary and tertiary gametophytes from the spore derived gametophytes of the fern *Phlebodium aureum*. But there is no report on the effect of plant growth regulators on spore derived gametophytes, differentiation of gametophytes, calli production and calli mediated sporophytes formation of *Phlebodium aureum*. To fulfill the lacuna the present study was initiated to study the impact of plant growth regulators on spore derived gametophytes, differentiation of gametophytes, calli initiation and direct and calli mediated sporophytes formation of *Phlebodium aureum*.

Materials and Methods

For morphogenesis, the *in vitro* spore derived gametophytes (Johnson and Manickam, 2011) were sub-cultured on different media Knudson C, Knops, MS, Mi and Knudson C modified medium with Nitsch's trace elements (KC (M) N) supplemented with 3% sucrose, 0.6% (w/v) agar (Hi-Media, Mumbai) and different concentration and combination of BAP, Kin, 2, 4, 5 –T, 2-iP, NAA, IAA and 2, 4 - D. The pH of the medium was adjusted to 5.8 before autoclaving at 121°C for 15min. The cultures were incubated at 25 ± 2°C under cool fluorescent light (1500 lux 14 hr photoperiod). The gametophyte derived callus was sub-cultured onto different media supplemented with different concentration and combinations of PGRs for sporophytes formation and calli proliferation. The *in vitro* raised gametophytes derived calli were sub – cultured on different concentration and combinations of PGRs supplemented MS medium for the organogenesis development. For rooting, the calli mediated sporophytes were inoculated onto half strength KC medium fortified with different concentration and combinations of auxins. For acclimatization, the plants with well developed roots (5-8 cms) were removed from culture tubes, washed in running tap water to remove the remnants of agar and each group was planted separately onto 10 cm diameter polycup filled with different potting mixtures river sand, garden soil and farm yard manures (1 : 1 : 1) and sand and soil (2 : 1). The plants were kept in mist chamber with a relative humidity of 70%. Plants were irrigated at 8h intervals for 3-4 weeks and establishment rate was recorded. The plantlets established in community pots were transferred to shade net house for 3-4 weeks and then repotted in larger pots (20 cm diameter) with one plant in each pot then transferred to its native habitat and also to the natural forest segment at KBG, Kodaikannal, Tamil Nadu, India.

Results and Discussion

In vitro raised gametophytes were transferred to Mi medium, KC (M) N medium and MS medium supplemented with various PGRs at different concentrations and none of them produced sporophytes. The PGRs supplemented media enhanced gametophyte multiplication. Highest percentage (75.8 ± 1.38) of gametophyte multiplication was observed in Mi + NAA 1.0 mg/l (Table 8.1). Media supplemented with PGRs not only enhanced gametophyte multiplication, but also induced callus induction and sporeling formation from the gametophyte derived callus. MS medium containing 2, 4- D induced more proliferation (60.7 ± 1.3 %) of callus than other PGRs (Fig. 8.1 A - E). The gametophyte-derived callus was subcultured on MS medium supplemented with various PGRs for sporophyte regeneration. Supplementation of Kin 1.0 mg/l and NAA 0.5 mg/l induced high percentage of sporophyte formation (73.8 ± 0.93) (Fig. 8.1 F; Table 8.2). *In vitro* derived shoots formed high percentage (73.1 ± 0.68) of roots

Table 8.1: Effect of PGRs on Gametophyte of *P.aureum*

Medium + PGRs (mg/l)	% of gametophyte multiplication ±S.D.	% of callus formation +S.D.
Basal	-	-
Mi + 2, 4-00.5	53.1 +1.28	-
Mi + 2, 4-D 1.0	59.3 +2.18	-
Mi +NAA 1.0	**75.8±1.38**	-
Mi + Kin 0.5	63.8 ±1.28	
Mi + Kin 1.0	65.6 ±1.43	-
Mi + BAP 0.5	59.1 ±1.28	-
Mi + 2, 4, 5-T 1.0	61.3 ±1.43	-
Mi + 2, 4, 5-T 0.5	58.1 ±1.43	43.8 ±1.4
Mi + 2, 4, 5-T 1.0	53.1 +1.28	48.3±1.6
MS + 2, 4-D 0.5	53.8 ±1.28	51.4 ±0.8
MS + 2, 4-D 1.0	40.7 ±1.31	**60.7 ±1.3**
MS +NAA 0.5	-	45.6 ± 0.83
MS + NAA 1.0	-	48.7 ± 0.72
MS + Kin 0.5.	42.1 ±1.28	48.1 +1.2
MS + Kin 1.0	48.1±1.38	54.3 ±l. 4
KC (M) N + 2, 4-D 0.5	54.3 ±1.24	50.3 ±1.3
KC (M) N + 2. 4-D 1.0	58.3 ±1.28	53.1 ±1.4
KC (M) N + 2, 4-D NAA 0.5	-	-
KC (M) N + 2, 4-D NAA 1.0	-	-
KC (M) N + 2, 4-D Kin 0.5	38.3±1.41	
KC (M) N + 2, 4-D Kin 1.0	43.8 :±.63	-

when they were cultured only on ½ MS medium supplemented with IBA, 1.0 and NAA 0.5 mg/l (Table 8.3). The frequency and number of shoot formation were varied between concentrations and combinations of PGR(s) tested. In addition to shoot formation, the PGRs had profound effect on gametophyte multiplication and also callus formation.

Table 8.2: Influence of PGR on gametophyte, sporophyte formation and callus induction on gametophyte derived calli of *P. aureum* cultured on MS medium

MS Medium + PGRs (mg/l)	% of Sporophyte formation ±S.D	% of Gametophyte formation ±S.D
Kin 1.0 + NAA 0.5	73.8 ±0.93	-
Kin 0.5	-	48.3 + 1.26
Kin 1.0	-	61.3 ±1.38
Kin 1.5	-	52.3+ 1.34
NAA 0.5	35.8 ±0.84	48.3 ±1.26
AA 1.0	-	58.7±1.24
BAP 0.5	-	52.3 ±1.28
BAP 1.0	-	60.8±1.43
2iP 0.5	-	52.6 ±1.31
2iP 1.0	43.1 ±1.38	58.4 ±1.61
2iP 1.5	32.8 + 1.43	48.6 +1.21

Pteridophytes, because of their distinct alternation of gametophytic and sporophytic generations offer an excellent material for morphogenetic studies. In the present study also we observed the alternation of gemetophytic and sporophytic generation in the endangered fern *P. aureum* with reference to the supplmentation of hormones on the cultured media. The completion of *in vitro* life cycle and raising whole plants in sterile culture have been attempted in few ferns (Cheema, 1984; Cheema and Kaur, 1986; Nigam *et al.*, 1991; Hegde and D' Souza, 1999; Sara et al., 1998, 2001; Vallinayagam 2003; Johnson and Manickam, 2010, 2011, 2012). Contributions on the developmental biology of ferns through *in vitro* spore culture are many. While majority of these works focus on spore germination and gametophyte development of ferns, few reports deals with *in vitro* sporophyte formation. Already a number of reviews on morphogenetic studies on ferns have been published (Bir and Anand, 1982; Cheema, 1997). In the middle of 20th century, fern spores and gametophytes were subjected to various experimental studies to understand the phenomena of growth and development. Bulk of morphogenetic and experimental investigations carried out deal with the reproductive and developmental physiology of spores and gametophyte (Bir, 1987; Hickok *et al.*, 1987).

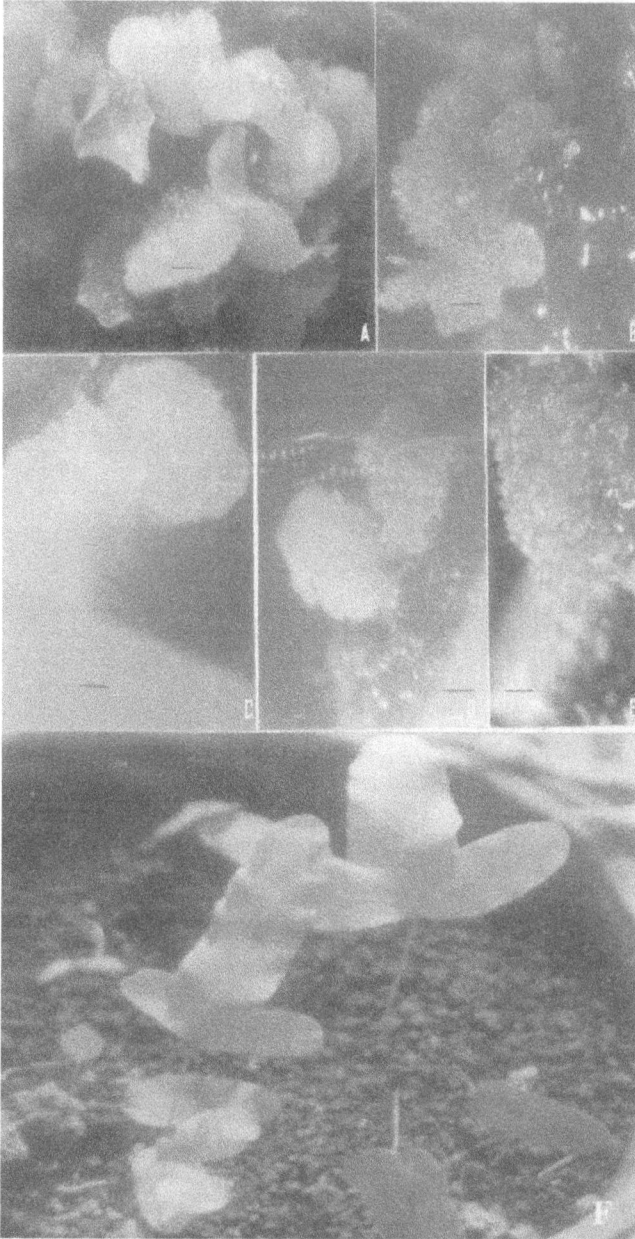

Fig. 8.1. Influence of PGR on the spore derived gameophytes of *P. aureum* L.

A	-	Gametophyte derived Yellow colour globular embryogenic calli
B	-	Gametophyte derived Green colour calli
C	-	Close up view of Yellow colour globular embryogenic calli
D & E	-	Gametophyte derived Green colour calli – Green colour calli
F	-	Sporophyte formation from the gametophyte derived calli

Table 8.3: Influence of PGR on root formation on the *in vitro* raised Sporophytes of *P. aureum* in 12 MS medium

1/2 MS Medium + PGRs (mg/l)	% of Root formation ±S.D.	Mean no. of roots ±S.D.
IBA 1.0	58.1 ±1.81	3.1 +0.83
IAA I.0	43.1 ±1.64	2.1±I.1
IBA and NAA 0.5	73.1 ±O.68	5.1 ±I.28

Normal plant growth and development will not take place without exceedingly small quantities of specific, internally induced plant hormone (Overbeek, 1968). Despite a vast role of hormones, in plant growth and development, information on hormonal control is largely arrived at through trial and error mechanism. The "mechanism of action" of auxins and cytokinin in organogenesis is still a mystery (Hicks, 1994). Even now almost nothing is known regarding the signal transduction pathway operating during events of embryogenesis, shoot bud formation and also during microscope visibility (Dey *et al.*, 1998). Cell differentiation involves activation of certain genes and repression of others, which control different basic metabolic pathways. Thus network of genes and their products play a crucial role in cell division and differentiation. Plant growth regulators have been used to induce callus formation, regeneration of shoots and roots (Sara, 2001; Manickam *et al.*, 2003). In the present study also the callus formation, sprophyte formation and gametophyte formation was stimulated by the plant growth regulators supplemented the cultured media. In most cases, growth regulators have been used over a wide range of concentrations and in numerous combinations, but quantitative data on their effects are scanty (Minocha, 1987). In the present study we witnesed the effect of hormononal combination, the MS medium augmented with Kin 1.0 + NAA 0.5 showed the sporophyte formation. Each type of plant growth regulators may have a wide range of physiological effects in different plants. These effects are determined by the type of the growth regulators, concentrations, presence or absence of other growth regulators, and by genetic make up and the physiological status of the target tissue (Einset, 1991). The results of the present study also supplemented and supported the Einset observations (Table 1-3). Regeneration may also occur spontaneously without exogenous addition of PGR in some such species as *Cyathea spinulosa* (Nigam *et al.*, 1991), *Histiopteris incisa, Pteris vittata* (Vallinayagam, 2003). Generally, a growth regulator, which elicits a positive response in a given tissue at a given concentration, may inhibit the same physiological response when used at higher concentration. Thus several important factors make it difficult to present a concise and critical review of the role of plant growth regulators in cell and tissue cultures. Several auxin regulated genes have been characterized and their possible roles in different cellular processes have been determined (Takahashi *et al.*, 1995). Still, it is not clear whether hormones primarily influence differentiation by activating early response genes or are involved at relatively late stages during morphogenesis (Dey *et al.*, 1998). The results of the present study may promote large-scale multiplication of the endangered fern *P. aureum* to compensate its depletion in nature.

Acknowledgement

The authors sincerely acknowledge the Ministry of Environment and Forests, New Delhi for their financial support.

References

Bir, S.S. 1987. Pteridophytic flora of India: Rare and endangered elements and their conservation. Indian Fern J. **4 (2):** 95-101.

Bir, S.S. and Anand, M. 1982. Morphogenetic studies on pteridophytes in India, In: Aspects of Plant Sciences 6, Today & Tomorrow's Printers & Publishers, New Delhi, pp. 105-118.

Cheema, H.K. 1983. *In vitro* studies on callus induction and differentiation of gametophytes and sporophytes in ferns in: National seminar on progress in Botanical Research (Eds.) Verma, S.C. and Sarein, T. S. Bot. Rept. Punjab Uni. Chd, pp. 44-46.

Cheema, H.K. 1984. *In vitro* studies on reproductive biology and regeneration of the fern *Ceratopteris pteridoides* (Hook) Hieron. Res. Bull. Punjab. Uni. Sci. **35:**13-17.

Cheema, H.K. 1997. Ferns as an excellent experimental system for morphogenesis: An overview. Indian Fern J. **14:** 1-9.

Cheema, H.K. and Kaur, M. 1986. *In vitro* regeneration and effects of growth regulators on aquatic heterosporous fern. Res. Bull Punjab. Uni. Sci. **36:** 35-37.

Cheema, H.K. and Sharma, M.B. 1994. Induction of multiple shoots from adventitious buds and leaf callus in *Ceratopteris thalictroides*. Indian Fern J. **11:** 63-67.

Dey, M., Kaila, S., Ghosh, S. and Mukherjee, S.G. 1998. Biochemical and molecular basis of differentiation in plant tissue culture. Curr. Sci. **74:** 591-596.

Einset, J.W. 1991. Woody plant micropropagation with cytokinins. In: Biotechnology in Agriculture and Forestry. Vol. 17. High-Tech and Micropropagation I. (Ed.) Bajaj, Y.P.S., Springer – Verlag, Berlin, Germany, pp. 190-201.

Fay, M.F. 1994. In what situations is *in vitro* culture appropriate to plant conservation? Biodiversity and Conservation **3:** 176-183.

Hegde, S. and D'Souza, L. 1999. Effect of season on contamination and *in vitro* germination of the spores (*Drynaria quercifolia*). Ind. Fern J. **16:** 30-34.

Hickok, L.G., Warne, T.R. and Slocum, M.K. 1987. *Ceratopteris richardii*: Applications for experimental plant biology. Amer. J. Bot. **59:** 458-465.

Hicks, G.S. 1994. Shoot induction and organogenesis *in vitro*: a developmental perspective. *In vitro* Cell. Dev. Biol. **30**: 10-15.

Johnson, M. and Manickam, V.S. 2010. Influence of Plant Growth Regulators on *in vitro* raised gametophytes of *Cheilanthes viridis* (Forssk.) Swartz. J. Basic & Applied Biol. **4(4)**: 18-23.

Johnson, M. and Manickam, V.S. 2011a. *In vitro* proliferation of secondary and tertiary gametophytes from the spore derived gametophytes of the fern *Phlebodium aureum* L. J. Basic & Applied Biol. **5(1&2)**: 352-356.

Johnson, M. and Manickam, V.S. 2011b. Influence of media, sucrose and plant growth regulators on *in vitro* apospory induction and developmental changes in *Sphaerostephanos unitus*. J. Basic & Applied Biol. **5(1&2)**: 344-351.

Johnson, M. and Manickam, V.S. 2012. *In vitro* organogenesis Studies of *Cheilanthes viridis* – An endangered fern. Proceedings of the National Conference on "Climate change, Biodiversity & Conservation". Rajkumar, S.D. Samuel, C. O. & Lal, J. K. (eds.) IJBT (2012) (Special Issue), Gayathri Technological Publication, Palayamkottai, India, pp. 76 – 88.

Kwa, S.H., Wee, Y.C., Lim, T.M. and Kumar, P.P. 1995. Establishment and physiological analyses of photoautotrophic callus cultures of the fern *Platycerium coronarium* (Koenig) Desv. under CO_2 enrichment. J. Exp. Bot. **46**: 1535-1542.

Kwa, S.H., Wee, Y.C., Lim, T.M. and Kumar, P.P. 1997. Morphogenetic plasticity of callus reinitiated from cell suspension cultures of the fern *Platycerium coronarium*. Plant Cell Tiss. and Org. Cult. **48**: 37-44.

Kwa, S.H., Wee, Y.C. and Loh, C.S. 1991. Production of aposporous gametophytes and calli from *Pteris vittata* L. pinnae strips cultured *in vitro*. Plant Cell Rep. **10(8)** 392-393.

Manickam, V.S., Vallinayagam, S. and Johnson, M. 2003. Micropropagation and conservation of rare and endangered ferns of Western Ghats through *in vitro* culture.. In: Subhash Chandra and M. Srivastava (eds.). Pteridology in the New Millennium, Kluwer Academic Publishers, Netherlands, pp: 497-504.

Minocha, S.C. 1987. Plant growth regulators and morphogenesis in cell and tissue culture of forest trees In: Cell and Tissue Culture in Forestry Vol. 1 (Eds.) Bonga, J.M. and Durzan D J, Martinus Nijhoff, The Hague, pp. 56-59.

Nigam, U.V., Agarwal, D.C., Raskar, S.V., Morwal, G.C. and Mascarenhas, A.F. 1991. *In vitro* studies on *Cyathea spinulosa* – an endangered tree fern. In: Bharadwaja, J.N. and Gena, E.B. (eds.) Aspects of plant sciences Vol. 13. Perspectives in Pteridology Present and Future. Today and Tomorrow printers and publishers, New Delhi, pp. 63-67.

Overbeek, J.V. 1968. The control of plant growth. Scientific American **219**: 75-81.

Padhya, M.A. 1985. Rapid propagation of *Adiantum trapeziformae* L. through tissue culture. Indian Fern J. **12**: 20-23.

Padhya, M.A. 1987. Mass propagation of ferns through tissue culture. Acta horticulture **212**: 645-648.

Sara, S.C. 2001. Conservation of selected rare and endangered ferns of the Western Ghats through micropropagation and restoration. Ph. D. thesis, Manonmaniam Sundaranar University, Tirunelveli, India.

Sara, S.C., Manickam, V.S. and Antonisamy, R. 1998. Regeneration in kinetin - treated gametophytes of *Nephrolepis multiflora* (Roxb.) Jarret in Morton. Curr. Sci.**75:** 503-508.

Takahashi, Y., Ishida, S. and Nagata, T. 1995. Auxin - regulated genes. Plant Cell Physiol. **36:** 383-390.

Vallinayagam, S. 2003. Micropropagation of rare and endangered ferns of Western Ghats. Ph.D. Thesis, Manonmaniam Sundaranar University, Tirunelveli, India.

Chapter 9

Antiulcer Activity of an Isolated Compound (KR–1) from *Kaempferia rotunda* Linn. Leaf in Rats

Prasanta Kumar Mitra, Tanaya Ghosh, Prasenjit Mitra and Gayatri Mitra

Introduction

An active compound (KR-1) was isolated from *Kaempferia rotunda* Linn. leaf and its antiulcer activity was studied against ethanol, hydrochloric acid, indomethacin, stress and pyloric ligation induced gastric ulceration in albino rats. A significant antiulcer activity of KR-1 was observed in all the models. KR-1 thus provides a scientific rationale for the use as antiulcer drug.

Sanyal *et al.* (1961) found that vegetable banana is efficacious not only for experimentally induced gastric ulcers in albino rats, mice, guinea pigs etc. but also for humans suffering from gastric ulcers. Akah and Nwafor (1999) demonstrated anti gastric ulcer activity of the herb *Cissampelos mucronata*. Shetty *et al.* (2000), Sairam *et al.* (2001), Maity *et al.* (1995, 2003) and Dharmani and Palit (2006) confirmed anti gastric ulcer activities of *Ginkgo biloba, Convolvulus pluricaulis,* tea root extract and *Vernonia lasiopus* respectively. We also reported anti gastric ulcer activities of few medicinal plants of this part of India in different experimental ulcer models (Mitra, 1980, 1981 , 1982, 1985, 2001; Mitra and Mitra 2005, 2008; Mitra *et al.* 2008).

Recently we have noted anti gastric ulcer activity of *Kaempferia rotunda* Linn. leaf against ethanol induced gastric ulcer in albino rats (Mitra *et al.* 2010). Tempted by the observation we had undertaken isolation studies of the active compound(s) from *Kaempferia rotunda* Linn. leaf for which the plant leaves showed anti gastric ulcer activity.

Kaempferia rotunda Linn.

Methodology

Kaempferia rotunda Linn. leaf was collected from the garden of medicinal plants of the University of North Bengal during September, 2009 and identified by the experts of the Department of Botany of the said University. Leaf was kept in the department with proper voucher number for future reference.

Isolation of active principle (KR-1) from *Kaempferia rotunda* Linn. leaf

Fresh plant leaves were shade dried at room temperature, ground into fine powder and then extracted (amount 100g) with 600 ml chloroform – water mixture (10 : 1, v/v) for 1 hour using soxhlet apparatus at room temperature. The extract was concentrated under reduced pressure by a rotary evaporator to a volume of about 10 ml.This was then subjected to column chromatography using alumina as adsorbent. Elution was done by 50% ethanol-chloroform mixture. Eluted material was evaporated to dryness and extracted with 10 ml ethyl acetate. The ethyl acetate extract was further subjected to column chromatography using silica gel mesh (200-400 size) as adsorbent. The fraction obtained after elution with 50% methanol - chloroform mixture was subjected to repeated crystallization when a compound was crystallized. The compound was given a trivial name KR-1. The compound was preserved for acute toxicity study as well as for anti gastric ulcer activity.

Experimental animals

Wistar strain albino rats of both sex were used for the study. The animals were housed in colony cages (4 rats/cage) and were kept for at least a week in the experimental wing of the animal house (room temperature 25–28 degree centigrade and humidity 60–65% with 12 h light and dark cycle) before experimentation. Animals were fed on laboratory diet with water *ad libitum*. For each set of experiment 10 animals were used. The animal experiment had approval of the institutional ethics committee.

Chemicals

Indomethacin (Torrent Research Centre, Gandhinagar), Ethanol (Baroda Chemical industries Ltd., Dabhoi), HCl LR (Thomas baker, Mumbai), omeprazole (Kopran Pharma Ltd. Mumbai).

Test drug

Isolated compound (KR-1) was used as the test drug.

Production of gastric ulcers

Ethanol induced gastric ulcer (Sairam *et al.* 2001)

Rats were fasted for 18 h when no food but water was supplied *ad libitum*. Gastric ulcers were induced by administering ethanol (95%, 1 mL/ 200 g body weight) orally through a feeding tube. 1h after administration of ethanol, animals were sacrificed by cervical dislocation and the stomach was taken out and incised along the greater curvature. Stomach was then examined for the presence of ulcers.

HCl induced gastric ulcer (Parmar and Desai, 1993)

0.6M HCl (1 mL/200 g body weight) was orally administered to all rats. Rest part is same to that of ethanol induced gastric ulcer group.

Indomethacin induced gastric ulcer (Parmar and Desai, 1993)

Indomethacin (10 mg/kg) was given orally to rats in two doses at an interval of 15 hour. Rest part is same to that of ethanol induced gastric ulcer group.

Stress induced gastric ulcer (Alder, 1984)

Rats were fasted for 24h when no food but water was supplied *ad libitum*. Stress ulcer was induced by forced swimming in the glass cylinder (height 45 cm, diameter 25 cm) containing water to the height of 35 cm maintained at 25 degree centigrade for 3h. Rats were then sacrificed. Rest part was same to that of ethanol induced gastric ulcer group.

Induction of gastric ulcer by pyloric ligation method (Parmar and Desai, 1993)

Rats were fasted for 24h when no food but water was supplied *ad libitum*. Under light ether anesthesia, abdomen was opened and the pylorus was ligated. The abdomen was then sutured. After 4h the rats were sacrificed with excess of anesthetic ether and the stomach was dissected out. Rest part was same to that of ethanol induced gastric ulcer group.

Acute oral toxicity study (Ghosh, 2005):

Acute toxicity studies were carried out on Swiss albino mice. Isolated compound (KR-1) from *Kaempferia rotunda* Linn. leaf was given orally at doses of 100, 200, 500, 1000 and 3000 mg/kg to five groups of mice, each group containing six animals. After administration of the compound, the animals were observed for the first three hours for any toxic symptoms followed by observation at regular intervals for 24 hours up to seven days. At the end of the study, the animals were also observed for general organ toxicity, morphological behaviour and mortality.

Anti gastric ulcer study

Rats were divided into 5 groups;

Group 1 : Control

Group 2 : Ulcerogenic drug or Method (Ethanol / HCl / Indomethacin / Stress / Pyloric ligation)

Group 3 : Ulcerogenic drug or method + KR-1 (5 mg/kg)

Group 4 : Ulcerogenic drug or method + KR-1 (10 mg/kg)

(KR-1 was given orally 30 minutes prior to administration of ulcerogenic drug or method)

Group 5: Ulcerogenic drug or method + Omeprazole (8 mg/kg orally 30 minutes prior to administration of ulcerogenic drug or method). Omeprazole was used as per the method of Malairajan *et al.* (2008).

Evaluation of ulcer index (Szelenyi and Thiemer, 1978)

Gastric lesions were counted and the mean ulcerative index was calculated as follows :

I - Presence of edema, hyperemia and single sub mucosal punctiform hemorrhage.

II – Presence of sub mucosal hemorrhagic lesions with small erosions.

III – Presence of deep ulcer with erosions and invasive lesions.

Ulcer index = (number of lesion I) x 1 + (number of lesion II) x 2 + (number of lesion III) x 3.

Statistical Analysis

The values were expressed as mean ± SEM and were analyzed using one-way analysis of variance (ANOVA) using Statistical Package for Social Sciences (SPSS) 10th version. Differences between means were tested employing Duncan's multiple comparison test and significance was set at p < 0.05.

Results

Acute toxicity studies

Acute toxicity studies revealed that KR-1 did not produce any toxic symptoms when administered orally to mice in doses of 100, 200, 500, 1000 and 3000 mg/kg. Animals were healthy, cheerful and behaved normal throughout the experimental period.

No death of animal was recorded during seven days of experiment.

Effect of KR-1 on ethanol induced gastric ulcer

Result is given in Table 9.1.

Ethanol produced massive gastric ulcers in all albino rats. Ulcers were mostly superficial. Bleeding of the stomach was followed by adhesion and dilatation. Ulcer index came 30.1 ± 1.91. Pretreatment of rats with KR-1 produced a dose dependent protection (12.95% and 46.51% for the doses of 5 mg/kg and 10 mg/kg of KR-1 respectively) from ethanol induced ulceration as compared to ethanol group. However, the protection was not statistically significant at 5 mg/kg dose. Omeprazole produced significant gastric ulcer protection (64.12%). Efficacy of KR-1 in the dose of 10 mg/kg was 67.08% when compared with that of omeprazole group.

Table 9.1: Effect of KR-1 on ethanol induced gastric ulcer

Group	Ulcer index (mean ± SEM)	% Ulcer protection
Control	Nil	--
Ethanol	30.1 ± 1.91	--
Ethanol+ KR-1 (5 mg/kg)	26.2 ± 1.52	12.95
Ethanol+ KR-1 (10 mg/kg)	16.1 ± 1.43**	46.51
Ethanol + Omeprazole (8mg/kg)	10.8 ± 1.12**	64.12

Effect of KR-1 on hydrochloric acid induced gastric ulcer

0.6M HCl when administered to rats orally produced massive ulcers in stomach of all rats. Adhesion and dilatation of the stomach were seen. Ulcer index was 28.1 ± 1.86. Pretreatment with KR-1 gave dose dependent protection (26.69% and 41.63% for the doses of 5 mg/kg and 10 mg/kg of KR-1 respectively) from HCl induced ulceration. Protection was statistically significant and comparable with omeprazole group where ulcer protection was 63.70% (Table 9.2).

Table 9.2 : Effect of KR-1 on hydrochloric acid induced gastric ulcer

Group	Ulcer index (mean ± SEM)	% Ulcer protection
Control	Nil	--
Ethanol	28.1 ± 1.86	--
Ethanol+ KR-1 (5 mg/kg)	20.6 ± 1.76*	26.69
Ethanol+ KR-1 (10 mg/kg)	16.4 ± 1.55**	41.63
Ethanol + Omeprazole (8mg/kg)	10.2 ± 1.00**	63.70

Effect of KR-1 on indomethacin induced gastric ulcer

Result is given in Table 9.3.

Indomethacin produced gastric ulcers in all albino rats. Ulcers were superficial in nature.There were adhesion, dilatation and bleeding in the stomach. Ulcer index came 29.2 ± 1.81. Pretreatment of rats with KR-1 produced dose dependent protection (22.26% and 43.49% for the doses of 5 mg/kg and 10 mg/kg of KR-1 respectively) from indomethacin induced ulceration as compared to indomethacin group. Protections were statistically significant. Omeprazole produced significant gastric ulcer protection (63.35%). In this model efficacy of KR-1 in the dose of 10 mg/kg was 68.65% when compared with that of omeprazole group.

Table 9.3: Effect of KR-1 on indomethacin induced gastric ulcer

Group	Ulcer index (mean ± SEM)	% Ulcer protection
Control	Nil	--
Ethanol	29.2 ± 1.81	--
Ethanol+ KR-1 (5 mg/kg)	22.7 ± 1.43*	22.26
Ethanol+ KR-1 (10 mg/kg)	16.5 ± 1.39**	43.49
Ethanol + Omeprazole (8mg/kg)	10.7 ± 1.23**	63.35

Effect of KR-1 on stress induced gastric ulcer

Swimming stress produced massive ulcers in stomach of all rats. Adhesion and dilatation of the stomach were seen. Ulcer index was 30.4 ± 1.99. Pretreatment with KR-1 gave dose dependent protection (25.65% and

51.64% for the doses of 5 mg/kg and 10 mg/kg of KR-1 respectively) from stress induced ulceration. Protection was statistically significant and comparable with omeprazole group. In omeprazole group ulcer protection was 65.13%. Results were shown in Table 9.4.

Table 9.4: Effect of KR-1 on stress induced gastric ulcer

Group	Ulcer index (mean ± SEM)	% Ulcer protection
Control	Nil	--
Ethanol	30.4 ± 1.99	--
Ethanol+ KR-1 (5 mg/kg)	22.6 ± 1.56*	25.65
Ethanol+ KR-1 (10 mg/kg)	14.7 ± 1.40**	51.64
Ethanol + Omeprazole (8mg/kg)	10.6 ± 1.21**	65.13

Effect of KR-1 on pyloric ligation induced gastric ulcer

Result is given in Table 9.5.

Pyloric ligation produced gastric ulcers in all albino rats. Ulcers were superficial in nature.There were adhesion, dilatation and bleeding in the stomach. Ulcer index came 26.3 ± 1.74. Pretreatment of rats with KR-1 produced dose dependent protection (53.61% and 59.31% for the doses of 5 mg/kg and 10 mg/kg of KR-1 respectively) from pyloric ligation induced ulceration as compared to pyloric ligation group. Protection was only statistically significant at 10 mg/kg dose of KR-1. Omeprazole produced significant gastric ulcer protection (62.35%).

Table 9.5: Effect of KR-1 on pyloric ligation induced gastric ulcer

Group	Ulcer index (mean ± SEM)	% Ulcer protection
Control	Nil	--
Ethanol	26.3 ± 1.74	--
Ethanol+ KR-1 (5 mg/kg)	12.2 ± 1.55	53.61
Ethanol+ KR-1 (10 mg/kg)	10.7 ± 1.31**	59.31
Ethanol + Omeprazole (8mg/kg)	9.9 ± 1.1**	62.35

Discussion

The term "Peptic ulcer" refers to an ulcer in the lower oesophagus, stomach or duodenum, in the jejunum after surgical anastomosis to the stomach or, rarely in the ileum adjacent to a Meckel's diverticulum. Ulcer in the stomach (gastric ulcer) may be acute or chronic. Quincke (1963) was probably the first to use the term 'Peptic ulcer'. Because of its frequency and worldwide distribution, peptic ulcer continues to be a subject of numerous investigations, both experimental and clinico pathological. In this respect peptic ulcer occupies a place secondary to carcinoma in the field of gastroenterology.

There is medicine to treat peptic ulcer (Tierrey *et al.*, 1978). In case, the ulcer is due to infection of *Helicobacter pylori (H. pylori)*, the different medications are usually prescribed. This is known as "Triple therapy". This includes a proton pump inhibitor viz. omeprazole to reduce acid production and two antibiotics to get rid of the organism. Sometimes, instead of one of the antibiotics, bismuth salicylate may be the third medication recommended. This drug, available over the counter, coats and soothes the stomach, protecting it from the damaging effects of acid. Two, rather than three, drug regimens are currently being developed. For non *H. pylori* ulcers number of drugs are now available for treatment. These drugs are broadly classified into two categories :

(1) Those that decrease or counter acid – pepsin secretion viz. ranitidine, famotidine etc. (H2 - blockers), pirenzepine, telenzepine etc. (M1 – blockers), omeprazole, lansaprazole etc. (proton pump inhibitors)

(2) Those that affect cytoprotection by virtue of their effects in mucosal defense factors like sucralfate , carbenoxolone etc. (Yeomans *et al.* 1998)

No doubt the above said drugs have brought about remarkable changes in peptic ulcer therapy, reports on clinical evaluation of these drugs show that there are incidences of relapses and adverse effects and danger of drug interactions during ulcer therapy. Hence, the search for an ideal anti – ulcer drug continues and has also been extended to medicinal plants / herbs in search for new and novel molecules, which afford better protection and decrease the incidence of relapse.

In Ayurvedic literature *Kaempferia rotunda* Linn. has been described as a medicinal plant of the family 'Zingiberaceae'. It has several names. In Nepali the plant is called 'Bhuichampa', in Hindi 'Sans', in Lepcha 'Ribirip' and in English it is known as 'Black horm'. *Kaempferia rotunda* Linn. is widely distributed at foothills of Himalayas from Kumaon to Sikkim, Bengal, Assam, middle and lower hill forests up to the height of 5000 ft. The plant has tuberous root stock with large and long erect leaves. Rainy season is the flowering time of the plant. *Kaempferia rotunda* Linn. has several medicinal uses. The whole plant in the form of powder and paste was found effective in healing fresh wounds and also removes coagulated blood or purulent matter from the body when taken internally. Roots of the plant was found good in swelling and are also stomachic. In some places of India specially in Bombay powdered root of this plant is popular in mumps (Chopra and Chopra,1958; Gurung, 2002).

Recently, we observed anti ulcer activity of *Kaempferia rotunda* Linn. leaf against ethanol induced gastric ulcer in albino rats (Mitra *et al.* 2010). Tempted by this observation we undertook studies on isolation of the active compound(s) present in *Kaempferia rotunda* Linn. leaf and the antiulcer activity of the isolated compound(s) against ethanol, hydrochloric acid, indomethacin, stress and pyloric ligation induced gastric ulceration in albino rats.

By different solvent extraction processes and chromatographic experiments an active compound was isolated from *Kaempferia rotunda* Linn. leaf. A trivial name of the compound was given as KR-1. Anti gastric ulcer activity of KR-1 was studied against ethanol, hydrochloric acid, indomethacin, stress and pyloric ligation induced gastric ulceration in albino rats. Two doses of KR-1 (5 mg/kg and 10 mg/kg) were used. Results were compared with omeprazole, a known anti gastric ulcer drug.

Significant anti gastric ulcer activity of KR-1 was observed in all the models employed. Results showed that pretreatment of rats with KR-1 produced dose dependent protection. The protections were statistically significant ($p<0.001$) and comparable to that of omeprazole group.

It is known that peptic ulcer is formed either through offensive mechanism (acid – peptic secretion) or through defensive mechanism (mucus secretion). Anti gastric ulcer activity of KR-1 may be related with any one of the two mechanisms. Work in this direction is now under progress.

References

Akah, P. A. and Nwafor, S. V. 1999. Studies on anti – ulcer properties of *Cissampelos mucronata* leaf extract. Indian J. Exp. Biol. **37**: 936 – 938.

Alder R, 1984. In breakdown in human adaptation to stress. Boston. Martinus Nihjihoff, p. 653.

Chopra, R.N. and Chopra, I. C. 1958. Indigenous drugs of India, U.N.Dhar and Sons Private Limted, Kolkata, p. 605.

Dharmani, P and Palit, G. 2006. Exploring Indian medicinal plants for anti ulcer activity. Indian J. Pharmacol. **38** : 95 – 99.

Ghosh, M.N. 2005. Toxicity studies in fundamentals of experimental pharmacology. Hilton and Company, Kolkata, pp. 190-7.

Gurung Bejoy, 2002. The medicinal plants of Sikkim Himalaya, Gangtok, Sikkkim, p. 271.

Maity, S., Vedasiromoni, J. R. and Ganguly, D. K. 1995. Anti – ulcer effect of the hot water extract of black tea (*Camellia sinensis*). J. Ethnopharmacol. **46**: 167 - 174.

Maity, S., Chaudhuri, T., Vedasiromoni, J. R. and Ganguly, D. K. 2003. Cytoprotection mediated anti ulcer effect of tea root extract. Indian J. Pharmacol. **35** : 213 – 219.

Malairajan P., Gopalakrishnan, Geetha., Narasimhan S. and Jessi Kala Vani, K. 2008. Evaluation of anti – ulcer activity of *Polyalthia longifolia* (Sonn.) Thwaites in experimental animals. Indian J. Pharmacol., **40**: 126 – 131.

Mitra, P. K. 1980. Anti gastric ulcer effect of vegetable banana . Am. J. Herbs. **1** : 11 – 12.

Mitra, P. K. 1981. Influence of a banana supplemented diet on restraint stress in rats. Br. J. Exp. Gastroenterol. **2** : 27 – 30.

Mitra, P. K. 1982 . Effect of *Piper longum* Linn. on aspirin induced gastric ulcer in albino rats. Bul. J. Herb Res. **1** : 61 – 64.

Mitra, P. K. 1985. Effect of *Emblica officinalis* Linn. on swimming stress induced gastric ulcer in mice. Arc. Nat. Prod. **1** : 81 – 86.

Mitra, P. K. 2001. In search of an anti ulcerogenic herbal preparation. Trans. Zool. Soc. East India. **5** : 59 – 64.

Mitra, P. and Mitra P.K. 2005. Biochemical studies of the anti ulcerogenic activity of Nirmali (*Strychnos potatorum* Linn.) in restraint induced gastric ulcers in rats. Trans. Zool. Soc. East. India. **9** : 39 – 42.

Mitra, P. and Mitra, P. K. 2008. Use of *Astilbe rivularis* Buch. – Ham. ex D. Don as anti – peptic ulcer agent. Pleione **2(1)** : 74 – 76.

Mitra, P. K.; Mitra, P and Debnath, P. K. 2008. Screening the efficacy of some East Himalayan medicinal plants against ethanol induced gastric ulcer in Albino rats. Pleione 2(2), 233 – 238.

Mitra, P. K., Mitra, P., Das, A.P., Ghosh, C., Sarkar, A. and Chowdhury, D. 2010. Screening the efficacy of some East Himalayan medicinal plants on ethanol induced gastric ulcer in Albino rats. Pleione **4(1):** 69 – 75

Parmar, N.S. and Desai, J.K. 1993. A review of the current methodology for the evaluation of gastric and duodenal antiulcer agents. Indian J. Phrmacol. **25** : 120-35.

Quincke H. 1963. Quoted from 'Pathophysiology of peptic ulcer' (ed.) S. C. Skoryna and H. L. Bockus, J. B. Lippman Cott Company, Philadelphia.

Sairam, K., Rao, Ch. V. and Goel, R. K. 2001. Effect of *Convolvulus pluricaulis Chois.* on gastric ulceration and secretion in rats. Indian J. Exp. Biol. **39** : 137 – 142.

Sanyal,A. K. ; Das, P. K., Sinha, S. and Sinha, Y. K. 1961. Banana and gastric secretion. J. Pharm. Pharmacol. **13** : 318 – 319.

Shetty, R., Vijay Kumar, Naidu, M. U. R. and Ratnakumar, K. S. 2000. Effect of *Ginkgo biloba* extract in ethanol induced gastric mucosal lesions in rats. Indian J. Pharmacol. **32** : 313 – 317.

Szelenyi, I. and Thiemer, K. 1978. Distension ulcer as a model for testing of drugs for ulcerogenic side effects. Arch. Toxicol. **41** : 99 – 105.

Tierney, L. M., Mephee S. J. and Papadakis M. A. 2001. In "Current Medical Diagnosis & Treatment". Mc Craw Hill, New York.

Yeomans N. *et al.* 1998. Sucralfate and carbenoxolone as mucosal defense factors. N. Eng. J. Med., **338** : 719 - 723.

Chapter 10

Comparative Assessment of *In vitro* Antioxidant and Antiproliferative Activity in *Morinda citrifolia* Fruit and Commercial Juice

K. Rajaram and P. Suresh Kumar

Introduction

Morinda citrifolia (noni) used as an important folk medicine to treat various diseases, is widely found in tropical countries. Comparative phytochemical screening in fruit and noni juice were attempted to check its efficiency. The alkaloids, flavonoids, saponins, steroids and terpenoids were present high in the fruit compared to commercial juice. Both fruit and juice were extracted with solvents (methanol, ethanol and water) and tested for *in vitro* antioxidant activity (DPPH scavenging assay). Maximum percentage of inhibition was observed only in the methanolic extract, fruit (IC_{50} 53.50±0.77 µg/ml) and in juice (IC_{50} 379.70 ± 1.03) respectively. Cell viability assay using normal cell line (RAW) and antiproliferative activity in liver cancer cell lines (Hep-G2) were tested using the methanolic extracts. The cell viability assay showed that the concentration up to 100µg/ml have lesser toxicity and gradually increase with higher dosages. So concentrations of 25, 50 and 75 µg/ml were selected for the antiproliferative assay. The fruit methanolic extracts (FME) exhibits significant (IC_{50} 19.94±0.90 µg/ml) antiproliferative effect than juice methanolic extracts JME (IC_{50} 50.69±3.79µg/ml). The outcome of this study indicated that baseline information for the bioactive potentiality of *M. citrifolia* dried fruit over the commercial juice.

Morinda citrifolia L. (Rubiaceae), commonly known as noni, is a small evergreen tree or large bush that grows widely across Polynesia and cultivated all over the world (Kamiya *et al.*, 2008). It has been traditionally used for its antibacterial, antiviral, antifungal, antitumor, anti-helminth, analgesic, hypotensive, anti-inflammatory, and immune enhancers effects (Wang *et al.*, 2002). Furthermore, noni juice, which is made from the ripe fruit, was very popular because of its many health benefits (Dixon *et al.*, 1999; McClatchey, 2002).

M. citrifolia (Rubiaceae, Noni) fruit juice is a well-known health drink traditionally used as a folk herbal remedy for the treatment of many diseases including diabetes, hypertension, and cancer. These effects have been attributed to its nutrient content, which yields antioxidative and anti-inflammatory effects (Hiranzumi *et al.*, 1996; Wang and Su, 2001; Kamiya *et al.*, 2004; Pawlus *et al.*, 2005). As previously reported, Noni contains many antioxidative and anti-inflammatory ingredients such as, neolignan, americanin A, vitamin C, vitamin E, scopoletin and xeronine (Wang *et al.*, 2002; Kamiya *et al.*, 2004). Renewed attention in recent years to natural therapies has stimulated a new wave of research interest in traditional practices. Herbs have become a target for the search for new anticancer drugs. About half of the drugs used in clinical practice come from natural products (Butler, 2005). In fact, many nutritive and nonnutritive phytochemicals with diversified pharmacological properties have shown promising responses for the prevention and treatment of various cancers, including colon cancer (Surh, 2003). In the last decade, several papers have reported the chemical constituents of noni and noni extracts, revealing the presence of anthraquinone (Kamiya *et al.*, 2004), iridoid (Kamiya *et al.*, 2004), flavones (Deng *et al.*, 2007), lignans (Deng *et al.*, 2007) and fatty acid derivatives (Daulatabad *et al.*, 1989) and also the biological activities such as hypotensive activity (Youngken *et al.*, 1960), anticancer activity (Wang and Su, 2001) and inhibition of copper-induced low-density lipoprotein oxidation (Kamiya *et al.*, 2004). The fruit of *Morinda citrifolia* was commercialized as juices and used for many treatments. Hence our present study designed to assess the importance of noni dried fruits and Noni commercial juice to study their efficiency in *in vitro* antioxidant and antiproliferative activity (Hepatocellular cancer cell line (Hep-G2)).

Materials and Methods

Plant collection

M. citrifolia fruits were procured from the local nursery and also noni juice purchased from store. The fruits were identified by Botanical Survey of India, Coimbatore, Tamil nadu, India (Authentication No.: BSI/SRC/5/ 23/09-10/Tech-1569).

Phytochemical analysis

Phytochemicals such as alkaloids, phenols, flavonoids, saponins, steroids, terpenoids and tannins were analyzed following the standard procedures (Harborne, 1973; Trease and Evans, 1996).

Preparation of Extracts

The fruits were chopped into pieces (~1cm) and air dried. The air dried fruits (20g) were powdered and extracted with 200ml solvents (methanol, ethanol and aqueous) separately using magnetic stirrer. Then the extracted solvents filtered using Whatman filter paper No.1. For commercial juice, 20ml of noni juice were taken in separating funnel and mixed with 200ml of solvents (Methanol, Ethanol and Aqueous) separately. The solvent phase was carefully removed and filtered using Whatman filter paper No.1. The methanol and ethanol extracts were concentrated using Rotary evaporator (Yamato, Japan) where as the aqueous extracts were lyophilized (Martin crest). Later the concentrated extracts were dissolved in DMSO (Dimethyl sulphoxide) and assayed for antioxidant and antiproliferative assays.

DPPH radical-scavenging effect

The radical scavenging activity of all extracts (Fruit Methanol Extract (FME), Fruit Ethanol Extract, Fruit Aqueous Extract, Juice Methanol Extract (JME), Juice Ethanol Extract and Juice Aqueous Extract) of *M. citrifolia* was evaluated using DPPH (Soares *et al.*, 1997). % Radical scavenging activity (DPPH) = (Control OD" sample OD/Control OD) ×100%. Each sample was assayed in triplicate for each concentration. Ascorbic acid was used as control.

Cell viability and anti- proliferative assays

RAW (cell viability) and Hep-G2 (anti- proliferative) cells were procured from National Center for Cell Science, Pune, India and it was cultured in multi-well plate at 37°C in a humidified atmosphere of 5% CO_2,(Sanyo) with DMEM that contained 10% fetal calf serum, 50 U/ml of penicillin, and 50 µg/ml streptomycin. For assay, 1×10^4 cells/ml was seeded in 96-well plates in the regular growth medium. After incubation for 24 h, the cells were maintained in the medium containing methanolic extracts of *M. citrifolia* fruit, commercial juice and standard for 24 h at 37°C (5% CO_2). After incubation, the spent medium was discarded and cells were washed once with DMEM and resuspended in fresh medium (200µl/well). 5 mg/ml MTT (10µL/well) was added to cells, followed by incubation for 4 h at 37°C (5% CO_2). The formazan crystals were solublized by incubating the cell with DMSO (150µg/well). The absorbance of the solution was measured at 570nm, using a microplate reader (Molecular devices, France). The percentage of live cells was determined using the formula, $([A_{control} - A_{experiment}]/A_{control})$ ×100%

Results

Phytochemical screening

Phytochemical analysis indicated the presence of abundant amount of alkaloids, saponins, steroids, terpenoids and flavonoids components in *M. citrifolia* fruits compared to commercial juice (Table 10.1.). This may be due to loss of some potential phytochemical components during juice processing.

Table 10.1. Preliminary phytochemical groups test for the methanolic extracts of commercial juices and *Morinda citrifolia* fruits

S.No.	Phytochemical tests	Fruit	Commercial juice
1	Alkaloids	+++	+
2	Steroids	+++	+
3	Terpenoids	+++	++
4	Flavonoids	+++	+
5	Phenols	+	+
6	Tannins	+	+
7	Saponins	++	++

(-): Absent, (+): Slightly present, (++): Fairly present, (+++): Abundant

Radical scavenging activity using DPPH method

There was a linear increase in the radical scavenging activity of all extracts with increment of the sample concentrations (Fig. 10.1). FME (IC_{50} 53.50±0.77 µg/ml) and JME (IC_{50} 379.70 ± 1.03 µg/ml) showed significant antioxidant activity than other fruit extracts tested (Table 10.2.). Hence, the methanol extract of both fruit and juice selected for anti-proliferative activity.

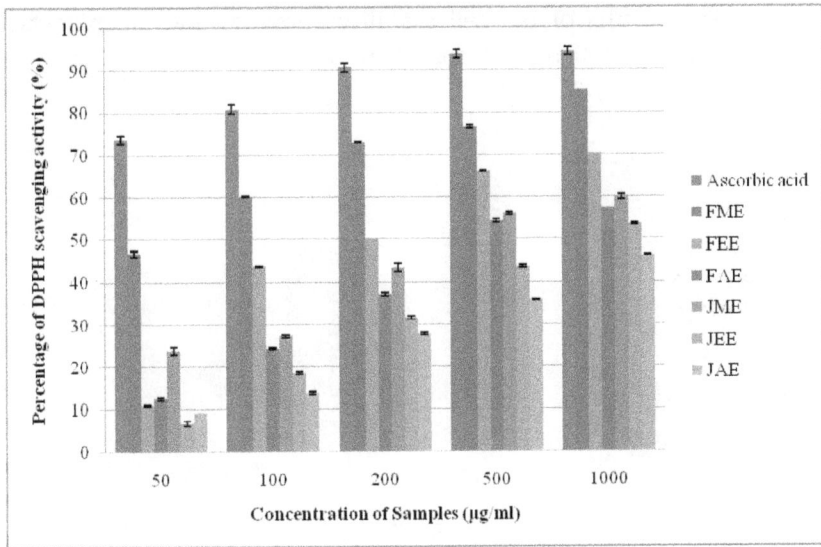

Fig. 10.1. DPPH Radical scavenging activity of fruit and Commercial juice extracts

FME -Fruit Methanol Extract; FEE - Fruit Ethanol Extract; FAE - Fruit Aqueous Extract
JME - Juice Methanol Extract; JEE - Juice Ethanol Extract; JAE - Juice Aqueous Extract
Data were presented as the mean ± SD (n=3).

Table 10.2. DPPH free radical scavenging activity of methanol, ethanol and Water extracts from commercial juices

S.No	Samples	DPPH scavenging activity (IC_{50}: µg/ml)
1	FME	53.50±0.77
2	FEE	199.50±0.25
3	FAE	459.50±3.51
4	JME	379.70±1.03
5	JEE	430.20±1.01
6	JAE	520.00±1.64
7	Ascorbic acid	33.93±0.19

Cell viability and anti-proliferative assays

The cell viability of FME and JME in RAW cell lines was analyzed with various concentrations (50-500 µg/ml), the MTT is reduced to purple formazan by the enzyme reductase which present in the living cells. The absorbances of the extracts were compared with the control (untreated

RAW cells). The results of the cell viability showed, that all extracts have lesser toxicity up to 100 µg/ml concentration (Fig. 10.2).

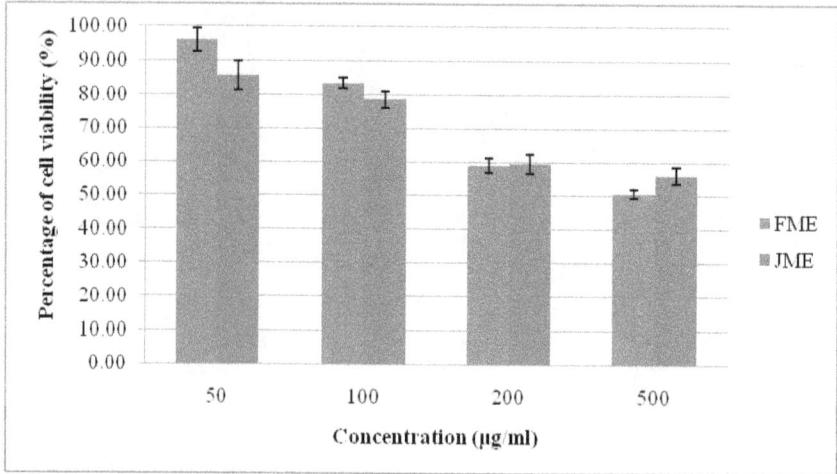

Fig. 10.2. Cell viability assay of methanol extracts (fruit and Commercial juice) in RAW cell line

FME -Fruit Methanol Extract ;JME - Juice Methanol Extract
Data were presented as the mean ± SD (n=3).

Anti-proliferative activity was performed in FME, JME and compared with standard. The IC_{50} of FME showed maximum activity (IC_{50} 19.94±0.90 µg/ml) and comparable with the standard (IC_{50} value 15.31±0.74µg/ml.). However, JME showed higher values. (IC_{50} 50.69±3.79µg/ml) (Fig. 10.3).

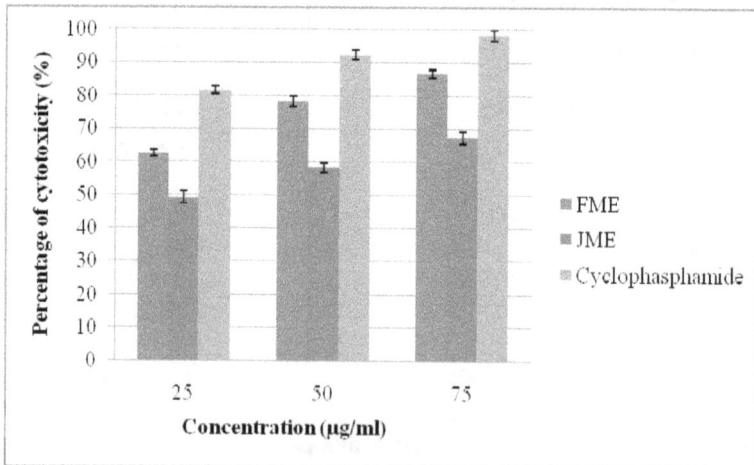

Fig. 10.3. Antiproliferative activity of methanol extracts (fruit and Commercial juice) in Hep-G2 cell line

FME -Fruit Methanol Extract ; JME - Juice Methanol Extract
Data were presented as the mean ± SD (n=3).

Discussion

Phytochemical analysis revealed the prominent differences between the phytochemical compositions present in the fruit and commercial juice. The presence of more phytochemicals in the fruit might be responsible for its significant antioxidant and anti-proliferative activity than commercial juice.

Antioxidants are the compounds which helps to delay or inhibit the oxidation of lipids and other molecules through the inhibition of either initiation or propagation of oxidative chain reactions (Jaleel *et al.*, 2009). Antioxidants can act as either reducing agents, or by free radical scavengers or singlet oxygen quenchers. Free radicals contribute to more than one hundred disorders in humans including atherosclerosis, arthritis, and ischemia and reperfusion injury of many tissues, central nervous system injury, gastritis, cancer and AIDS (Cook and Samman, 1996; Kumpulainen and Salonen, 1997). Free radicals are often generated as byproducts of biological reactions or from exogenous factors. The involvement of free radicals in the pathogenesis of a large number of diseases are well documented by Pourmorad *et al.* (2006). Medicinal plants can protect against harmful effects of ionizing radiation. Natural plant extracts or pure compounds are safe ingredients, which do not have any higher toxic effects. Plant extracts can be characterized by polyvalent formulations and interpreted as additive, or, in some cases, potentiating. First, the therapeutic benefit of medicinal plants is usually attributed to their antioxidant properties and oxidative stress is a prominent feature of these diseases (Feher *et al.*, 2003; Aboutwerat *et al.*, 2003).

DPPH is a stable free radical that accepts an electron or hydrogen radical and becomes a stable diamagnetic molecule (Soares *et al.*, 1997; Loo *et al.*, 2007). A deep purple color with an absorption maximum at 517 nm is formed from DPPH solution, but it generally fades when some antioxidants are present in the solution. In addition, antioxidant activity had a linear relationship with the total phenolic or anthocyanin content in some plants (Kalt *et al.*, 1999), which can be validated by the following research. Our result clearly indicated that the FME have higher antioxidant activity among all other extracts. A strong positive correlation has been reported between total polyphenol content and DPPH free radical scavenging activity (Oki *et al.*, 2002; Siriwardhana and Shahidi, 2002). The antioxidant activity was derived from some flavonoid-type compounds, which are one of the most diverse and widespread groups of natural phenolic (Cakir *et al.*, 2003). Therefore, the presence of flavonoids, alkaloids, steroids, saponins, terpenoids content in methanolic extract of fruit might account for the better scavenging effect on DPPH. Phenolic compounds which originate from plant resources have multiple biological effects such as antioxidant activity and mainly attribute antioxidant activity to their redox properties (Osawa, 1994; Siriwardhana *et al.*, 2003).

The present study demonstrates the methanolic extract of *M. citrifolia* fruits possessed significant antioxidant and antiproliferative activity when compared to commercial juice. This may be due to the presence of significant amount of phytochemicals such as alkaloids, saponins, steroids, terpenoids, present in the fruit is responsible for the momentous antioxidant and antiproliferative activity. Some of the phytochemicals may be lost during commercial juice processing and which is very well correlated with reduced bioactivity.

Acknowledgement

The authors thank DST-SERC, New Delhi, India for providing the financial assistance to carry out this work.

References

Aboutwerat, A., Pemberton, P.W., Smith, A., Burrows, P.C. and McMahon, R.F.T. 2003. Oxidant stress is a significant feature of primary biliary cirrhosis. Biochimica et Biophysica Acta: Molecular Basis of Disease **1637**: 142-150.

Butler, M.S. 2005. Natural products to drugs: natural product derived compounds in clinical trials. Nature Product Reports **22**: 62-195.

Cakir, A., Mavi, A., Yildirim, A., Duru, M.E., Harmandar, M. and Kazaz, C. 2003. Isolation and characterization of antioxidant phenolic compounds from the aerial parts of *Hypericum hyssopifolium* L. by activity-guided fractionation. J. Ethnopharmacol. **87**: 73-83.

Cook, N.C. and Samman, S. 1996. Flavonoids chemistry, metabolism, cardio protective effects, and dietary sources. Nutritional Biochemistry **7**: 66-76.

Daulatabad, C.D., Mulla, G.M. and Mirajkar, A.M. 1989. Ricinoleic acid in *Morinda citrifolia* seed oil. J. Oil Technologists Association of India **21**: 26-27.

Deng, S., Palu, A.K., West, B.J., Su, C.X., Zhou, B.N. and Jensen, J.C. 2007. Lipooxygenase inhibitory constituents of the fruits of noni (*Morinda citrifolia*) collected in Tahiti. J. Nat. Prod. **70**: 859-862.

Dixon, A.R., McMillan, H. and Atkin, N.L. 1999. Ferment this: the transformation of noni, a traditional Polynesian medicine (*Morinda citrifolia*, Rubiaceae). Economic Botany **53**: 51-68.

Feher, J., Lengyel, G. and Blazovics, A. 1998. Oxidative stress in the liver and biliary tract diseases. Scandinavian Journal of Gastroenterology - Supplement **228**: 38-46.

Harbone, J.B. 1973. Phytochemical methods. Chapman and Hall, Ltd, London, pp 49-188.

Hirazumi, A., Furusawa, E., Chou, S.C., and Hokama, Y. 1996. Immunomodulation contributes to the anticancer activity of *Morinda citrifolia* (Noni) fruit juice. Proc. Western Pharmacol. Soc. **39**: 25-27.

Jaleel, C.A., Gopi, R., Manivannan, P., Sankar, B. and Kishorekumar, A. 2009. Antioxidant potentials and ajmalicine accumulation in *Catharanthus roseus* after treatment with giberellic acid. J. Biotech. **5(11)**: 1142-1145.

Kalt, W., McDonald, J.E., Ricker, R.D. and Lu, X. 1999. Anthocyanin content and profile within and among blueberry species. Canadian J. Plant Sci. **79**: 617-623.

Kamiya, K., Hamabe, W., Harada, S., Murakami, R., Tokuyama, S. and Satake, T. 2008. Chemical constituents of *Morinda citrifolia* roots exhibit hypoglycemic effects in streptozotocin-induced diabetic mice. Biological and Pharmaceutical Bulletin, **31**:935-938

Kamiya, K., Tanaka, Y., Endang, H., Umar, M. and Satake, T. 2004. Chemical constituents of *Morinda citrifolia* fruits inhibits copper induced low-density lipoprotein oxidation. J. Agric. Food Chem. **52**: 5843-5848.

Kumpulainen, J.T. and Salonen, J.T. 1997. Natural antioxidants and anticarcinogens in Nutrition, health and disease. The Royal Society of Chemistry 178-187.

Loo, A.Y., Jain, K. and Darah, I. 2007. Antioxidant and radical scavenging activities of the pyroligneous acid from a mangrove plant *Rhizophora apiculata*. Food Chem. **104**: 300-307.

McClatchey, W. 2002. From Polynesian healers to health food stores: changing perspectives of *Morinda citrifolia* (Rubiaceae). Integrative Cancer Therapies **1**:110-120.

Oki, T., Masuda, M., Furuta, S., Nishibia, Y., Terahara, N. and Suda, I. 2002. Involvement of anthocyanins and other phenolic compounds in radical scavenging activity of purple-fleshed sweet potato cultivars. Food Chem. **67**: 1752-1756.

Osawa, T. 1994. Novel natural antioxidants for utilization in food and biological systems. In: Uritani, I., Garcia, V.V. and Mendoza, E.M. (Eds.) Post harvest biochemistry of plant food materials in the tropics. Japan Scientific Societies Press, Japan, pp241-251.

Pawlus, A.D., Su, B.N., Jung, H.A., Keller, W.J., McLaughlin, J.L. and Kinghorn, A.D. 2005. Chemical constituents of the fruits of *Morinda citrifolia* (Noni) and their antioxidant activity. J. Nat. Prod. **68**: 592-595.

Pourmorad, F., Hosseinimehr, S.J. and Shahabimajd, N. 2006. Antioxidant activity, phenol and flavonoid contents of some selected Iranian medicinal plants. African J. Biotech. **5(11)**: 1142-1145.

Siriwardhana, N., Lee, K.W., Kim, S.H., Ha, J.W., Jeon, Y.J. 2003. Antioxidant activity of *Hizikia fusiformis* on reactive oxygen species scavenging and lipid peroxidation inhibition. Food Science and Technology International **9**: 339-346.

Siriwardhana, S.S.K.W. and Shahidi, F. 2002. Antiradical activity of extracts of almond and its byproducts. J. American Oil Chemists' Society **79**: 903-908.

Soares, J.R., Dins, T.C.P., Cunha, A.P. and Ameida, L.M. 1997. Antioxidant activity of some extracts of *Thymus zygis*. Free Radical Research **26**: 469-478.

Surh, Y.J. 2003. Cancer chemoprevention with dietary phytochemicals. Nature Reviews Cancer **3**: 768-80.

Trease, G.E. and Evans, W.C. 1996. Phenols and phenolic glycosides. In: Trease and Evans Pharmacognosy. Bailliere Tindall Publishers, London.

Wang, M.Y., Nowicki, D. and Anderson, G. 2002. Protective effect of *Morinda citrifolia* on hepatic injury induced by a liver carcinogen. The Proceedings of 93rd Annual Meeting of American Association for Cancer Research **43**: 477.

Wang, M.Y. and Su, C. 2001. Cancer preventive effect of *Morinda citrifolia* (Noni). Annals of the New York Academy of Sciences **952**:161-168..

Youngken, H.W., Jenkins, H.J., and Butler, C.L. 1960. Studies on *Morinda citrifolia* L. II. J. Amer. Pharmacists Association **49**: 271-273.

Chapter 11

Remediation of Uranium Contaminated Soils: Conventional and Emerging Technologies

Sravani Konduru, Chandra Obul Reddy Puli*, Chandra Sekhar Akila , Varakumar Pandit, Krishna Kumar Guduru and Jayanna Naik Banavath

Introduction

Uranium (U) is present in most continental earth's crust soils as a natural trace element. Mining and milling of U ores are the most important sources of U contamination of the environment. They generate large quantities of waste materials stored up as heaps of mining rock debris or dumps of mill wastes after U processing. Like other heavy metals U is chemically toxic, and it can also pose radiological hazards. Uranium contaminated soils and water pose a major environmental and human health hazards. Therefore, it is necessary that together with permanent monitoring of environmental contamination, selection of cost effective remediation technology appropriate for large areas such as contaminated water and soil. Conventional remediation techniques: excavation, treatment (soil washing, chelating), conditioning and disposal as low-level radioactive waste are necessary for heavy contaminated sites. Though this problem is taken up by physical and chemical remediation techniques to solve, these days

Phytoremediation technology proved to be promising, eco-friendly and economically viable option. This cost-effective plant based approach to Uranium remediation takes advantage of the remarkable ability of plants to concentrate uranium and other radio nuclides from the environment and to metabolize these into their tissues. The search for the accumulator plants, development of omics platform and knowledge of the physiological and molecular mechanisms of Phytoremediation began to emerge in hand with biological and engineering strategies designed to optimize the feasibility of using plants to remediate uranium contaminated soils. Advancements in the field of genetic engineering to develop transgenic plants and recognition of role of microbial association with the plants to enhance remediation has opened new avenues in Phytoremediation .

Contamination of soil, water, and sediments by radionuclides and other toxic metals has become a worldwide problem, affecting crop yields, soil biomass, fertility, and contributing to bioaccumulation in the food chain causing serious health hazards. Increased application of nuclear energy and the use of radionuclides for industrial, medical and research purposes have caused significant contamination of certain sites and their surrounding environment. Till today, no efficient remediation methodology was developed that could be feasible and affordable over a vast areas and decontamination of considerable amounts of radioactive wastes, except for the close environment facility, where decontamination activities may be effectively applied.

Uranium, thorium and potassium are the main elements contributing to natural terrestrial radioactivity of which Uranium (U) is reported as one of the most frequent pollutants of groundwater and surface soils (Riley *et al.,* 1992). It is the 49th most abundant element in the earth crust, rocks and soils and heaviest natural element known with atomic number 92, first found by Henry Becquerel to possess radioactivity in 1896 (Hopkins, 1923). Uranium exists in three isotopic forms in nature, viz., ^{238}U, ^{235}U and ^{234}U, that exhibit radioactivity and release radiation in the form of alpha, beta and gamma rays. In its natural state, uranium consists of a mixture of U-238 (99.27%), U-235 (0.72%), and U-234 (0.006%), with half-lives of 4.5 billion, 7.13 million, and 0.247 million years, respectively. Uranium is fast acquiring notoriety as a radiological hazard. In fact, its radiotoxicity is known to be low. However, its chemical toxicity cannot be ignored in dissolved form.

Uranium is present in the soil primarily (80–90%) in the + 6 oxidation state as the uranyl (UO_2^{+2}) cation (Ebbs *et al.,* 1998; Allard *et al.,* 1984) and Uranium (VI) is the most mobile form of it. Uranium is a very reactive element which can readily form a variety of complexes with many other metals. It exists in solution predominantly as the stable linear ion UO_2^{+2} and as soluble carbonate complexes, $(UO_2)_2\,CO_3\,(OH)^{"3}$, UO_2CO_3, $UO_2(CO_3)_2^{"2}$, $UO_2(CO_3)_4^{"3}$, and possibly $(UO_2)_3$, $(CO_3)_6^{"6}$ (Duff and Amrhein, 1996). In general, in nature U is present in most of soils at a low concentration, so the mere fact that a soil contains U does not mean that it has been artificially contaminated by U. Typical concentration range of uranium in non-contaminated soils ranges from 0.40 to 6.00 mg kg^{-1} (Shacklette and Boerngen, 1984).

1. Major Sources of Uranium Contamination

Sources of uranium contamination include both natural and manmade operations.

1.1 Nuclear power plants: Nuclear energy is the one of the main source of uranium contamination and especially UO_2. Uranium dioxide (UO_2) is the chemical form most often used for nuclear reactor fuel. Uranium fluoride compounds- uranium hexafluoride and uranium tetra fluoride are also common in uranium processing. Most reactors are powered by uranium fuel rods, which are slightly radioactive at the beginning. However, when the fuel rod is 'spent,' or used, it is both highly radioactive and thermally hot. Intensity of radioactivity decreases with time called half-life time, which varies from one form of radioactive material to other. The half-life time is the time required for reducing the activity to half of its initial value. Radioactive half-life times can span from fractions of a second to millions of years. So, the waste products generated from the nuclear power plants are continuously rising and become a serious problem that need to be solved carefully. Nuclear weapons programme: Uranium from different kind of weapons is presented like depleted uranium (DU). Depleted uranium is the waste product of the process to enrich uranium ore for use in nuclear weapons and reactors. A typical example for DU effect is a 1991 Gulf War. The specific activity of DU is 21.100 Bq/g. Almost 2 million kilograms of radioactive uranium was dropped on Iraq in 2003 alone.

1.2 Nuclear accidents: Occur when a nuclear chain reaction is accidentally allowed to occur in fissile material, such as enriched uranium or plutonium. The Chernobyl accident is an example of a nuclear accident. This accident destroyed a reactor at the plant and left a large geographic area uninhabitable. In a smaller scale accident at Sarov a technician working with highly enriched uranium was irradiated while preparing an experiment involving a sphere of fissile material. The Sarov accident is interesting because the system remained critical for many days before it could be stopped, though safely located in a shielded experimental hall. This is an example of a limited scope accident where only a few people can be harmed, while no release of radioactivity into the environment occurred. Recently the Fukushima Daiichi nuclear disaster was due to a series of equipment failures, nuclear meltdowns and releases of radioactive materials at the Fukushima-I Nuclear Power Plant, following the Tôhoku earthquake and tsunami on 11 March 2011. It is the largest nuclear disaster since the Chernobyl disaster of 1986 and only the second disaster (along with Chernobyl) to measure Level 7 on the International Nuclear Event Scale.

1.3 Research and Development programmes: Artificial radionuclides may also be released into the environment from non – nuclear cycle activities in industry and research, and from usage in diagnostic and therapeutic medicine. The contamination with uranium and waste products from medical and scientific uses is relatively low but nevertheless, these products must be handled with care.

1.4 Uranium mining and milling process: Typically uranium concentrations can be as low as 0.1 to 0.2% in mined ore, meaning that well over 99% of what is mined is rejected after processing. Once mined, ore must be milled to produce useful uranium concentrate. Milling is the process of grinding the ore and adding chemicals, usually sulfuric acid, to extract the uranium it contains. During milling, other constituents of the ore are released as well, including toxins like arsenic and lead. The byproduct of milling is a toxic sludge of tailings. Because of the low concentration of uranium in ore, nearly as much sludge is produced as ore is mined. This leftover sludge contains a high amount of radioactivity – as much as 85% of the initial radioactivity of the ore. The tailings contain low-grade radioactivity but can be dangerous because of the very large quantities that are stored in rather small areas. Additionally, ground or surface water that is pumped away from the site during mining operations can also contain low levels of radiation and therefore contaminate local rivers, lakes and other drinking water sources.

The mining of phosphate rock, which contains relatively high concentrations of uranium, may release radionuclides in mining and processing effluent, as may the use of phosphate rock products (e.g., fertilizer). Erosion of agricultural soils, for example, may input the ^{238}U decay radionuclides into drinking water supplies in areas with heavy fertilizer usage.

2. Effects of Uranium contamination

Uranium is a radioactive and chemotoxic heavy metal. It is highly toxic to a broad range of organisms, particularly mammals in terrestrial ecosystems (Fellows *et al.*, 1998). Uranium ore is relatively harmless, as long as it remains outside of the body, because it only contains a little pure uranium. But through the mechanical extraction of uranium ore from the rock around it, miners are exposed not only to fine particles of uranium but also to radon, a by-product of uranium in the form of radioactive gas, which they breath in. The inhalation of uranium particles and radon gas can cause variety of cancers, particularly bronchial and lung cancer in mine workers. The EPA estimates that radon causes 21,000 lung-cancer deaths a year (MARYANN BATLLE Cronkite News, 2012). Uranium in soil does not often present a radiological hazard to humans, but toxicity to plants could lead to prescribe cleanup and assessment criteria, for industrial activities (Sheppard *et al.*, 1992). The dangers arising from the biochemical toxicity of U as a heavy metal are considered to be about six orders of magnitude higher than those from its radioactivity. Compared to other heavy metals its chemical toxicity lies between mercury and nickel (Schnug *et al.*, 2005). Kidneys are considered to be the most sensitive target organ for chemical toxicity of U and long-term ingestion of this by humans leads to progressive kidney injury. Precipitation of uranium in bodies is in the form of uranyl-carbonate complexes, after long exposures which lead to renal failure (ATSDR, 1999). During the gulf war in 1991, the wounded American soldiers were reported to have twice raised levels of depleted uranium in the body fluids (Hooper *et al.*, 1999). Uranium in the human body is derived mostly from U in food, especially from vegetables, and cereals (Fisenne *et al.* 1987). Different types of cancers *i.e.*, leukemia, breast, bladder, colon, liver, lung, esophagus, ovarian, and stomach cancers, neurological disorders, birth defects and infertility were reported to caused due to uranium toxicity (Salem, 2000; John *et al.*, 2012).

3. Remediation methods of contaminated Uranium

Remediation of U-contaminated soils has thus become an urgent need because of many adverse effects on living organisms, especially on humans. To reduce the radioactive contamination in soil, effective strategies are to be employed to contain, to accumulate at specific sites, so as to prevent their way to food chain. Remediation technologies of radionuclides are of two types depending on the site of operation. If remediation operated at the site of generation of radionuclides, known as *in situ* and away from the site by transferring to some other site referred to *ex situ*. Depending on the operational modes and technologies involved remediation of uranium are classified into three types.

   (i) Physical remediation technologies
   (ii) Chemical remediation technologies

(iii) Biological remediation technologies: By Microbes-Bioremediation

By Plants –Phytoremediation

3.1 Physical remediation technologies: Physical processes are an attempt to contain U-contaminated soils to prevent their entry to food chain. In this method of remediation the contaminated soil at *in situ* or *ex situ*, operated without an attempt to change the state of contaminant. In physical remediation technologies contaminant is contained rather than its destruction. The following are the types of physical remediation technologies to remediate U contaminated soils.

3.1.1 Excavation: In this approach the uranium contaminated waste is moved from the site of generation to some other site where it is contained and stored to allow further processes to remediate radionuclide soil. This method is of least significance since no efforts are done to isolate contaminants and is expensive. Excavation of contaminated soil and replacement with clean fill is the traditional remedial option for metal contaminated soil; however, it's cost as high as $100 to $500 per cubic yard make this approach prohibitive for large soil volumes (Glass, 1999).

3.1.2 Soil–flushing/Soil-washing: Soil washing and *in situ* flushing involve the addition of water with or without additives including organic and inorganic acids, sodium hydroxide which can dissolve soil organic matter, water soluble solvents such as methanol, nontoxic cations complexing agents such as ethylene-diamine-tetraacetic acid (EDTA), acids in combination with complexation agents or oxidizing/reducing agents. Bio-surfactants, biologically produced surfactants may be promising agents for enhancing removal of metals from contaminated soils and sediments (Mulligan *et al.*, 2001). Virtually all soil-washing or soil-flushing systems are designed to treat soils where the majority of the contaminants are concentrated in the finer-grained materials or on the surfaces of the larger soil particles. Many soil-washing processes are simply screening processes that separate the fine, contaminated particles from the bulk of the soil. The large particle fraction, which constitutes the bulk of many soils, is then clean and does not need further treatment before it can be placed back onsite.

3.1.3 Solidification and stabilization: This process allows binding of contaminated soil waste into a solid matrix made of cement or silicates which limit the solubility and mobility of radionuclides by reducing its leaching potential. The radionuclides are made to bind solid matrix mechanically, thus generates blocks of waste. This is the process of immobilizing toxic contaminants so that it does not have any effect temporally and spatially. Stabilization-solidification (SS) is performed in single step or in two steps. In single step, the polluted soil is mixed with a special binder so that polluted soil is fixed and rendered insoluble. In two step process, the polluted soil is first made insoluble and non-reactive and in the second step it is solidified. SS process is mostly justified for highly toxic pollutants. In-situ SS process is mostly influenced by the transmissivity characteristics of the soil, viscosity and setting time of the binder. Well compacted soil, high clay and organic content do not favour in-situ SS. In ex-situ methods, polluted soil is first grinded, dispersed, and then mixed with binder material. The resultant SS material need to be disposed in a well contained landfill. It is essential that the resultant SS product does not undergo leaching. The common binders used in practice include cement, lime, fly ash, clays, zeolites, pozzolonic products etc. Organic binders include bitumen, polyethylene, epoxy and resins. These organic binders are used for soil contaminated with organic pollutants.

3.1.4 Electro-kinetic methods: This method allows *in situ* use of electric fields to mobilise and remove contaminants (Wood, 1997). When electric field is applied to contaminated soil and treated with water (EPA, 1997), the contaminants separate based on the charge further allowing to enter into porous housings at the electrodes thus facilitates specific separation of contaminants. The procedure is more effective for granular type of soils. Two metal electrodes are inserted into the soil mass which act as anode and cathode. An electric field is established across these electrodes that produces electronic conduction as well as charge transfer between electrodes and solids in the soil-water system. This is achieved by applying a low intensity direct current across electrode pairs which are positioned on each side of the contaminated soil. The electric current results in electrosmosis and ion migration resulting in the movement of contaminants from one electrode to the other. Contaminants in the soil water or those which are desorbed from the soil surface are transported to the electrodes depending upon their charges. Contaminants are then collected by a recovery system or deposited at the electrodes. Sometimes, surfactants and complexing agents are used to facilitate the process of contaminant movement. This method is commercially used for the removal of heavy metals such as uranium, mercury etc., from the soil.

3.1.5 Permeable reactive barriers (PRBs): Permeable reactive barrier (PRB) is one of the most promising technologies that has the potential to treat subsurface U (VI) plumes (Fuller *et al.*, 2003). USEPA defined PRB as: a zone of reactive material which extends below the water table to intercept and treat contaminated groundwater (http://www.epa.gov/ahaazuuc/topics/prb.html). A trench arranged downstream of the contaminant source and filled with reactive material allows the treatment of contaminated groundwater passing slowly through (Navratil *et al.*,2001; Boyd *et al.*, 2002;

Conca and Wright, 2003; Barton *et al.*,2004; Agrawal, 2006). They are in fact permeable walls that are installed across the flow path of a contaminant plume. The wall is designed to be at least as permeable as the surrounding aquifer material. The PRBs contain a zone of reactive material that is designed to act as an in situ treatment zone for specific contaminants as groundwater flows through it (USEPA, 1998; Bronstein, 2005; England, 2006) The feasibility of using low-cost organic materials within a permeable reactive barrier to treat uranium contaminated environmental components was being investigated. Various types of sorbent materials have been used within the PRB to remove U (VI) from groundwater include activated carbon, zero-valent iron (ZVI), zeolites, phosphate rocks, and hydroxyapatites (HAs) (Han *et al.*, 2007; Phillips *et al.*, 2008; Raicevic *et al.*, 2006; Saxena *et al.*, 2006).

4. Chemical remediation technologies

The first step in chemical remediation process is to understand the nature of bonding between contaminant and the soil surface. A suitable extractant need to be selected for selective sequential extraction (SSE) of uranium from the soil mass. The extractants include electrolytes, weak acids, complexing agents, oxidizing and reducing agents, strong acids etc. The use of these extractants in single or in combination will depend upon the concentration of contaminant and nature of the soil mass. A variety of chemical remediation techniques are available for remediation of radionuclide-contaminated soils that can be grouped as

- Chemical conversion into a water-soluble form;
- Chemical immobilization.

Chemicals used in the extraction of U from contaminated soils include - Carbonates, Citric acid, Sodium peroxide and Chelating agents like a) organic chelating agents and b) Inorganic chelating agents

4.1 Carbonate Extractions: Carbonate extraction of soils is an attractive procedure because carbonate and uranium (IV) form a very stable complex and also generates a waste stream with lower concentrations of secondary soil constituents (i.e., iron, aluminum, calcium, and silica) than acid extractions do (Phillips *et al.*, 1995; Kulpa *et al.*, 2001; Zhou and Gu, 2005). Sodium bicarbonate has been used in the mining industry to extract U from carbonate bearing ore material. The bicarbonate ion forms strong aqueous complexes with U (VI) according to reactions and enhances the dissolution of $UO_2^{2+.}$ Mason *et al.*, (1997) used $NaHCO_3$ solution as an alkaline treatment for U contaminated soils from a processing facility in OH, USA. Using the 0.5 M sodium bicarbonate as the dominant reagent, they were able to achieve uranium removals of 75-90%, which corresponding approximately to the percentage of uranium in the oxidized state. Further the aluminum and silicon concentrations in the leachate are low in the effluents suggest that carbonate leach is selective for removal of U, and leaves the soil material relatively unaffected.

4.2 Citric Acid Extractions: Weak organic acids or their salts can be used as environmentally compatible compounds. Citrate is used as a complexing agent to mobilize sorbed and precipitated uranium in both *in situ* and *ex situ* extraction of soils and nuclear reactor components. Various researches revealed that citric acid is highly effective in uranium mobilization and the efficiency of extraction from contaminated soils increased with the acid concentration (Stumm and Morgan, 1996; Gramss *et al.*, 2004). However, care should be taken with the quantity of citric acid used in such systems, because additional quantities may result in uranium migration which contaminates groundwater. The enhanced U(VI) desorption in the presence of citrate may be explained through several processes, including the complexation of U(VI) with citrate and extraction of secondary coatings (e.g., Fe), together with the liberation of Fe–citrate complexes into solution (Francis *et al.*, 1993; Francis *et al.*, 1999; Ebbs *et al.*, 2001; Kantar *et al.*, 2006). An important advantage of U–citrate complexes for remediation accomplishments is their biodegradability. This may depend on the pH of the system, initial *U:citrate* molar ratios, contact time. At pH 6–7, a significant amount of uranium is also observed to associate with biomass, whereas only a negligible amount is observed at pH 8 - 9.

4.3 Sodium peroxide: Use of sodium peroxide (oxidizing agent), improved uranium removal due to oxidation of U(IV), enhancing the solubility of the uranium (AbdEl-Sabour *et al.*, 2007). The oxidation of U(IV) to U(VI) occurred by a two electron transfer from U(IV) to H_2O_2 (Shoesmith, 2007). The resulting uranyl ion (UO_2^{2+}) was then available for subsequent complexation with HCO_3 ions by reaction (1). A 10:1 molar ratio of oxidant to U enhanced the extraction of DU by 20% (Hossain, 2006).

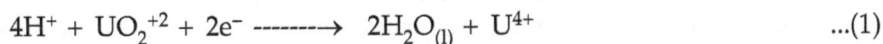

$$4H^+ + UO_2^{+2} + 2e^- \longrightarrow 2H_2O_{(l)} + U^{4+} \qquad ...(1)$$

4.4 Extractions by Complex Organic Chelating Agents: The best synthetic organic chelating agents appeared to be Tiron (0.012 M at pH 7) and catechol (0.012 M at pH 6 4 , which removed between 6 an 8% of the uranium after 48 h. Although these chelators did enhance the aqueous dissolution of UO, at neutral pH values, the rates were disappointing, and as a consequence, Tiron was the only one of this group used in the soil extraction studies. Screening tests were used to compare the extraction effectiveness of two amino carboxylate chelators [1, 2-diaminocyclo-hexanetetraacetic acid (CDTA) and diethylene triamine pentaacetic acid (DTPA)], Tiron, and potassium carbonate. Carbonate and Tiron were the most effective in extracting uranium from the Fernald soils.

4.5 Inorganic chelating agents: A number of inorganic chelators have been investigated for remediation and some of complexation constants for those chelators are now available (Langmuir.1978; Waite *et al.*, 1994; Pabalan *et al.*, 1997; Pabalan *et al.*, 1998; Fjeld *et al.*, 2000; Winde, 2002; Gupta and Singh, 2005; Chao *et al.*, 1998). Polyphosphates are considered as the most efficient inorganic chelators. Their annual consumption is higher than that of organic chelating agents (Yousfi *et al.*, 1999; Chao *et al.*, 1998; Stumm and Morgan, 1996).

Traditional remediation approaches for U-contaminated soils include physical and chemical methods (Francis and Dodge, 1998; Zhu and Shaw, 2000) are often very costly, and may significantly reduce soil quality and damage the local ecosystem.

5. Biological remediation technologies

Bioremediation is the process of application of microbial agents and their products to clean up hazardous chemicals. Microorganisms are being used to remove organic matter and toxic chemicals from domestic and manufacturing waste effluents for many years. Microorganisms are ubiquitous, abundant and possess inherent capacity to adapt to extreme climatic conditions including heavy metal toxicity. Microorganisms can absorb heavy metals, but they cannot metabolize heavy metals, since heavy metals (especially Uranium) are not biodegradable, unlike hydrocarbons. Bioremediation of U involves the detoxification by immobilization, thus preventing their entry to food chain or by accumulation inside the cell, thus reducing the levels in the contaminated area. The microbes indulged in bioremediation process of U, includes bacteria- *Shewanella putrifaciens* (Baiget *et al.*, 2013), *Geobacter uraniireducens* (Holmes *et al.*, 2013), *Deinococcus radiodurans* (Misra *et al.*, 2012). The other biological organisms include eukaryotic algae- *Nostoc linckia, Porphyridium cruentum, Spirulina platensis* (Cecal *et al.*, 2012), *Synechococcus elongatus* (Acharya *et al.*, 2009), *Chara fragilis* (Dakovic *et al.*, 2008); yeast- *Rhodotorula glutinis* (Bai *et al.*, 2012) and fungi- *Trichoderma harzianum* (Akhtar *et al.*, 2000), *Clostridium acetobutylicum* (Vecchia *et al.*, 2010). Now-a-days serious attempts were made to explore indigenous microbial bioreduction capabilities merged with technological applications to transform U for effective and affordable environmental restoration.

Bioremediation process based on its operation mode is of two types: On-site and off-site remediation. The on-site bioremediation method ensures minimal exposure to public or site personnel's, minimal disruption to the site, immediate processing of contaminated soil and is cost effective. On the other side it is exposed to environmental factors, seasonal variation of microbial activity, supplying nutrients is time consuming and tedious process. The off-site bioremediation is carried when duration is less, contaminated area being small, process can be carried out with desired microbes in a controlled manner. The disadvantages include high disruption of site, highly expensive and disposal of treated waste. The site operations of bioremediation are at the interest of regulatory authorities and area of contamination.

A successful, cost-effective microbial dependent bioremediation programme is dependent on hydrogeologic conditions, the contaminant, microbial ecology, and other spatial/temporal factors that vary widely. Properly executed, microbal bioremediation can cost-effectively and expeditiously destroy or immobilize contaminants in a manner that fosters regulatory compliance and is protective of human health and the environment (Roane and Kellog, 1996; Dua *et al.*, 2002; Wagner-Döbler, 2003). Compared to physical and chemical methods this technology proved to be affordable and gained significance in remediating uranium contaminated soils, at the laboratory stage since manipulating microbes rather easy compared to plants.

Based on contact, uptake and its fate after entry of uranium with the microbe and their interactions bioremediation is of different types:

5.1 Biotransformation: Radionuclides, including uranium (U), technetium (Tc), and chromium (Cr), have been subjected to enzymatic detoxification by microbes. The process by which transformation of U occurs is reduction hence the process is referred as bioreduction, where in the oxidised form of U (VI) is converted to reduced form U (IV) by the microbial activity. The Oxidized forms of U is highly soluble in aqueous media and carried along with ground water, whereas reduced forms are insoluble and often precipitale into the solution (Humphries and Macaskie, 2002; Istok *et al.*, 2004). Carbon sources of microbes such as acetate, lactate or molasses and vegetable oils are the physiological electron donors to stimulate U (VI) reduction by microbial populations, native to contaminated aquifers (Anderson *et al.*, 2003; Istok *et al.*, 2004; Wall and Krumholz, 2006; Barlett *et al.*, 2012; Watson *et al.*, 2013). The enzyme system responsible for U(VI) reduction, tetraheme cytochrome c3, was characterized in the cells of *D. vulgaris* (Payne *et al.*, 2002). Because of the insoluble nature of U(IV) oxide, the site of its deposition in the cells should give an indication of the location of reductase. TEM analysis indicated that the precipitated uraninite has been located in the periplasm and outside of both Gram-negative and Gram-positive bacterial cells (Lovley and Phillips, 1992; Xu *et al.*, 2000; Liu and Fang, 2002) suggesting that U(VI) complexes do not generally have access to intracellular enzymes (Wall and Krumholz, 2006). For effective long-term immobilization of uranium through bioreduction, there should be a low probability of abiotic or biotic reoxidation of the insoluble U(IV) (Wall and Krumholz, 2006). DiSpirito and Tuovinen (1982) reported for the first time bacteria mediated aerobic U(IV) oxidation by *Acidithiobacillus ferroxidans* at acidic conditions (pH 1.5), where the energy is conserved and used for carbon dioxide fixation. This enzymatic U(IV) oxidation is not linked to energy conservation required for the cell growth (Finneran *et al.*, 2002; Beller, 2005; Senko *et al.*, 2005), but is coupled to nitrate reduction, a common contaminant at uranium contaminated sites (Riley and Zachara, 1992).

5.2 Biosorption: Biosorption describes the association of soluble substances with the cell surface. It encompasses both adsorption and absorption (Lloyd and Macaskie, 2002). This reactivity arises from the presence of a wide array of ionisable groups, such as carboxylate and phosphate, present in the lipopolysaccharides (LPS) of a Gram-negative (Beveridge and Koval, 1981), and the peptidoglycan, teichuronic acids, and teichoic acids of a Gram-positive bacterial cell wall (Frankel and Bazylinski, 2003). The bacterial cell wall may be over layered by a number of surface structures, which can also interact with metal ions. These may be composed primarily of carbohydrate polymers (capsules) or proteinaceous surface layers (S-layers) (Beveridge, 1994). Biosorbents used for biosorption of metals include exopolysaccharides, living cultures and non-living biomass. Cell surface sites capable for metal binding are sites containing carboxyl groups (pKb4.27–4.37), sites enriched with phosphate groups (pKH≈7), and sites rich on hydroxyl and amine groups (pKN8). The S-layer protein may function as a molecular sieve, ion trap, or protective shell (Sleytr, 1997). Because of the ability of S-layer to replace the "older" S-layer sheets on the cell surface, one can speculate about the mechanism of its protective function against uranium and other toxic metals. The saturation with metals (e.g., uranium) may lead to denaturation of the S-layer lattice, which is then replaced by freshly synthesized protein monomers (Merroun *et al.*, 2005). Recently a modified yeast *Rhodotorula glutinis* proved as a potential sorbent for U waste water treatment (Bai *et al.*, 2012).

5.3 Bioaccumulation: It is a metabolism-independent process in which, solutes are transported from the outside of the microbial cell into the cytoplasm through the cellular membrane. Various microbes, such as *Sphingomonas sp*. S15-S1 (Merroun *et al.*, 2006) *Micrococcus luteus, Arthrobactor nicotianae, Bacillus megaterium*, and *Citrobacter sp*. N14 ((Merroun *et al.*, 2006), have been reported to be used in the bioremediation of radioactive waste materials. Bacterial cells have developed several intracellular mechanisms to immobilize the accumulated U. One of the well studied processes is uranium chelation by polyphosphate bodies. Polyphosphate bodies have been observed in different bacterial strains such as *A. ferroxidans* (Merroun et al., 2003) and *Sphingomonas* sp. *S15-S1* (Merroun et al., 2006). It is reported that at pH value ranging from 4.5 to 7 uranium precipitates in the bacterial cells as Hautunite (Macaskie et al., 2000), and autunite/meta-autunite mineral phases (Jroundi et al., 2007; Martinez et al., 2007; Nedelkova et al., 2007). At physiological conditions (pH 6.9), HUO_2PO_4 and $NaUO_2PO_4$ phases have been precipitated around the cells of *Citrobacter* sp. (Macaskie et al., 1992, 2000). At acidic conditions (pH 4–5) which are the characteristics for the uranium mining wastes, indigenous acidic phosphatase activity of naturally occurring strains belonging to the genera *Bacillus* and *Rahnella* isolated from radionuclide- and metal contaminated soils have been involved in the precipitation of uranium (Martinez et al., 2007; Beazley et al., 2007).

5.4 Biostimulation / Bioaugumentation: One of the promising strategies for *in situ* remediation of U(VI) is the biostimulation of U(VI) immobilization. Biostimulation can be defined as addition of nutrients (carbon and other nutrient sources) that serves to increase the number or activity of indigenous microflora available for bioremediation activity (North et al., 2004). Uranium U(VI) is the most common radionuclide contaminant at the sites of nuclear complexes. Soluble U(VI) can be biologically reduced to U(IV), which is insoluble, thus immobilizing the radionuclide and posing less threat to drinking water resources located in the near areas of contamination. At the DOE Field Research Center (FRC) in Oak Ridge, TN, where groundwater contains >130 mM nitrate and micromolar concentrations of uranium, addition of a biodegradable electron donor results in denitrification as the primary terminal electron-accepting process (Istok et al., 2004). Because nitrate serves as a more energetically favourable electron acceptor, uranium reduction has been shown to occur only after nitrate has been depleted to low levels (Elias et al., 2003; Finneran et al., 2002, Michalsen et al., 2006, Senko et al., 2005). Thus, at sites such as the FRC, denitrifying bacteria are likely to play a critical role in uranium bioremediation. A recent phylogenetic survey of sediment from the FRC revealed several potential nitrate-reducing bacteria (Akob et al., 2007), but it remains unclear which species are involved in nitrate removal upon biostimulation.

5.5 Biofilms: Biofilms are defined, as single or multiple populations of microbes, growing by attaching to abiotic or biotic surfaces through extracellular polymeric substances. In the past, hexavalent uranium U(VI) has been immobilized using biofilms of the sulfate-reducing bacterium *Desulfovibrio desulfuricans* (Beyenal et al., 2004). The available evidences suggested that biofilm populations present in the spent nuclear fuel are directly involved in the accumulation of radionuclides, especially ^{60}Co from the contaminated water (Sarro *et al.*, 2005). Most of the radionuclides detected in the biofilm could be accumulated by biosorption processes. Bioremediation can be accelerated by genetic engineering and gene transfer methods for improved chemotactic ability of microbial strains among the biofilm organisms (Singh *et al.*, 2006).

5.6 Recent developments in Bioremediation process: The strategy and outcome of bioremediation in open systems or confined environments depend on a variety of physico-chemical and biological factors that need to be assessed and monitored. In particular, microorganisms are key players in bioremediation applications, yet their catabolic potential and their dynamics *in situ* remain poorly characterized. There is a need for development of more sensitive, precise and automated methodologies for large scale screening of structural and functional changes in microbial communities involved in bioremediation processes. Over the last few years, the scientific literature has revealed the progressive emergence of genomic high-throughput technologies in environmental microbiology and biotechnology.

Day to day technological advancements are available, even to monitor *in situ* bioremediation process to understand complex mechanisms like terminal electron acceptors, electron donors, etc., Further, other inventions like enzyme probe that can measure functional activity in the environment, functional genomic microarrays, phylogenetic microarrays, metabolomics, proteomics, and quantitative PCR are few that can provide unprecedented insights into the key microbial reactions employed in bioremediation. In some cases, a change in redox state is the simplest tool to bring about detoxification of hazardous metals and organic compounds. This is particularly true for metals and other radionuclides like U(VI), Cr(VI), and Tc(VII). While these cannot be degraded, they can be biotransformed decreasing their bioavailability, mobility and thus toxicity (Desjardin *et al.*, 2002; Elias *et al.*, 2003). Microbes can directly mediate such immobilization and detoxification by changing the valence states, utilizing them as electron acceptors (Finneran *et al.*, 2002). Proteogenomic analysis during U(VI) reduction field studies have been able to identify and track Geobacter-specific biomarker peptide citrate synthase (Wilkins *et al.*, 2011) during the process of biotransformation. Using qPCR as a technique for detection of phylogenetic and catabolic genes as indication of microbially mediated remediation is a popular and successful approach for monitoring detoxification in metal. Examples include monitoring *Anaeromyxobacter* strains involved in reduction of U(VI) (Thomas *et al.*, 2010). Recently, high throughput microarrays like the PhyloChip and the GeoChip have been extensively used in metal and organic bioremediation studies in order to quickly characterize microbial community and function (Van Nostrand *et al.*, 2009).

Previously using the 16S clone library based analysis at the area 3 FRC site in Oak Ridge which is contaminated with Uranium (up to 200mM) Cardenas *et al.* (2008) identified the bacterial species *Desulfovibrio, Geobacter, Anaeromyxobacter, Desulfosporosinus, Acidovorax,* and *Geothrix* spp. present concomitant with U(VI) reduction. Geo-Chip analysis of several groundwater monitoring wells reported widespread diversity of dsrAB genes (He *et al.*, 2007; Waldron *et al.*, 2009), which showed that sulfate-reducing bacteria were key players in U(VI) reduction. During the U(VI) reoxidation phase as studied in a sediment column with samples from FRC, observed decrease in biomass, but increase in microbial activity (Brodie *et al.*, 2006).

Using the PhyloChip, the study showed no decline in Geobacter or Geothrix spp. during the reoxidation phase, but members of Actinobacteria, Firmi-cutes, Acidobacteria, and Desulfovibrionaceae exhibited increased abundance (Brodie et al., 2006). GeoChip analysis during the reoxidation phase from field samples showed a decline in dsr genes but reoxidation did not appear to effect microbial functional diversity (Van Nostrand et al., 2009) suggesting that the microbial community was able to recover and continue to reduce U (VI) in the post oxidation phase. Recently, GeoChip 2.0 was used to characterize microbial communities under Fe-reducing conditions and the shift from Fe-reducing to sulfate-reducing conditions during *in situ* uranium Bioreduction. The results indicated that functional microbial communities altered with a shift in the dominant metabolic process as indicated with abundance of dsrAB genes (dissimilatory sulfite reductase genes) and methane generation-related mcr genes (methyl coenzyme M reductase coding genes) increased when redox conditions shifted from Fe-reducing to sulfate reducing conditions. The cytochrome genes detected were primarily from *Geobacter* sp. and decreased with lower subsurface redox conditions (Liang *et al.*, 2009). Recently Giloteaux *et al.*, (2013) developed a transcriptomic approach using with quantitative reverse transcription-PCR and analyzed the expression levels of arsenic respiration during *in situ* uranium bioremediation. The results demonstrated that subsurface *Geobacter* species can tightly regulate their physiological response to changes in ground water arsenic concentrations.

In addition to genomics technology, proteomics technology has been proven effective in studying proteins involved in radionuclide bioremediation. Proteomic analysis of membrane proteins from a radioresistant and moderate thermophilic bacterium *Deinococcus geothermalis* revealed a total of 552 differentially regulated proteins, including a cytochrome bd ubiquinol oxidase that were associated with radioresistant mechanisms of the organism (Tian *et al.*, 2010). Comparative proteomics analysis of wild type (R1) and a *pprI* knock-out strain (YR1) of a extremely radioresistant bacterium *Deinococcus radiodurans* under ionizing irradiation conditions shows significant up-regulation of 31 proteins involved in DNA replication and repair, in the wild type when compared to its knock-out counterpart (Lu *et al.*, 2008). Very recently environmental metaproteomics approach has identified the microbial community members of the Betaproteobacteria (i.e., *Dechloromonas, Ralstonia, Rhodoferax, Polaromonas, Delftia, Chromobacterium*) following biostimulation and also elucidated active pathways responses to biostimulation at a uranium- and nitrate-contaminated site (Chourey *et al.*, 2013).

Advanced technologies are very much required to identify the specific microbes and to understand their specific mechanisms of tolerance by studying pathways and underlying genes and proteins involved in the remediation of uranium.

5.7 Role of genetic engineering in uranium Bioremediation: Genetic engineering has the potential to improve or redesign microorganisms, where biological metal-sequestering systems will have a higher intrinsic capability as well as specificity and greater resistance to environmental conditions (Bae *et al.*, 2000; Majare and Bulow, 2001). Naturally occurring bacteria highly resistant to radiation are ideal metabolic engineering candidates for enhanced radionuclide cleanup (Brim *et al.*, 2003). First reports had come from the use of an extremely radiation-resistant and thermophilic bacterium *Deinococcus geothermalis* by expressing the *mer* operon from *E. coli* coding for Hg^{2+} reduction. The engineered bacteria were capable of reducing mercury at higher temperature and ionizing radiation. The engineered bacteria also had the capability to reduce other metals like Fe (III), U(VI), and Cr(VI). This demonstrates the possibility of utilizing such engineered microorganisms for mixed radioactive wastes and at higher temperatures. Polyphosphates could also be used in *P. aeruginosa* for radionuclide precipitation as metal phosphate by overexpressing polyphosphate kinase and exopolyphosphatases (Renninger *et al.*, 2004). Recently a non-specific phosphatase phoN was expressed in a radiation-resistant bacterium *D. radiodurans* leading to bioprecipitation of uranium from dilute nuclear waste (Appukuttan *et al.*, 2006). Lyophilized cells of recombinant strains of *Deinococcus* retained viability and PhoN activity for up to six months of storage at room temperature and could efficiently precipitate uranium from aqueous solutions. The precipitated uranyl phosphate remained tightly associated with the cell surface, thus facilitating easy recovery (Appukuttan *et al.*, 2011). These lyophilized cells could able to remove 70% of uranium when immobilized in polyacrylamide gels (Mishra *et al.*, 2012). Directed evolution has also been used to improve the enzymatic efficiency of chromate and uranyl reductases; engineered enzymes expressed in *E. coli* and *P. putida* showed improved enzymatic kinetics (Barak *et al.*, 2006). It is clear that these improved enzymes when expressed in radiation-resistant microorganism could further enhance the radionuclide precipitation efficiency. Although, genetically engineered microbes hold considerable promises, their potential use in field bioremediation will require additional studies to develop a safe environmental cleanup.

6. Plant based remediation methods

Phytoremediation is an integrated approach that combines the disciplines of plant physiology, soil chemistry, and soil microbiology to cleanup the contaminated soils. This technique takes advantage of the natural abilities of plants to take up (absorb) and accumulate metals and radionuclides (McIntyre, 2003). These plants could be used in an efficient way if they are adapted to a wide range of environmental conditions. Plants for phytoremediation are tolerant plants, having heavy metal hyper accumulation potential, which could be beneficial in phytoremediation for cleanup of soil and water. On the other hand, tolerant food crops, if exposed to heavy metals in their growth medium, may be dangerous as carriers of these toxic metals into the food chain leading to food toxicity (Gavrilescu *et al.*, 2009).

Plants for phytoremediation of U-contaminated soils could be selected by using a mathematical model related to plant characteristics (e.g., biomass and planting density) to predict a long-term U-removal rate from the contaminated soil (Hashimoto *et al.*, 2005).

Plant-assisted remediation of soil can generally occur through one or more of the following mechanisms viz.,

Phytoextraction is the use of plants for removal of pollutants.

Phytostimulation is representing the stimulation of microbial degradation by the effect of plant exudates (Dushenkov *et al.*, 1999; Gavrilescu *et al.*, 2009).

Rhizofiltration is the exploitation of plant roots for absorption or adsorption of pollutants, especially metals, from water. From many tested plants, e.g., hydroponically grown sunflowers proved to be very effective in removal of radionuclides from surface waters around Chernobyl (Meagher, 2000).

Phytostabilisation is a method that uses plants to reduce the mobility and availability of the pollutants in the environment.

Phytovolatilisation is uptake and transfer of some pollutants into gas phase by plants.

6.1 Phytoextraction of Uranium contaminated soils: Among all the methods discussed above Phytoextraction is an environmentally friendly and cost effective technique that has been proposed by many researchers to extract U and various other inorganic contaminants from soils (Salt *et al.*, 1995; Cunningham *et al.*, 1995; Dushenkov *et al.*, 1997; Ebbs *et al.*, 1998; Huang *et al.*, 1998). This technology involves the extraction of metals by plant roots and the translocation thereof to shoots. The shoots are subsequently harvested to remove the contaminants from the soil. Salt *et al.*, (1995) reported that the costs involved in phytoextraction would be more than ten times less per hectare compared to conventional soil remediation techniques. Phytoextraction also has environmental benefits because it is considered a low impact technology. Furthermore, during the phytoextraction procedure, plants cover the soil erosion and leaching will thus be reduced. With successive cropping and harvesting, the levels of contaminants in the soil can be reduced (Vandenhove *et al.*, 2001). Harvested biomass can be incinerated to reduce volume (Chaney *et al.*, 1997) and stored as hazardous waste or the metals can be recycled and sold (termed phytomining; Anderson *et al.*, 1999). To remove sufficient amounts of heavy metals with this technique, plants have to be highly efficient in metal uptake and translocation into their above ground vegetative parts. The phytoextraction process is, however, limited because of low bioavailability of metals in soils.

There are two general approaches to phytoextraction: Natural hyperaccumulation and induced phytoextraction.

6.1.1 Natural hyperaccumulation: The first approach uses naturally hyperaccumulating plants with the ability to accumulate an exceptionally high metal content in the shoots (Lentan, 2006). There could be several problems identified with applications of U hypraccumulator plants: (1) plants take up the more available metal fraction, but less available fractions cannot be extracted, and (2) hyperaccumulators often have low biomass, which results in a low amount of metal extracted from the site.

Researchers attributed several traits for the ideal natural hyperaccumulator plant species for phytoremediation of radioactive metals. First, the plants should have either a low biomass with a high metal capacity or a high biomass plant with an enhanced metal uptake potential. Specifically, the plant should have a sufficient capacity to accumulate the metal of concern within the harvestable biomass at a level greater than 1% (for some metals, greater than 1000 mg kg^{-1}). Furthermore, the plant should have a sufficient capacity to tolerate the site conditions and accumulate multiple metal contaminants. Finally, the species should be fast growing and have a suitable plant phenotype for easy harvest, treatment, and disposal (McIntyre, 2003).

In the investigations of Shahandeh and Hosssner (2002), thirty four plant species were screened for uranium (U) accumulation, among the plants screened sunflower and Indian mustard had the highest capability of uranium accumulation from U contaminated soil. According to the PHYTOREM data base, sunflowers are recognized as hyperaccumulators of uranium. PHYTOREM was developed by the Environment of Canada and these databases consist of 775 plants with capabilities to accumulate or hyperaccumulate one to many of 19 key metallic elements. Species were considered as hyperaccumulators if they took up greater than 1,000 mg/kg^{-1} dry weight of most metals. Sunflowers had a content of uranium of more than 15,000 mg kg^{-1} dry weight. Plant hyperaccumulators like sunflowers (*Helianthus annuus*) have the highest phytoremediation potential since there are also crop plants with well established cultivation methods (McIntyre, 2003). The index of tolerance and the bioaccumulation coefficient were two indices used for screening plants and evaluating metal uptake and phytotoxicity effects (Dushenkov *et al.*, 1995; Nanda-Kumar *et al.*, 1995). *Uncinia leptostachya* and *Coprosma arborea* were considered unusual U accumulators, whose U contents were around 3 mg kg^{-1} a.w. (Peterson, 1971). Furthermore, the leaves of black spruce (*Picea mariana*) were reported to contain U in excess of 1,000 mg kg^{-1} dry weight (Chang *et al.*, 2005). Recently to screen the candidate plant species for phytoremediation of uranium mill tailings, Li *et al.*, (2011) were collected 15 dominant plant species belonging to 9 different families from the uranium mill tailings repository in South China. Among the plants screened, based on the phytoremediation factor (PF) *Phragmites australis* was found to have the greatest removal capabilities for uranium (820 μg). Although the concentration of a target element in a plant does not satisfy the criteria for a hyperaccumulator, the plant may also be considered as the candidate for phytoremediation if it has relatively high biomass. Hence, based on the hyperaccumulator criteria *Cyperus iria* was identified as a hyperaccumulator for uranium (36.4 μg/g), and could be the candidates for phytoremediation of uranium.

So far, plant species used for extraction of U(VI) from contaminated soils were dicotyledonous and monocotyledonous plants: (field crops, cool and warm season grasses, and the *Brassica* family). Plant species selection was based on the agronomic importance of the crop, dry matter production, and apparent tolerance to heavy metals. Research findings found a significant difference in accumulation between plant species, and among the plants studied sunflower and Indian mustard plants showed the highest uranium accumulation.

6.1.2 Induced phytoextraction: This method involves the addition of metal-chelates (Shahandeh *et al.*, 2002), microbial activities (de Souza *et al.*, 1999), and genetic engineering approaches (Dhankher *et al.*, 2002) could be enhance the efficiency of phytoextraction.

6.1.2.1 Metal –Chelater mediated induced phytoextraction: This soil remediation technique makes use of fast growing high biomass accumulating crop plants, which differ from natural hyperaccumulating plants in that they are not capable of accumulating and translocating sufficient amounts of metals without the addition of amendments (Blaylock *et al.*, 1997; Huang *et al.*, 1998; Vandenhove *et al.*, 2001). Chelates bind metals in the soil and/ or acidify the soil solution, which increases bioavailability and aid in the translocation of metals from root to shoot (Blaylock *et al.* 1997). Amendments could be organic compounds such as synthetic chelating agents (ethylenediaminetetraacetic acid (EDTA), N-hydroxyethyl-ethylenediamine-N,N′,N′-triacetic acid (HEDTA), diethylenetrinitrilopentacetic acid (DTPA)), natural fulvic acid, humic acid, and more natural low molecular weight organic acids (citric, malic, oxalic, and acetic acid). In soils, complexation of heavy metals with various complexing agents may follow the order, EDTA and related synthetic chelators > nitrilotriacetic acid (NTA) > citric acid (CA) > oxalic acid (OA) > acetic acid, as was shown by several comparative experiments (Krishnamurti *et al.*, 1998; Evangelou *et al.*, 2007). The most frequently used is EDTA, which has been reported as more effective than other synthetic chelators for several heavy metals. Although EDTA is much effective but EDTA-heavy metal complexes are toxic to plants and soil microorganisms because they can persist in the environment due to their lower rate of biodegradation (Grcvman *et al.*, 2001; Quartacci *et al.*, 2005). However, recently Jagetiya and Sharma (2013) observed that minimum growth inhibition produced by chelators occurred in NTA which was followed by OA, moderate in CA and maximum was traced in EDTA applications. Chelator strengthened U uptake in the present study follows the order: CA > EDTA > OA > NTA.

Several researchers (Qualls and Haines, 1992; Jones and Darrah, 1994; Jones *et al.*, 1996; Huang *et al.*, 1998) have reported that citric acid is a more environmentally friendly chelate to use in phytoextraction due to the rapid degradation rate of citric acid. Huang *et al.*, (1998) reported that *Brassica juncea* achieved maximum shoot-U concentrations after three days of citric acid addition where after the concentration curve reached a steady state. This *in situ* decontamination strategy is also more environmentally friendly and cost effective than the conventional soil remediation techniques (Cunningham *et al.*, 1995), which include soil excavation and metal leaching.

6.1.2.2 Microbial mediated induced phytoextraction: A promising alternative to chemical amendments could be the application of microbe-mediated processes, in which the microbial metabolites/processes in the rhizosphere effect plant metal uptake by altering the mobility and bioavailabity (Aafi *et al.*, 2012; Glick, 2010; Ma *et al.*, 2011; Miransari, 2011; Rajkumar *et al.*, 2010; Wenzel, 2009; Yang *et al.*, 2012). Microbial activities in the root/rhizosphere soils enhance the effectiveness of phytoremediation processes in metal contaminated soil by two complementary ways: (i) Direct promotion of phytoremedation in which plant associated microbes enhance metal translocation (facilitate phytoextraction) or reduce the mobility/ availability of metal contaminants in the rhizosphere (phytostabilization) and (ii) Indirect promotion of phytoremediation in which the microbes confer plant metal tolerance and/or enhance the plant biomass production in order to remove/arrest the pollutants.

Among the microorganisms involved in heavy metal phytoremediation, the rhizosphere bacteria deserve special attention because they can directly improve the phytoremediation process by changing the metal bioavailability through altering soil pH, release of chelators (e.g., organic acids, siderophores), oxidation/reduction reactions (Gadd, 2000; Khan *et al.*, 2009; Kidd *et al.*, 2009; Ma *et al.*, 2011; Rajkumar *et al.*, 2010; Uroz *et al.*, 2009; Wenzel, 2009). Similarly the metal tolerant mycorrhizal fungi have also been frequently reported in hyperaccumulators growing in metal polluted soils indicating that these fungi have evolved a heavy metal-tolerance and that they may play important role in the phytoremediation of the site (Gohre and Paszkowski, 2006; Miransari, 2011; Or³owska *et al.*, 2011; Zarei *et al.*, 2010). Among soil microorganisms, mycorrhizal fungi are the only ones providing a direct link between soil and plant. Arbuscular mycorrhizal (AM) fungi form association through root symbiosis with 80- 90% of all seed plant species (Harrison, 1997). However, mycorrhizas can also lead to reduced metal uptake and increased plant growth in metal-contaminated soils (Leyval and Joner, 2001). Chen *et al.*, (2005) showed that mycorrhiza decreased U translocation from roots to shoots from soil containing 2% phosphate rock. In another set of studies by Rufyikiri *et al.*, (2002, 2003) actual uptake and translocation of U from extra radical AM fungus mycelium to roots were observed in an *in vitro* culture by utilising a separate compartment for the U where only the mycelium had access. The uptake of U by the roots was largely influenced by the presence or absence of AM fungus. They also observed that the concentration of U in hyphae was about 5-10 times the concentration in mycorrhizal and non-mycorrhizal roots, while the concentration was only 1.8 times larger for mycorrhizal roots than for non-mycorrhizal roots. The high specific U concentration in the fungal mycelium combined with an observed higher cation exchange.

Most plant-associated bacteria and fungi can produce iron chelators called siderophores in response to low iron levels in the rhizosphere. Siderophores are low-molecular mass (400–1,000 Daltons) compounds with high association constants for complexing iron, but they can also form stable complexes with other metals, such as Al, Cd, Cu, Ga, In, Pb and Zn (Glick and Bashan, 1997; Schalk *et al.*, 2011). Although siderophores contain other functional groups, they are broadly classified into three main groups based on the chemical nature of themoieties donating the oxygen ligands for Fe(III) coordination, which are either of the catecholates (enterobactin), hydroxamates (desferrioxamines), or (á-hydroxy-)carboxylates (aerobactin). Since siderophores solubilize unavailable forms of heavy metal bearing minerals by complexation reaction, siderophores producing microbes that inhabit the rhizosphere soils are believed to play an important role in heavy metal phytoextraction (Braud *et al.*, 2009; Dimkpa *et al.*, 2009; Rajkumar *et al.*, 2010).

6.1.2.3 Genetic Engineering mediated induced phytoextraction: The plant species currently being developed for phytoremediation seem capable of effective bioaccumulation of targeted contaminant, but efficiency might be improved through the use of transgenic (genetically engineered) plants. In general, any dicotyledon plant species can be genetically engineered using the *Agrobacterium* vector system, while most monocotyledon plants can be transformed using particle gun or electroporation techniques. Some promising transgenics that show higher tolerance, accumulation, and /or degradation capacity for various pollutants have been developed. This was achieved either by overproducing metal chelating molecules such as citrate (de la Fuente *et al.*, 1997), PCs (Zhu *et al.*, 1999 a,b), MTs (Evans *et al.*, 1992; Hasegawa *et al.*, 1997) or ferritin (Goto *et al.*, 1999) or by overexpression of metal transporter proteins (Samuelsen *et al.*,1998; Arazi *et al.*, 1999; Van der Zaal *et al.*, 1999; Curie *et al.*, 2001). Naturally occurring plant species that can be genetically engineered for improved phytoremediation include *Brassica juncea* for phytoremediation of heavy metals from soil (Dushenkov *et al.*, 1995), *Helianthus annuus* (Dushenkov *et al.*, 1995) and *Chenopodium amaranticolor* (Eapen *et al.*, 2003) for rhizofiltration of uranium.

The *A. rhizogenes*-induced hairy root system has a wide range of applications, from altering plant architecture to phytoremediation, and also in a broad spectrum of biotechnological research (Doran, 1997; Shanks and Mo rgan, 1999; Eapen and Mitr a, 2001). Plant species belonging to Cruciferae (mustard, *Thlaspi*), Chenopodiaceae (*Chenopodium*, spinach), and Compositae (sunflower) have been shown to possess genetic potential to extract heavy metals from soil or water and accumulate them in plant parts (Koboi *et al.,* 1986; Baker and Brooks, 1989). Eapen *et al.,* (2003) demonstrated the potentiality of hairy root cultures of *Brassica* and *Chenopodium,* the two species endowed with genetic potential of extracting heavy metals from aqueous solutions. The results indicated that the hairy roots could remove uranium from the aqueous solution within a short period of incubation. *B. juncea* could take up 20–23% of uranium from the solution containing up to 5000mM, when calculated on g/g dry weight basis. *C. amaranticolor* showed a slow and steady trend in taking up uranium, with 13% uptake from the solution of 5000mM concentration. Root growth was not affected up to 500mM of uranium nitrate over a period of 10 days.

Certain plants have developed the ability to acquire plant nutrients such as iron through the production of root exudates (organic acids and siderophores) that chelate iron in the rhizosphere. The genetic manipulation of plants to enhance the production of specific chelates or organic acids would provide a promising means to enhance the natural uptake of heavy metals and minimize or eliminate the need for soil applied chelating agents.

The increase in metal accumulation as the result of these genetic engineering approaches is typically two to threefold more metal per plant, which potentially enhances phytoremediation efficiency by the same factor. It is not yet clear how applicable these transgenics are for environmental cleanup, since no field studies have been reported except one using transgenics. Indian mustard plant that overexpresses enzymes involved in sulfate/ selenate reduction. (Pilon Smits *et al.,* 1999; Zhu *et al.,* 1999). Potential environmental impacts of transgenics such as competitiveness of transgenic to wild type, effect on birds, insects etc., that might feed on plant biomass containing high concentration of toxic metals and possibility of gene transfer to other plants by pollination require continuous monitoring. Genetic engineering of the chloroplast genome offers a novel way to obtain high expression without the risk of spreading the transgene via pollen (Ruiz *et al.,* 2003). In future, as more data on field trials and associated risk assessment would be available, transgenics will play an important role in commercial phytoremediation.

7. Assets of Phytoremediation of Uranium

Phytoremediation of radionuclides has many advantages over the traditional treatments. First, in phytoremediation the soil is treated *in situ,* which does not cause further disruption to the soil dynamics. Secondly, once plants are established, they remain for consecutive harvests to continually remove the contaminants. In a single growing season, it is possible to grow and harvest multiple crops (Huang *et al.* 1998). Plants also stabilize the soil by reducing wind and water erosion. Thirdly, phytoremediation reduces the time workers are exposed to the radionuclides (Negri and Hinchman 2000). Finally, phytoremediation can be used as a long term treatment that can provide an affordable way to restore radionuclide contaminated areas (Dushenkov *et al.* 1998).

8. Criteria to be met by plants for successful Phytoremediation

- The most important is that the radionuclides be spread through-out a large area and present in low-level concentrations.
- The radionuclides must be bioavailable in water or soil solution in order to take them up into their roots by plants.
- The plants themselves must also be tolerant of these radionuclides when they are accumulated into their biomass.
- The best plants for phytoremediation are those that have an extensive root system and adequate above-ground biomass

Conclusions and future prospects

There is an increasing trend of uranium accumulating in soils due to a number of deliberate or wrong practices. Public and political pressure to solve a problem situation of this nature occurs when critical toxic levels are reached. As a consequence, there would be a risk for ecosystems, agro-systems and health. It is suggested that knowledge of the mechanisms that control the behaviour of such heavy metals must be improved and can be used for risk assessment and proposition of remediation treatments. To limit U dispersion in the environment, different remedial actions including physio-chemical and biological methods have been proposed. Physio chemical technologies, such as PRB techniques, provide the possibility of efficient clean-up of the contaminated groundwater. According to the bioremediation data available to date, it seems that biosorption might be a promising approach to remediate uranium contaminated environments, since some bacteria and microbial assemblages have shown to have the capability to adsorb uranium from aqueous phase. However, additional efforts need to be paid, towards a better understanding of principles behind biosorption, especially in uranium biosorption by biofilm, a promising but complex microbial system for uranium bioremediation. Very limited data is available, regarding microbial bioremediation of uranium, at a systems biology level. This could be done, by combining two and more of the newly developed metagenomic tools, e.g., 454-pyrosequencing, functional gene array, proteomics, transcriptomics and metabolomics.

For remediation of uranium-contaminated soils, or ecological restoration of uranium mine tailings, phytoremediation turns out to be one of the best choices, although further research is still necessary to search for ideal uranium hyperaccumulators and also to understand molecular mechanisms of hyperaccumulation, for more efficient phytoextraction. Hyperaccumulator plants possess genes that regulate the amount of metals taken up from the soil by roots and sequestration within the plant. These genes govern processes that can increase the solubility of metals in the soil surrounding the roots as well as to mobilize the metals into root cells. From there, the metals enter the plant's vascular system for further deposit in the plant. Secretion of organic acids and organic ligands by plants is a universal phenomenon that has a profound influence on several abiotic and biotic interactions in soil. Plants produce and secrete a variety of organic acids including citrate, malate and oxalate, among which, citrate is gaining importance as the most efficient chelators of heavy metals in the soil. Therefore, the identification of novel genes involved in the acquisition and the homeostasis of toxic compound, as well as an understanding the way they are regulated, will encourage real improvement in phytoremediation systems.

Acknowledgments

This work was supported by financial grant (No.BT/PR-10531/BCE/08/660/2008) from Department of Biotechnology (DBT), New Delhi, India.

References

Aafi, N.E., Brhada, F,. Dary, M., Maltouf, A.F. and Pajuelo E. 2012. Rhizostabilization of metals in soils using *Lupinus luteus* inoculated with the metal resistant rhizobacterium *Serratia* sp. MSMC 541. International Journal of Phytoremediation,**14**:261–74.

AbdEl-Sabour, M.F. 2007. Remediation and bioremediation of uranium contaminatedsoils, Electronic Journal of Environmental Agricultural Food Chemistry, **6** : 2009–2023.

Acharya, C., Joseph, D. and Apte, S.K. 2009. Uranium sequestration by a marine cyanobacterium, *Synechococcus elongates* strain BDU/75042. Bioresource Technology, **100(7):** 2176-81.

Agrawal, Y.K., 2006. Selective supercritical fluid extraction of uranium(VI) using *N*-phenyl-(1,2- methano-fullerene C60)61-formohydroxamic acid and simultaneous on-line determination by inductively-coupled plasma-mass spectrometry (ICP-MS), fullerenes, Nanotubes Carbon Nanostructure, **14** : 621–639.

Akhtar, K., Khalid, A., Akhtar, M., and Ghauri, M. 2009. Removal and recovery of uranium from aqueous solutions by Ca-alginate immobilized *Trichoderma harzianum*. Bioresource Technology, **100 (20):** 4551–4558.

Akob, D. M., H. J. Mills, L. Kerkhof, and J. E. Kostka. 2007. Determination of the metabolically active microbial groups in contaminated ORFRC subsurface sediments using stable isotope probing (SIP). DOE 2nd Annual ERSP Spring PI Meeting,

Allard, B., Olofsson, U. and Torstenfelt, B. 1984. Environmental actinide chemistry. Inorganica Chimica Acta, **94:** 205–221.

Anderson, C. W. N., Brooks, R. R., Chiarucci, A., LaCoste, C. J., Leblanc, M., Robinson, B. H., Simcock, R. and Stewart, R. B. 1999. Phytomining for nickel, thallium and gold. Journal of Geochemical Exploration, 67:407-415.

Anderson, R.T., Vrionis, H.A., Ortiz-Bernad, I., Resch, C.T., Long, P.E., Dayvault, R., *et al.* 2003. Stimulating the *in situ* activity of *Geobacter* species to remove uranium from the groundwater of a uranium-contaminated aquifer. Applied Environmental Microbiology, **69:** 5884–5891.

Appukuttan, D., Rao, A.S. and Apte, S.K. 2006. Engineering of *Deinococcus radiodurans* R1 for bioprecipitation of uranium from dilute nuclear waste. Applied Environmental Microbiology, **72(12):** 7873-8.

Appukuttan, D., Seetharam, C., Padma, N., Rao, A.S. and Apte, S.K. 2011. PhoN-expressing, lyophilized, recombinant *Deinococcus radiodurans* cells for uranium precipitation. Journal of Biotechnology, 20; **154(4):** 285-290.

Arazi, T., Sunkar, R., Kaplan, B. and Fromm,H. 1999. A tobacco plasma membrane calmodulin binding transporter confers Ni^{2+} tolerance and Pb^{2+} hypersensitivity in transgenic plants. Plant Journal, **20:** 171-82.

ATSDR (Agency for Toxic Substances and Disease Registry), 1999: US Public HealthService, Department of Health & Human Services. Toxicological Profile for Uranium, Atlanta, GA

Bae, W., Chen, W., Mulchandani, A. and Mehra, R.K.2000. Enhanced bioaccumulation of heavy metals by bacterial cells displaying synthetic phytochelatins. Biotechnology Bioengineering, **70:** 518–24.

Bai, J. Wu, X.,Fan, F., Tian,W.,Yin, X., Zhao, L.,Tian,L., Qin, Z. and Guo, J. 2012. Biosorption of uranium by magnetically modified *Rhodotorula glutinis*. Enzyme Microbe Technology, 10;**51(6-7):** 382-7.

Baiget , M., Constanti, M., Lopez, M.T. and Medina , F. 2013 . Uranium removal rom a contaminated effluent using a combined microbial and nanoparticle system. Nature Biotechnology ,S1871-6784 (13) 00064-2.

Baker, A.J.M. and Brooks, R.R. 1989. Terrestrial higher plants which hyperaccumulate metallic elements—a review of their distribution, ecology and phytochemistry. Biorecovery, **1:** 81-126.

Barak, Y., Ackerley, D.F., Dodge, C.J., Banwari, L., Alex, C., Francis, A.J. and Matin, A. 2006. Analysis of novel soluble chromate and uranyl reductases and generation of an improved enzyme by directed evolution. Applied Environmental Microbiology.**72(11):** 7074-82.

Barlett, M., Moon, H.S., Peacock, A.A., Hedrick, D.B.,Williams, K.H., Long, P.E., Lovely, D. and Jaffe, P.R. 2012. Uranium reduction and microbial community development in response to stimulation with different electron donors. Biodegradation,

Barton ,C.S., Stewart D.I., Morris, K.. and Bryant, D.E. 2004. Performance of three resin based materials for treating uranium-contaminated groundwater within a PRB, Journal of Hazardous Materials, B116: 191–204.

Beazley, M.J., Martinez, R.J., Sobecky, P.A., Webb, S.M. and Taillefert, M. 2007. Uranium biomineralization as a result of bacterial phosphatase activity: insight from bacterial isolates from a contaminated subsurface. Environmental Science and Technology, 41: 5701–5707.

Beller, H. 2005. Anaerobic, nitrate-dependent oxidation of U(IV) oxide minerals by the chemolithoautotrophic bacterium *Thiobacillus denitrificans*. Applied Environmental Microbiology. 71: 2170–2174.

Beveridge, T.J., 1994. Bacterial S-layers. Current Opinions in Structural Bioliogy, 4: 204–212.

Beveridge, T.J. and Koval, S.F., 1981. Binding of metals to cell envelopes of *Escherichia coli* K12. Applied Environmental Microbiology, 42: 325–335.

Beyenal, H., Sani, R.K., Peyton, B.M., Dohnalkova, A.C., Amonette, J.E. and Lewandowski, Z. 2004. Uranium immobilization by sulfate-reducing biolms. Environmental Science & Technology, 38 (7): 2067–2074.

Blaylock, M.J., D.E. Salt, S. Dushenkov, O. Zakharova, C. Gussman, Y. Kapulnic, B.D. Ensley and I. Raskin. 1997. Enhanced accumulation of Pb in Indian mustard by soil-applied chelating agents. Environmental Science and Technology 31: 860-865.

Boyd, G. and Hirsch, R.M. 2002. Handbook of groundwater remediation using permeable reactive barriers. Academic Press.

Braud, A, Jézéquel ,K, Bazot, S. and Lebeau, T. 2009. Enhanced phytoextraction of an agricultural Cr,Hg- and Pb-contaminated soil by bioaugmentation with siderophore producing bacteria. Chemosphere,74: 280–6.

Brim, H., Venkateshwaran, A., Kostandarithes, H.M., Fredrickson, J.K and, Daly, M.J. 2003. Engineering *Deinococcus geothermalis* for bioremediation of high-temperature radioactive waste environments. Applled Environmental Microbiology,69: 4575-4582.

Brodie, E.L., DeSantis, T.Z., Joyner, D.C., Baek, S.M., Larsen, J.T., Andersen, G.L., Hazen, T.C., Richardson, P.M., Herman , D.J., Tokunaga, T.K. et al. 2006. Application of a high-density oligonucleotide microarray approach to study bacterial population dynamics during uranium reduction and reoxidation. Applied Environmental Microbiology , 72:6288-6298.

Bronstein, K. 2005. Permeable reactive barriers for inorganic and radionuclide contamination, U.S. Environmental Protection Agency Office of Solid Waste and Emergency Response Office of Superfund Remediation and Technology Innovation, Washington, DC, Onlineat:http://cluin.org/download/studentpapers/brnsteinprbpaper.pdf.

Cardenas, E., Wu, W.M., Leigh, M.B., Carley, J., Carroll, S., Gentry, T., Luo, J., Watson, D., Gu, B., Ginder-Vogel, M. et al. 2008. Microbial communities in contaminated sediments, associated with bioremediation of uranium to submicromolar levels. Applied Environmental Microbiology, 74: 3718-3729.

Cecal, A., Humelnicu, D., Rudic, V., Cepoi, L., Ganju,D. and Cojocari, A. 2012. Uptake of uranyl ions from uranium ores and sludges by means of *Spirulina platensis, Porphyridium cruentum* and *Nostoc linckia* alga. Bioresource and Technology, **118:** 19-23.

Chaney, R.L., Malik, M., Li, Y.M., Brown, S.L., Brewer, E.P., Angle, J.S. and Baker, A.J.M. 1997. Phytoremediation of soil metals. Current Opinions in Biotechnology, **8:** 279-284.

Chang, P., Kim, K.W., Yoshida, S. and Kim, S.Y. 2005. Uranium accumulation of crop plants enhanced by citric acid. Environmental Geochemistry and Health, **27:** 529–538.

Chao, J.C., Hong, P.T., Okey, R.W. and Peters, R.W. 1998. Selection of chelating agents for remediation of radionuclide contaminated soil, in: Proceedings of the 1998 Conference on hazardous waste research, Bridging gaps in technology and culture, Snowbird, Utah,:18–21, 142–160.

Chen, B., Roos, P., Borggard, O.K., Zhu, Y.G. and Jakobsen, I. 2005. Mycorrhiza and root hairs in barley enhance acquisition of phosphorus and uranium from phosphate rock but mycorrhiza decreases root to shoot uranium transfer. New Phytologist, **16:** 591-598.

Chourey, K., Nissen, S., Vishnivetskaya, T., Shah, M., Pfiffner, S., Hettich, R.L. and Löffler, F.E. 2013. Environmental proteomics reveals early microbial community responses to biostimulation at a uranium- and nitrate-contaminated site. Proteomics. 2013 Jul 26. doi: 10.1002/pmic.201300155.

Conca, J.L. and Wright, J. 2003. Apatite II to remediate soil or groundwater containing uranium or plutonium, in: Radiochemistry Conference, Carlsbad, On line at: www.clu in.org/conf/itrc/prb/pu=apatite.pdf.

Cunningham, S. and Ow, O.W. 1996. Promises and prospects of phytoremediation. Plant Physiology, **110:** 715-719.

Curie, C., Panaviene, Z., Loulergue, C., Dellaporta, S.L., Briat, J.F. and Walker, E.L. 2001. Maize yellow stripe 1 encodes a membrane protein directly involved in Fe(III) uptake. Nature, **409:** 346-349.

Dakovic, M., Kovacevic, M., Andjus, P. and Bacic, G. 2008 On the mechanism of uranium binding to cell wall of *Chara fragilis*. European Biophysics Journal, **37:** 1111-1117.

De la Fuente, J.M., Ramírez-Rodríguez, V., Cabrera-Ponce,J.L. and Herrera-Estrella, L. 1997. Aluminum tolerance in transgenic plants by alteration of citrate synthesis. Science **276:** 1566–1568.

de Souza, M.P., Huang, C.P., Chee, N. and Terry, N. 1999. Rhizosphere bacteria enhance the accumulation of selenium and mercury in wetland plants. Planta, **209(2):** 259–263.

Desjardin, V., Bayard, R., Huck, N., Manceau, A. and Gourdon, R. 2002. Effect of microbial activity on the mobility of chromium in soils. Waste Management, **22:** 195-200.

Dhankher, O.P. L., Rosen, Y.B.P., Shi, J., Salt, D., Senecoff, J.F., Sashti, N.A. and Meagher, R.B. 2002. Engineering tolerance and hyperaccumulation of arsenic in plants by combining arsenate reductase and gamma-glutamylcysteine synthetase expression. Nature Biotechnology, **20:** 1140–1145.

Dimkpa, C.O., Merten, D., Svatos, A., Büchel, G. and Kothe, E. 2009. Siderophores mediate reduced and increased uptake of cadmium by *Streptomyces tendae* F4 and sunflower (*Helianthus annuus*), respectively. Journal of Applied Microbiology, **107:** 1687–96.

DiSpirito, A.A. and Tuovinen, O.H. 1982. Uranous ion oxidation and carbon dioxide fixation by *Thiobacillus ferrooxidans*. Archives of. Microbiology, **133:** 28–32.

Doran, P.M. 1997. Hairy Roots: Culture and Applications. Harwood Academic, Australia.

Dua, M., Singh, A., Sethunathan, N. and Johri, A.K. 2002. Biotechnology and bioremediation: successes and limitations. Applied Microbiology and Biotechnology, **59:** 143–15.

Duff, M. C. and Amrhein, C. 1996. Uranium(VI) adsorption on goethite and soil in carbonate Solutions. Soil Science Society of American Journal, **60:** 1393-1400.

Dushenkov, S., Vasudev, D., Kapulnik, Y., Gleba, D.,Fleisher, D.,Ting K.C. and Ensley. B. 1997. Removal of uranium from water using terrestrial plants. Environmental Science and Technology, **31 (12):** 3468-3474.

Dushenkov, S.,Mikheev, A., Prokhnevsky, A., Ruchko, M. and Sorochinsky, B. 1999. Phytoremediation of radiocesium-contaminated soil in the vicinity of Chernobyl, Ukraine, Environmental Science and Technology, **33:** 469–475.

Dushenkov, V., Nanda Kumar, P. B. A., Motto, H. and Raskin, I. 1995. Rhizofiltration: the use of plants to remove heavy metals from aqueous streams. Environmental Science and Technology, **29:** 1239 - 1245.

Eapen, S., Suseelan, K.N., Tivarekar, S., Kotwal S.A. and Mitra R.2003. Potential for rhizofiltration of uranium using hairy root cultures of *Brassica juncea* and *Chenopodium amaranticolor*. Environmental Research, **91(2):** 127-133.

Eapen, S. and Mitra, R. 2001. Plant hairy root cultures: prospects and limitations. Proceedings of Indian National Science Academy, **304:** 107–120.

Koboi, T., Noguchi, A. and Yazaki, J. 1986. Family dependent cadmium accumulation characteristics in higher plants. Plant & Soil, **92:** 405–415.

Ebbs, S.D., Brady, D. Norvell, W.A., and Kochian, L. 2001. Uranium speciation, plant uptake, and phytoremediation. Practice Periodical of Hazardous, Toxic and Radioactive Waste Management, 5:130–135.

Ebbs, S.D., Norvell, W.A. and Kochian, L.V. 1998. The effect of acidification and chelating agents on the solubilization of uranium from contaminated soil. Journal of Environmental Quality, **27:** 1486–1494.

Elias, D.A., Krumholz, L.R., Wong, D., Long ,P.E. and Suflita, J.M. 2003. Characterization of microbial activities and U reduction in a shallow aquifer contaminated by uranium mill tailings. Microbial Ecology, **46:** 83-91.

England, E.C., 2006. Treatment of uranium-contaminated waters using organic based permeable reactive barriers, Federal Facilities Environmental Journal, **17:** 19–35.

EPA (Environmental Protection Agency).1997. Electrokinetic laboratory and field processes applicable to radioactive and hazardous mixed waste in soil and groundwater. EPA 402/R- 97/006. Washington, DC.

Evangelou, M.W.H., Ebel, M. and Schaeffer, A. 2007. Chelate assisted phytoextraction of heavy metals from soil. Chemosphere, **68**: 989-1003.

Evans, K.M., Gatehouse, J.A., Lindsay,W.J., Shi, J., Tommey, A.M. and Robinson, N.J. 1992. Expression of the pea metallothionein like gene PsMTA in *Escherichia coli* and *Arabidopsis thaliana* and analysis of trace metal ion accumulation: implications for gene PsMTA function. Plant Molecular Biology, **20**: 1019-1028.

Fellows, R.J., Ainsworth, C.C., Driver, C.J., and Catoldo, D.A.1998. Dynamics and transformation of radionuclides in soils and ecosystem health. In: Soil Chemistry and Ecosystem Health. Soil Science Society of American Journal, Madison, WI, USA, pp. 85–131.

Finneran, K.T., Anderson, R.T, Nevin, K.P. and Lovley, D.R. 2002. Potential for bioremediation of uranium-contaminated aquifers with microbial U(VI) reduction. Soil Sediment Contamination, **11**: 339-357.

Fisenne, I.M., Perry, P.M., Decker, K.M. and Keller, H.K. 1987. The daily intake of, 234, 235, 238U, 228, 230, 232 Th, and 226, 228 Ra by New York City Residents. Health Physics **53**: 357–363.

Fjeld, R.A., Coates, J.T. and Elzerman, A.W. 2000. Column tests to study the transport of plutonium and other radionuclides in sedimentary interbed at INEEL, Final Report Submitted to the Idaho National Engineering and Environmental Laboratory, Idaho Falls.

Francis, C.W., Mattus, A.J., Elless, M.P. and Timpson, M.E. 1993. Carbonate- and citratebased selective leaching of uranium from uranium-contaminated soils, in: Removal of Uranium from Uranium-contaminated soils, Phase I: Bench-Scale Testing, ORNL-6762, Oak Ridge National Laboratory, Oak Ridge, TN.

Francis, C.W., Timpson, M.E. and Wilson, J.H. 1999. Bench- and pilot-scale studies relating to the removal of uranium from uranium-contaminated soils using carbonate and citrate lixiviants, Journal of Hazardous Materials, **66** : 67–87.

Frankel, R.B. and Bazylinski, D.A., 2003. Biologically induced mineralization by bacteria. Reviews of Mineralology & Geochemistry, **54**: 95–114.

Fuller, C., Bargar, J. and Davis, J. 2003. Molecular-scale characterization of U(VI) sorption by bone apatite materials for a permeable reactive barrier demonstration. Environmental Science and Technology, **37**: 4642–4649.

Gadd, G.M. 2000. Bioremedial potential of microbial mechanisms of metal mobilization and immobilization. Current Opinion in Biotechnology, **11**: 271–9.

Gavrilescu, M., Pavel, L. V. and Cretescu, I. 2009. Characterization and remediation of soils contaminated with uranium, Journal of Hazardous Materials, **163**: 475–510.

Giloteaux, L., Holmes, D.E., Williams, K.H., Wrighton, K.C., Wilkins, M.J., Montgomery, A.P.and Lovley, D.R. 2013. Characterization and transcription of arsenic respiration and resistance genes during in situ uranium bioremediation. The ISME Journal, **7**: 370-383.

Glass, D.J. 1999. U.S. and international markets for phytoremediation. Needham, Mass., D. Glass Associates, p. 266.

Glick, B.R. and Bashan, Y .1997. Genetic manipulation of plant growth-promoting bacteria to enhance biocontrol of phytopathogens. Biotechnology Advances, **15**: 353–78.

Glick, B.R. 2010. Using soil bacteria to facilitate phytoremediation. Biotechnology Advances, **28**: 367–74.

Gohre, V. and Paszkowski ,U. 2006. Contribution of the arbuscular mycorrhizal symbiosis to heavy metal phytoremediation. Planta , **223**:1115–22.

Goto, F., Yoshihara, T., Shigemoto, N., Toki, S., and Takaiwa, F. 1999. Iron fortification of rice seed by the soybean ferritin gene. Nature Biotechnology, **17**: 282–286.

Gramss, G., Voigt, K..D. and Bergmann, H. 2004. Plant availability and leaching of (heavy) metals from ammonium-, calcium-, carbohydrate-, and citric acid-treated uranium-mine-dump soil. Journal of Plant Nutrition and Soil Science, **167**: 417–427.

Grcman, H., Velikonja Bolta, S. Vodnik, D., Kos, B. and Leš tan, D. 2001. EDTA enhanced heavy metal Phytoextraction of heavy metals phytoextraction: metal accumulation. leaching, and toxicity. Plant and Soil **235**: 105-114.

Gupta, D.C. and Singh, H. 2005. Uranium resource proceecessing: secondary resources, developments in uranium resources, production, demand and the environment,in: Proceedings of a Technical Committee Meeting Held in Vienna, June 15–18, 1999, IAEA-TECDOC-1425.

Han, R., Zou,W.,Wang, Y. and Zhu, L. 2007. Removal of U(VI) (VI) from aqueous solutions by manganese oxide coated zeolite: discussion of adsorption isotherms and pH effect. Journal of Environmental Radioactivity, **93**: 127–143.

Harrison, M.J. 1997. The arbuscular mycorrhizal symbiosis: An underground association. Trends in Plant Science, **2**: 54–56.

Hasegawa, I.,Terada,E., Sunairi, M., Wakita, H. and Shinmachi, F.1997. Genetic improvement of heavy metal tolerance in plants by transfer of the yeast metallothionein gene (CUP1). Plant Soil, **196**: 277-281.

Hashimoto, Y. Lester, B.G. Ulery, A.L. and Tajima, K. 2005. Screening of potential phytoremediation plants for uranium-contaminated soils: a sand culture method and metal-removal model, Journal of Environmental Chemistry, **15**: 771–781.

He, Z., Gentry, T.J., Schadt, C.W., Wu, L., Liebich, J., Chong, S.C., Huang, Z., Wu, W., Gu, B., Jardine, P. et al. 2007. GeoChip: a comprehensive microarray for investigating biogeochemical, ecological and environmental processes. The ISME Journal, **1**: 67-77.

Holmes,D.E., Giloteaux, L.Barlett, M., Chavan, M.A., Smith, J.A., Williams,K.H.,Wilkins,M., Long, P. and Lovely, D.R. 2013. Molecular analysis of the *in situ* growth rates of subsurface *Geobacter* species. Applied Environmental Microbiology, **79(5)**: 1646-53.

Hooper, F.J., Squibb, K.S., Siegel, E.L., McPhaul, K. and Keogh, J.P. 1999. Elevated urine uranium excretion by soldiers with retained uranium shrapnel. Health Physics **77:** 512-519.

Hopkins, B.S.1923. Chemistry of the Rarer Elements, D.C. Heath and Company, Boston/New York/Chicago/London, Online at: http://www.sciencemad ness.org/library/books/chemistry of the rarer elements.pdf.

Hossain, M.M. 2006. Effects of HCO3" and ionic strength on the oxidation and dissolution of UO_2, Thesis, School of Chemical Science and Engineering Nuclear Chemistry Royal Institute of Technology Stockholm, Sweden.

Huang,W.J., Blaylock, J.M., Kapulnik, Y. and Ensley B.D. 1998. Phytoremediation of Uranium-contaminated soils: Role of organic acids in triggering Uranium hyperaccumulation in plants, Environmental Science and Technology, **32(13):** 2004- 2008.

Humphries, A.C. and Macaskie, L.E. 2002. Reduction of Cr(VI) by *Desulfovibrio vulgaris* and *Microbacterium* sp. Biotechnology Letters, **24:** 1261-1267 .

Istok, J.D., Senko, J.M., Krumholz, L.R., Watson, D., Bogle, M.A., Peacock, A., *et al.* 2004. *In situ* bioreduction of technetium and uranium in a nitrate-contaminated aquifer. Environmental Science Technology, **38:** 468–475.

Jagetiya, B. and Sharma, A. 2013. Optimization of chelators to enhance uranium uptake from tailings for phytoremediation. Chemosphere, **91(5):** 692-6.

John, E.,Ten Hoevea and Mark Jacobson, Z. 2012. Worldwide health effects of the Fukushima Daiichi nuclear accident.DOI: 10.1039/c2ee22019a.

Jones, D.L., Prabowo, A.M. and Kochian, L.V. 1996. Kinetics of malate transport and decomposition in acid soils and isolated bacterial populations: the effect of microorganisms on root exudation of malate under Al stress. Plant and Soil, **182:** 239-247.

Jones, D.L. and Darrah, P.R. 1994. Role of root derived organic acids in the mobilization of nutrients from the rhizosphere. Plant and Soil, **166:** 247-257.

Jroundi, F., Merroun, M.L., Arias, J.M., Rossberg, A., Selenska Pobell, S. and Gonzalez-Munoz, M.T. 2007. Spectroscopic and microscopic characterization of uranium biomineralization in *Myxococcus xanthus*. Geomicrobiology Journal, **24:** 441–449.

Kantar, C. and Honeyman, B.D. 2006. Citric acid enhanced remediation of soils contaminated with uranium by soil flushing and soil washing, Journal of Environmental Engineering, **132** : 247–255.

Khan, M.S., Zaidi, A., Wani, P.A. and Oves, M. 2009. Role of plant growth promoting rhizobacteria in the remediation of metal contaminated soils. Environmental Chemical Letters,**7:** 1-19.

Kidd,P., Barcelo, J., Bernal, M.P., Navari-Izzo, F., Poschenrieder, C., Shilev, S., et al. 2009. Trace element behaviour at the root–soil interface: implications in phytoremediation. Environmental Experimental Botany, **67:** 243–59.

Krishnamurti, G.S.R., Cielinski, G., Huang, P.M. and van Rees, K.C.J., 1998. Kinetics of cadmium release from soils as influenced by organic acid: implementation in cadmium availability. Journal of Environmental Quality, **26:** 271–277.

Kulpa, J.P. and Hughes, J.E. 2001. Deployment of chemical extraction soil treatment on uranium contaminatedsoil in:WM'01Conference,Online at: http://www.wmsym. org/Abstracts/2001/54/54-5.pdf.

Langmuir, D.1978.Uranium solution-mineral equilibria at low temperature with applications to sedimentary ore-deposits, Geochimica et Cosmochimica Acta, **42**: 547–569.

Lentan, D.2006. Enhanced heavy metal phytoextraction, In: Mackova M *et al.* (eds.) Phytoremediation Rhizoremediation. Springer the Netherlands; 115 - 132.

Leyval, C. and Joner, E.J. 2001. Bioavailability of heavy metals in the mycorrhizosphere. In: Gobran, G.R., Wenzel, W.W., Lombi, E. (Eds.), Trace Elements in the Rhizosphere. CRC Press LLC, London, pp :165-185.

Li, G.Y., Hu N., Ding D.X., Zheng, J.F., Liu, Y.L., Wang, Y.D. and Nie, X.Q. 2011. Screening of plant species for phytoremediation of uranium, thorium, barium, nickel, strontium and lead contaminated soils from a uranium mill tailings repository in South China. Bulletin of Environmental Contamination and Toxicology, **6(6)**: 646-52.

Liang, Y., Li, G., Van Nostrand, J.D., He, Z., Wu, L., Deng, Y., Zhang, X. and Zhou, J. 2009. Microarray-based analysis of microbial functional diversity along an oil contamination gradient in oil field. FEMS Microbiology & Ecology, **70**: 324-333.

Liu, H. and Fang, H.H.P. 2002. Characterization of electrostatic binding sites of extracellular polymers by linear programming analysis of titration data. Biotechnolology and Bioengineering, **80**: 806–811.

Lloyd, J.R. and Macaskie, L.E. 2002. Biochemical basis of microbe–radionuclide interactions. In: Keith-Roach, M., Livens, F. (eds.), Interactions of Microorganisms with Radionuclides. Elsevier Sciences, Oxford, UK,: 313–342.

Lovley, D.R., Phillips, E.J.P., 1992. Reduction of uranium by *Desulfovibrio desulfuricans*. Applied Environmental Microbiology, **58**: 850–856.

Lu, H., Gao, G., Xu G, Fan, L., Yin, L., Shen, B. and Hua, Y. 2009. *Deinococcus radiodurans* PprI switches on DNA damage response and cellular survival networks after radiation damage. Molecular Cell Proteomics, **8(3)**:481-94

Ma, Y., Prasad, M.N.V., Rajkumar, M. and Freitas, H. 2011a.Plant growth promoting rhizobacteria and endophytes accelerate phytoremediation of metalliferous soils. Biotechnology Advances, **29**: 248–58

Macaskie, L.E., Bonthrone, K.M., Yong, P. and Goddard, D.T., 2000. Enzymically mediated bioprecipitation of uranium by a *Citrobacter* sp. A concerted role for exocellular lipopolysaccharide and associated phosphatase in biomineral formation. Microbiology, **146**: 1855–1867.

Majáre, M. and Bülow, L. 2001. Metal-binding proteins and peptides in bioremediation and phytoremediation of heavy metals. TIBTECH, **19**: 67–73.

Macaskie, L.E., Empson, R.M., Cheetham, A.K., Grey, C.P. and Skarnulis, A.J. 1992. Uranium bioaccumulation by a *Citrobacter* sp. as a result of enzymically mediated growth of polycrystalline HUO_2PO_4. Science, **257**:782–784.

Martinez, R.J.M., Beazley, J., Teillefert,M., Arakaki, A.K., Skolnick, J. and Sobecky, P.A. 2007. Aerobic uranium (VI) bioprecipitation by metal-resistant bacteria isolated from radionuclide- and metal-contaminated subsurface soils. Environmental Microbiology, 9: 3122–3133.

Mason, C.V.F. Turney, W.R.J.R.,Thomson, B.M., Lu N., Longmire, P.A. and Chrisholm-Brause, C.J. 1997. Carbonate leaching of uranium from contaminated soils, Environmental Science and Technology, 31: 2707–2711.

McIntyre, T. 2003. Phytoremediation of Heavy Metals from Soils In: Scheper, T., Tsao, D.T. (eds.),Advances in Biochemical Engineering/Biotechnology, 78. New York: Springer-Verlag Berlin Heidelberg, 97-123.

Merroun, M.L., Hennig, C., Rossberg, A., Reich, T. and Selenska-Pobell, S. 2003. Characterization of U (VI)-*Acidithiobacillus ferrooxidans* complexes by using EXAFS, transmission electron microscopy and energy-dispersive X-ray analysis. Radiochimica Acta, 91: 583–591.

Merroun, M.L., Nedelkova, M., Rossberg, A., Hennig, C. and Selenska-Pobell, S. 2006. Interaction mechanisms of uranium with bacterial strains isolated from extreme habitats. Radiochimica Acta, 94: 723–729.

Merroun, M.L. and Selenska Pobell, S. 2007. Transmission Electron Microscope Analysis of Eu(III) accumulated by *Bacillus sphaericus JG-A12*, Report FZD-459: 44.

Michalsen, M.M., Goodman, B.A., Kelly, S.D., Kemner, K.M., McKinley, J.P., Stucki, J.W. and Istok, J.D. 2006. Uranium and technetium bio-immobilization in intermediate-scale physical models of an in situ bio-barrier. Environmental Science and Technolology, 40: 7048–7053.

Miransari, M. 2011. Hyperaccumulators, arbuscular mycorrhizal fungi and stress of heavy metals. Biotechnoogyl Advances, 29: 645–53.

Misra, C.S., Appukuttan, D., Kantamreddi, V.S., Rao, A.S. and Apte, S.K. 2012. Recombinant D.radiodurans cells for bioremediation of heavy metals from acidic/neutral aqueous wastes. Bioeng Bugs, 1;3(1): 44-8.

Mulligan, C.N., Yong, R.N. and Gibbs, B.F. 2001. Remediation technologies for metal-contaminated soils and groundwater: an evaluation. Engineering Geology, 60: 19-207.

Nanda Kumar, P. B. A., Dushenkov, V., Motto, H. and Raskin, I. 1995. Phytoremediation: the use of plants to remove heavy metals from soils. Environmental Science and Technology, 29: 1232-1238.

Navratil, J.D. 2001. Advances in treatment methods for uranium contaminated soil and water, Archives of Oncology, 9: 257–260.

Nedelkova, M., Merroun, M.L., Rossberg, A., Hennig, C. and Selenska Pobell, S. 2007. Microbacterium isolates from the vicinity of a radioactive waste depository and their interactions with uranium. Fems Microbiology and Ecology, 59: 694–705.

Negri, M.C. and Hinchman, R.R. 2000. The use of plants for the treatment of radionuclides. In: Phytoremediation of toxic metals using plants to clean up the environment: I. Raskin and B.D. Ensley (eds.). Wiley, New York: 107-132.

North, N.N., Dollhopf, S.L., Petrie, L., Istok, J.D., Balkwill, D.L. and Kostka, J.E. 2004. Change in bacterial community structure during *in situ* biostimulation of subsurface sediment co contaminated with uranium and nitrate. Applied Environmental Microbiology. **70:** 4911–4920.

Or³owska, E., Przyby³owicz, W., Orlowski, D., Turnau, K. and Mesjasz Przyby³owicz, J. 2011. The effect of mycorrhiza on the growth and elemental composition of Ni-hyper accumulating plant *Berkheya coddii* Roessler. Environmental Pollution. **159:** 3730–8.

Pabalan, R.T. and Turner, D.R. 1997. Uranium (6+) sorption on montmorillonite: experimental and surface complexation modelling study, Aquatic Geochemistry. 2: 203–226.

Pabalan, R.T., Turner, D.R., Nertetti, F.P. and Prikryl, J.D. 1998. Uranium (VI) sorption onto selected mineral surfaces, key geochemical parameters, in: E.A. Jenne (ed.), Adsorption of Metals, Geomedia, Academic Press. San Diego, CA.

Payne, R.B., Gentry, D.M., Rapp-Giles, B.J., Casalot, L., and Wall, J.D. 2002. Uranium reduction by *Desulfovibrio desulfuricans* strain G20 and a cytochrome c3 mutant. Applied Environmental Microbiology. **68:** 3129–3132.

Peterson, P.J. 1971. Unusual accumulations of elements by plants and animals. Science Progress, 59: 505–526.

Phillips, D., Gu, B., Watson, D. and Parmele, C. 2008. U(VI)removal from contaminated groundwater by synthetic resins.Water Research, **42:** 260–268.

Phillips, E.J.P., Landa E.R. and Lovley, D.R. 1995. Remediation of uranium contaminated soils with bicarbonate extraction and microbial U (VI) reduction. Journal of Indian Microbiology. **14:** 203–207.

Pilon-Smits, E.A.H., De Souza, M.P., Hong, G., Amini, A. and Bravo, R.C. 1999. Selenium volatilization and accumulation by twenty aquatic plant species. Journal of Environmental Quality. **28:** 1011-1017.

Qualls, R.G. and Haines, B.L. 1992. Biodegradability of dissolved organic matter in forest through fall, soil solution, and stream water. Soil Science Society of American journal. **56:** 578-586.

Quartacci, M.F., Baker, A.J.M. and Navari Izzo, F. 2005. Nitrilotriacetate- and citric acid-assisted phytoremediation of cadmium by Indian mustard (*Brassica juncea* (L.) Czern, Brassicaceae). Chemosphere **59:** 1249-1255.

Raicevic, S., Wright, J.V., Veljkovic, V., and Conca, J. L. 2006. Theoretical stability assessment of uranyl phosphates and apatites: selection of amendments for in situ remediation of U (VI). Science of the Total Environment. **355:** 13–24.

Rajkumar, M., Ae, N., Prasad, M.N.V. and Freitas, H. 2010. Potential of siderophore-producing bacteria for improving heavy metal phytoextraction. Trends in Biotechnology. **28:** 142–9.

Renninger, N., Knopp, R., Nitsche, H., Clark, D. and Keasling, J. 2004. Uranyl precipitation by *Pseudomonas aeruginosa* via controlled polyphosphate metabolism. Applied Environmental Microbiology. **70:** 7404–7412.

Riley, R.G. and Zachara, J.M. 1992. Nature of chemical contamination on DOE lands and identification of representative contaminant mixtures for basic subsurface science research. Subsurface Science Program, Office of Energy Research, U.S. Department of Energy, Washington, D.C.

Roane, T.M. and Kellogg, S.T. 1996. Characterization of bacterial communities in heavy metal contaminated soils. Canadian Journal of Microbiology, **42**: 593-603.

Rufyikiri, G., Thiry, Y. and Declerck, S. 2003. Contribution of hyphae and roots to uranium uptake and translocation by arbuscular mycorrhizal carrot roots under root-organ culture conditions. New Phytologist, **158 (2)**: 391-399.

Rufyikiri, G., Thiry, Y., Wang, L., Delvaux, D. and Declerck, S. 2002. Uranium uptake and translocation by the arbuscular mycorrhizal fungus *Glomus intraradices* under root-organ culture conditions. New Phytologist, **156 (2)**: 275-281.

Ruiz, O.N., Hussein, H.S., Terry, N. and Daniell, H. 2003. Phytoremediation of organomercurial compounds via chloroplast genetic engineering. Plant Physiology, **132**: 1344-1352.

Salem,H.M. 2000. Uranium ores and the environmental impact on human health risks.. ICEHM, Cairo University, Egypt,: 580- 585.

Salt, D.E., Blaylock, M., Kumar, N.P.B.A., Dushenkov, V., Ensley, D., Chet, I. and Raskin, I. 1995. Phytoremediation: a novel strategy for the removal of toxic metals from the environment using plants. Biotechnology, **13**: 468-474.

Samuelsen, A.I., Martin, R.C., Mok, D.W.S. and Machteld, C.M. 1998. Expression of the yeast FRE genes in transgenic tobacco. Plant Physiology. **118**: 51-58.

Sarro I.M, García Ana, M. and Moreno Diego A. 2005. Biofilm formation in spent nuclear fuel pools and bioremediation of radioactive water. International Microbiology, **8**: 223-230.

Saxena, S., Prasad, M., and D'Souza, S. F. 2006. Radionuclide sorption on to low-cost mineral adsorbent. Industrial and Engineering Chemistry Research, **45**: 9122–9128.

Schalk, I.J., Hannauer, M. and Braud, A. 2011. New roles for bacterial siderophores in metal transport and tolerance. Environmental Microbiology, **13**: 2844–54.

Schnug, E., Steckel, H. and Haneklaus, S. 2005. Contribution of uranium in drinking waters to the daily uranium intake of humans - a case study from Northern Germany. Landbauforschung Voelkenrode, **55**: 227-236.

Senko, J.M., Mohamed, Y., Dewers, T.A. and Krumholz, L.R. 2005. Role for Fe(III) minerals in nitrate-dependent microbial U(IV) oxidation. Environmental Science and Technology, **39**: 2529–2536.

Shacklette, H.T. and Boerngen, J.G. 1984. Element concentrations in soils and other surficial materials of the conterminous United States. U.S. Geological Survey Professional Paper, 1270. United States Printing Office, Washington.

Shahandeh, H. and Hossner, L. R. 2002. Role of soil properties in phytoaccumulation of Uranium. Water, Air, and Soil Pollution, **141**: 165–180.

Shanks, J.V. and Morgan, J. 1999. Plant hairy root culture. Current Opinion in Biotechnology, **10**: 156–159.

Sheppard, S. C., Evenden, W. G. and Anderson, A. J. 1992. Multiple assays of uranium toxicity in soil. Environmental Toxicology and Water Quality, **7**: 275-294.

Shoesmith, D.W. 2007. Used fuel and uranium dioxide dissolution studies - a review, Report No. NWMO TR-2007-03, The University of Western Ontario, Toronto, Ontario,Canada,Onlineat:http://www.nwmo.ca/adx/asp/adxGetMedia.asp?DocID=1607,1554,1,Documents&MediaID=3320&Filename=NWMO-TR-2007-03 Uranium Review R0a.pdf.

Singh, R.,Paul, D. and Rakesh, K. Jain. 2006. Biofilms: implications in bioremediation. Trends in Microbiology, **14(9)**: 389–3.

Sleytr, U.B., 1997. I. Basic and applied S-layer research: an overview. FEMS Microbiology. Reviews, **20**: 5–12.

STUK-A169. Removal of Radionuclides from Private Well Water with Granular Activated Carbon (GAC): Removal of U, Ra, Pb and Po.

Stumm, W., and Morgan, M.M. 1996. Aquatic Chemistry, 3rd ed., John Wiley & Sons Inc., New York, : 252–424.

Thomas, S. H., Sanford, R. A., Amos, B. K., Leigh, M. B., Cardenas, E. and Loeffler, F. E. 2010. Unique Ecophysiology among U(VI)-Reducing Bacteria Revealed by Evaluation of Oxygen Metabolism in *Anaeromyxobacter dehalogenans* Strain 2CP-C. Applied and Environmental Microbiology, **76(1)**: 176-183.

Tian, B., Wang, H., Ma, X., Hu, Y., Sun, Z., Shen, S., Wang, F. and Hua, Y. 2010. Proteomic analysis of membrane proteins from a radioresistant and moderate thermophilic bacterium *Deinococcus geothermalis*. Molecular Biosystematics, **10**: 2068–2077.

USEPA 1998. NATO/ CCMS Pilot Study: Evaluation of demonstrated and emerging technologies for the treatment of contaminated land and groundwater (Phase III), Special session: Treatment walls and permeable reactive barriers,EPA 542-R-98-003, No.229,p.108

Van der Zaal, B.J., Neuteboom, L.W., Pinas, J.E., Chardonnens, A.N. and Schat, H. 1999. Overexpression of a novel *Arabidopsis* gene related to putative zinc-transporter genes from animals can lead to enhanced zinc resistance and accumulation. Plant Physiology, **119**: 1047-1055.

Van Nostrand, J.D., Wu, W.M., Wu, L., Deng, Y., Carley, J., Carroll, S., He,. Z, Gu, B., Luo J., Criddle, C.S. et al. 2009. GeoChip-based analysis of functional microbial communities during the reoxidation of a bioreduced uranium-contaminated aquifer. Microbiology, Environmental **11**: 2611-2626.

Vandenhove, H. Van Hees, M. and Van Winckel, S. 2001. Feasibility of phytoextraction to clean up low-level uranium-contaminated soil. International Journal of Phytoremediation, **3**: 301–320.

Vecchia, E.D., Veeramani, H., Suvorova, E.I., Wigginton, N.S., Bargar, J.R. and Bernier-Latmani, R. 2010. U(VI) reduction by spores of *Clostridium acetobutylicum*. Research in Microbiology, **161(9)**: 765-71.

Wagner-Dobler, I. 2003. Mini review: Pilot plant for bioremediation of mercury contaminated industrial waste water. Applied Microbiology and Biotechnology. **62:** 124-13

Wagner-Dobler, I., H.F. von Canstein, Y. Li, J. Leonhäuser, and W.-D. Deckwer. 2003. Process-Integrated microbial mercury removal from wastewater of chlor-alkali. Electrolysis Plants. Engineering Life Sciences, **3** (4):177-181.

Waite, T.D., Davis, J.A., Payne,T.E., Waychunas,G.A. and Xu, N.1994. Uranium(VI) adsorption to ferrihydrite. Application of a surface complexation model. Geochimica et Cosmochimica Acta, **58:** 5465–5478.

Waldron, P.J., Wu, L., Nostrand, J.D.V., Schadt, C.W., He, Z., Watson, D.B., Jardine, P.M., Palumbo, A.V., Hazen, T.C. and Zhou, J. 2009. Functional gene array-based analysis of microbial community structure in ground waters with a gradient of contaminant levels. Environmental Science and Technology, **43:** 3529-3534.

Wall, J. D., and Krumholz, L. R. 2006. Uranium Reduction. Annual Review of Microbiology, 149-166.

Watson, D.B., Wu,W.M., Mehlhorn, T., Tang , g., Earles, J., Lowe ,K., Gihring, T. M., Zang , G., Phillips, J., Boyanov, M.I., Spalding , B.P., Schadt , C., Kemner , K.M., Criddle, C.S, Jardine , P.M. and Brooks S.C. 2013. *In situ* bioremediation of uranium with emulsified vegetable oil as the electron donar. Environmental Science & Technology, **47 (12):** 6440-8.

Wenzel, W.W. 2009. Rhizosphere processes and management in plant-assisted bioremediation (phytoremediation) of soils. Plant Soil, **321:** 385–408.

Wilkins, M.J., Callister, S.J., Miletto, M., Williams, K.H., Nicora, C.D., Lovley, D.R., Long, P.E. and Lipton, M.S. 2011. Development of a biomarker for Geobacter activity and strain composition; Proteogenomic analysis of the citrate synthase protein during bioremediation of U(VI). Microbial Biotechnology, **4:** 55-63

Winde, F. 2002. Uranium contamination of fluvial systems-mechanisms and processes, Part III, Cuadernos de Investigacion Geografica, **28:** 75–100.

Wood, P.A. 1997. Remediation methods for contaminated sites. p. 47-72. In R.E. Hester and R.M. Harrison (eds.) Issues in environmental science and technology: contaminated land and its reclamation. The Royal Society of Chemistry, Letchworth, U.K.

Xu, H., Barton, L.L., Zhang, P. and Wang, Y. 2000. TEM investigation of U(VI) and Re(VII) reduction by *Desulfovibrio desulfuricans*, a sulfate reducing bacterium. Material Research Society Symposium Proceedings, **608:** 299–304.

Yousfi, I., Bole, J. and Geiss,O. 1999. Chemical extractions applied to the determination of radium speciation in uranium mill tailing: study of different reagents, Journal of Radioanal Nuclear Chemistry, **240:** 835–840.

Zarei,M., Wubet,T., Schäfer, S.H., Savaghebi, G.R., Jouzani, G.S., Nekouei, M.K., *et al.* 2010. Molecular diversity of arbuscular mycorrhizal fungi in relation to soil chemical properties and heavy metal contamination. Environmental Pollution, **158:** 2757–65.

Zhou, P. and Gu, B. 2005. Extraction of oxidized and reduced forms of uranium from contaminated soils: effects of carbonate concentration and pH, Environmental Science and Technology, **39**: 4435–4440.

Zhu, Y., Pilon Smits, E.A.H., Jouanin, L. and Terry, N. 1999a. Overexpression of glutathione synthetase in *Brassica juncea* enhances cadmium tolerance and accumulation. Plant Physiology, **119**: 73-79.

Zhu, Y., Pilon Smits, E.A.H., Tarun, A. Weber, S.U., Jouanin, L. and Terry,N. 1999b. Cadmium tolerance and accumulation in Indian mustard is enhanced by overexpressing glutamylcysteine synthetase. Plant Physiology, **121**: 1169-1177.

Zhu, Y.G. and Shaw, G. 2000. Soil contamination with radionuclides and potential remediation. Chemosphere, **41**: 121-128.

Subject Index

Aconitumferox 145

heterophyllum 144

Aegle marmelos 144

AFLP 32

Allergy 7

Amaranthus 59

Amla 144

Andrographis paniculata 145, 150

Apple 163

Artemisia annua 150

Ascorbate peroxidase isozyme 26

Ashoka 144

Ashwagandha 144

Asparagus racemosus 145, 150

Atis 144

Atropa belladonna 150

Azadirachta indica 150

Bacopa monnieri 144, 149

Bael 144

Banana 163

Berberis aristata 144

Bhuiamalaki 144

Bifldobacterium 2

Adolescentis 5

animalis 5

bifidum 5

breve 5

lactis 5

Bioaccumulation 223

Bioaugumentation 223

Biological remediation 220

Bioremediation 224

Biosorpation 222

Biotransformation 221

Bone disorders 10

Brahmi 144

CAPS 40

Cassia angustifolia 145

Cattleya 97

Centella asiatica 149, 150

Chandan 144

Chirata 144

Chlorophytum borivilianum 145, 149, 150

Coleusforskohlii 145

Commiphora wightii 144

Curcuma longa 150

DArT 40

Daru haridra 144

Database 72

Dendrobium 55, 69, 85

Dental caries 10

Diabetes 9

Digitalis lanata 150

DNA based molecular techniques 57

DNA Markers 27

E. coli 3

Effect of banana I 14

Effect of coconut water 113

Effect of light 120

Effect of nitrogen 108

Effect of peptones 112

Effect of seed maturity 120

Effect of surface adsorbants 118

Effect of temperature 120

Effect of vitamins 110

Effect of yeast extract 113

Embelia ribes 145

Emblica officinalis 144

Erycina pusilla 73

ESTs38

EST-SSRs 39

Functional markers 37

Galeola 106

Garcinia indica 145

Geodorum 107

Gi/oe 144

Gloriosa superba 144, 149

Glycyrrhiza glabra 145

Gymnema sylvestre 144

Goodyera 105

Gudmar 144

Guggu/144

Gut dysbiosis 3

Gut microflora 2

Habenaria 107

Hybridization-based markers 57

Hyperlipidaemia 8

Infections 8

In-silico 55, 69, 74

lsabgol 144

ISSR 34

ITS-microarray 69

Jatamansi 144

Kaempferia rotunda 189

Kal ihali 144

Kalmegh 145

Kesar 145

Kokum 145

Kuth 145

Kutki 145

Lactobacillus acidophilus 5, 7, 9

casei 5,7,9

fermentus 9

gasseri 5

hilgardii 7

johnsonii 5, 7

lactis 9

paracasei 5

plantarum 5, 7

reuteri 5, 7, 10

rhamnosus 5, 7

Makoi 145

Mango 163

Medicinal plants 141

Microsatellite analysis 65

Molecular markers 21

Molecular markers types 24

Morinda citrifolia 201

Morphological markers 85

Mulethi 145

Nardostachys jatamansi 144

Neottia nidus-aris 73
Nuclear accidents 213
utrient medium 93
Ocimum 145
Oncidium 74
Ophrys 64
Orchid genome 73
Orchids 83
Orchid seeds 95, 97
Papaya 163
Patharchur 145
peR based markers 30, 58
Pediococcus acidilactici 5
Phalaenopsis 69, 74
bellina 74
equestris 74
Phlebodium aureum 179
Phyllanthus niruri 144
Phytoextraction 228
Phytoremediation 233
Pippali 145
Plantago ovata 144
Pogostemon patchouli 149
Potato 163
Probiotics 1
Pronephrium triphyllum 169
Protein markers 25
Protocorm 92
QTL 38
RAPD 30
Rauwolfia serpentina 150

RFLP28
Rhizanthella gardneri 74
Saccharomyces boulardii 6
Safed musli 145
Santalum album 144
Saraca indica 144, 149, 150
Sarpagandha 145
SCAR 36
Seed dispersal 96
Seed structure 91
Seed viability 90
Senna 145
Sequence-based markers 60
Shatavari 145
SNPs 39
Somaclona variation 157
Soybean 163
SSRs 33
Stevia rebaudiana 150
Strawberry 163
STS 36
Swertia chirata 144
Tinospora cordifolia 144
Tomato 163
Tulsi 145
Uranium contamination 213, 214
Uranium mining 214
Uranium remediation 215
Val vidangi 145
Vatsanabh 145
Wheat 163
Withania somnifera 149, 150

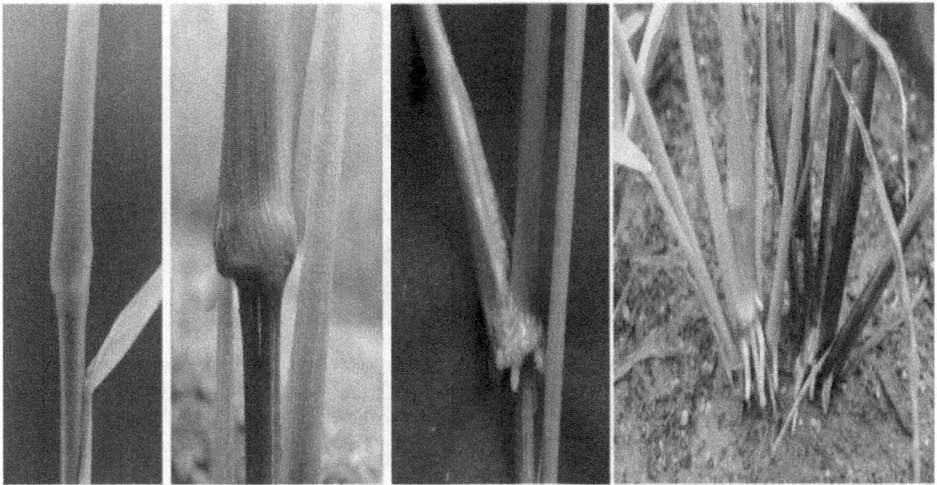

(a)   (b)   (c)   (d)

Fig. 2.1: Foxtail Millet Internode Colour and Leaf Sheath Colour inheritance.
(a) Female Parent (b) Male parent (c) F1 Plant and (d) F2 Segregating population in field **(Page-25)**

Fig. 4.1: Fruits of Orchids **(Page-89)**

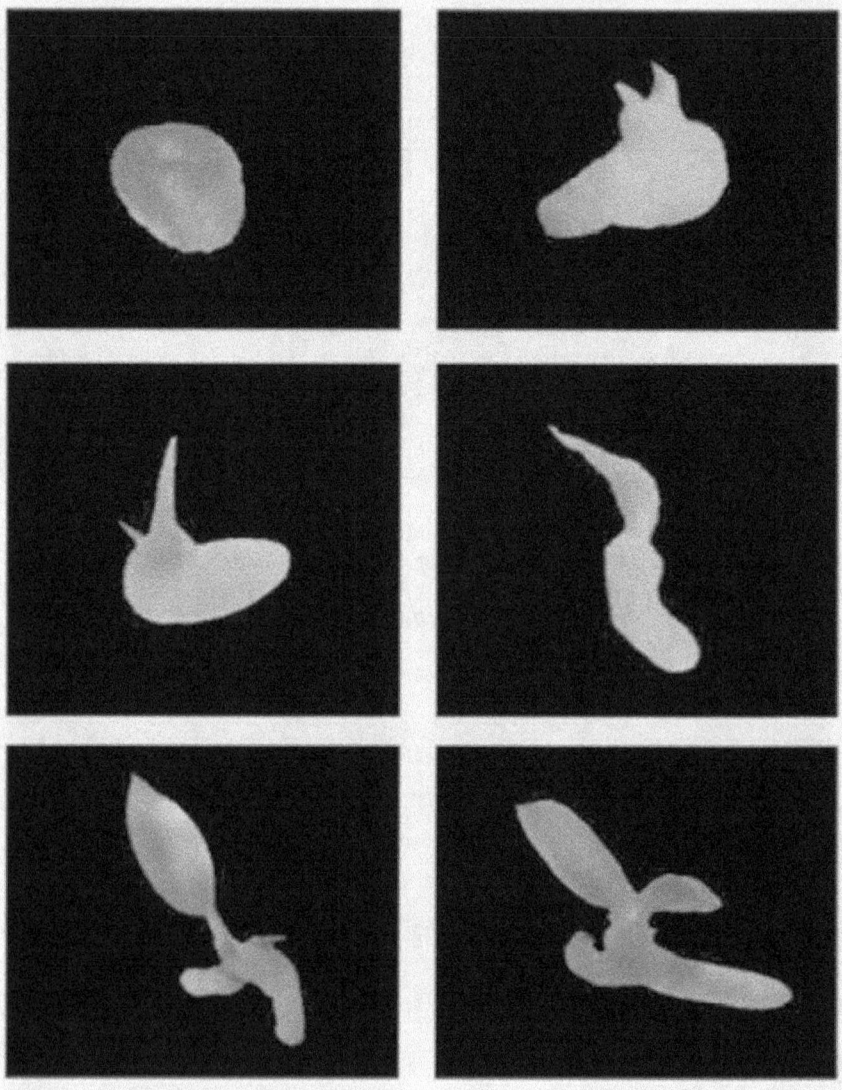

Fig. 4.5. Developmental stages of Protocorms after Germination of Seeds of Orchids
(Page-90)

Fig 7.1. *In vitro* multiplication of *Pronephrium triphyllum* (Sw.) Holttum

A - Aposporous gametophyte proliferation from crozier derived calli; B - Sporophyte formation from crozier derived calli; C-F - Different stages of sporophyte formation from crozier derived calli; G - Aposporous gametophyte formation from crozier derived calli; H - Rhizoid formation from the crozier deirved calli; I - Sporophyte formation from aposporous gametophyte **(Page-94)**

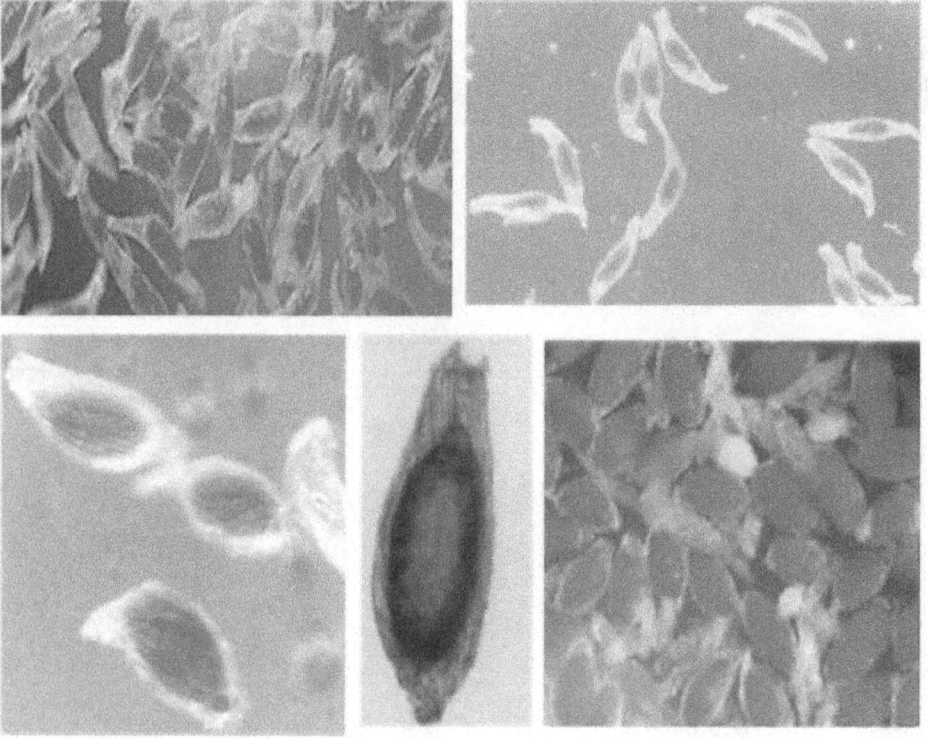

Fig. 4.2. A-E. Qualitative testing of viability through TTC porofile for seeds of : A. *Dendrobium nobile* (10x); B. D. *chrysanthum* (10x) C. *D.* farmeri (10x); D. *D. parishii* (10x); E. *Hygrochita parishii* (10x) **(Page-174)**

www.ingramcontent.com/pod-product-compliance
Lightning Source LLC
Chambersburg PA
CBHW021436180326
41458CB00001B/294